For Patricia Olmstead,
with best wishes,
Roger Pasquier

Birds in Winter

Surviving the Most Challenging Season

Roger F. Pasquier

Illustrations by Margaret La Farge

Princeton University Press

Princeton and Oxford

Copyright © 2019 by Princeton University Press

Published by Princeton University Press
41 William Street, Princeton, New Jersey 08540
6 Oxford Street, Woodstock, Oxfordshire OX20 1TR

press.princeton.edu

LCCN 2018958180
ISBN 978-069-117855-4

British Library Cataloging-in-Publication Data is available

Editorial: Robert Kirk and Kristin Zodrow
Production Editorial: Ellen Foos
Text Design: D & N Publishing, Wiltshire, UK
Jacket Design: Lorraine Doneker
Jacket image: Markus Varesvuo/Agami Picture Library
Production: Steven Sears
Publicity: Matthew Taylor and Julia Hall
Copyeditor: Laurel Anderton

This book has been composed in Baskerville MT for the main text and Museo Slab for
the captions and headings

Printed on acid-free paper. ∞

Printed in the United States of America

10 9 8 7 6 5 4 3 2 1

For

Phoebe Goodhue Milliken

CONTENTS

Male Black-billed Capercaillies in Siberia eat larch twigs through the winter, picking whatever is above the snow. This increases the number of shoots that will grow from the stem the next spring, providing food the following winter.

INTRODUCTION

IN 1976, WHEN I was writing *Watching Birds: An Introduction to Ornithology*, I had an abundance of books and papers in scientific journals to draw from as I described the breeding behavior and migrations of birds. Conspicuously missing, however, was a comparable level of information on the remaining portion of the annual cycle—winter—which to me seemed no less interesting, and no less vital to understanding the lives of birds. I believe the chapter on winter I wrote for *Watching Birds* was then the first unified, albeit modest, account of the ecology and behavior of how both permanent residents and migrants deal with that season. I was sure there was much more to be researched and brought together, and I immediately proposed the idea to my editor as my next book. "Too narrow a topic!" was the response then, as well as later in a succession of intervals, each of about fifteen years, when I proposed it again to subsequent editors.

Finally, in 2015, after completing *Painting Central Park*, another book I had long wanted to write, I decided to take the plunge, research the topic of birds in winter in all its aspects, and determine for myself whether there was a book in it. Over those nearly 40 intervening years, winter had also come to interest a new generation of ornithologists. The Smithsonian Institution's 1977 and 1989 symposia on the ecology and conservation of migrant birds in the Neotropics, for example, brought together tremendous amounts of essential information, inspired even more research, and produced publications that are still primary references. The years around the turn of this century saw expanded interest in austral migration and winter in Australia and South America—where, in fact, W. H. Hudson had vividly depicted it in *Far Away and Long Ago* (1918) based on his boyhood memories of Argentina in the 1870s.

Over these decades, ornithologists in Finland and Scandinavia were publishing a steady stream of papers describing the social and ecological winter dynamics of boreal forest birds, while from Britain and the Netherlands were coming accounts of how shorebirds spend that season in northerly latitudes, permitting useful comparisons with the North American and Eurasian shorebirds that winter far south of the equator. Similarly, in North America intensive long-term studies of chickadees and juncos have revealed the complexities of their winter lives. (Sadly, there is still very little information about the birds wintering in Southeast Asia, where forests are being cut at an alarming rate, or from much of Africa, where more frequent and intense droughts are transforming the landscape.) The recent rapid advances in satellite tracking technology and stable isotope analysis have enabled scientists to follow individual birds day by day and practically meal by meal all through the winter when these birds are continents and oceans away. And today the growing concern about conservation, especially the loss of wintering habitat, and the impacts of climate change have added new perspectives and new urgency to understanding the entire annual cycle of birds.

So, in 2015 I was glad to return to a seat at the same table in the library of the Department of Ornithology in the American Museum of Natural History where I had written *Watching Birds*. The perfect place to research and write. I've never had more fun!

When discussing the feathered toes of the Snowy Owl, the differences in bill lengths of North American and Caribbean kingbirds and vireos, or a species like the Tibetan Ground Tit that I did not know at all, I could look at the specimens. It was hard to say *basta!* to research when every newest issue of one of more than 20 ornithological, behavioral, ecological, and conservation journals I was following might have something amazing and irresistible to include. I drew the line at the last of the mid-2018 publications—and hope I will not soon regret it.

Of course, when I described my project, some people thought my topic was simply the familiar birds of northern latitudes that stay where it is cold. But no, winter is a global phenomenon, and it influences birds all over the world, including many of those that never leave the tropics but must share their habitat during parts of the year with birds from the Northern or Southern Hemisphere, or from both. We routinely say "Blackburnian Warblers winter in Colombia" and "White Storks winter in Africa," so there seemed every reason to follow them there to learn the challenges of that season and how the birds fit into those ecosystems. (I have kept the discussion of how migrants reach their winter destinations to a minimum, since today there are more books on migration, dazzling in their comprehensiveness, than ever before.) And, since life is a continual cycle, what happens to birds during the winter affects their subsequent breeding season, often in unexpected ways, as new technology is revealing. Finally, since one of the greatest pleasures of watching birds is knowing how far some have traveled and how their physical and behavioral adaptations have enabled them to survive in what may be widely different environments, one surely wants to include the birds that "winter" where it may not seem like winter at all.

Another of the pleasures of watching birds is understanding more of what's really going on among the most familiar species that we may see all year or through the winter when many birds are so much more visible. In the midlatitudes, where in winter there may be no concealing foliage, it is much easier to follow birds through the day. Many also seem tamer, perhaps because of hunger and, in extreme cold, their tendency to reduce movement to save energy. Naive young birds living through their first winter sometimes give us glimpses experienced adults do not allow. I have seen a first-year Red-tailed Hawk bathing in a stream and a young Northern Goshawk carelessly bold.

Showing readers how much there is to see through winter, when so many birders are least active, has been one of my goals. On my daily walks through Central Park to and from the museum as I was writing this book, I observed with new eyes some of what I was researching and writing—and smiled as I thought, "I know now what *you're* doing." I saw woodpeckers, jays, and tits beginning to store food while it still seemed like summer, and interactions at feeders starting in November as different species made others give way and as individuals asserted their rank in a flock. The robins and grackles that used to be harbingers of spring in New York were now commonly foraging successfully on the lawns through our mostly snowless winters. As the males of each of the duck species that winters on the lakes and reservoir in the park completed their molt, I saw the beginnings of their courtship. By February, when most of them were paired, I looked with sympathy at the extra males that had had no luck. And also in February the titmice and cardinals began singing, weeks before either would actually start nesting.

I hope, therefore, readers will look with new insight, just as I did, at the behavior of the birds around them through the winter and will consider how many of the physical and ecological adaptations of all birds result from selective pressure during the season when survival is the priority wherever birds are. The consequences of molt and migration schedules, of rank due to age, gender, and relative body size, winter separations of competing populations and species by latitude and habitat, seasonal rainfall shifts, incidents of extreme weather, and other factors all interweave to make winter at any latitude more challenging for most birds than their breeding season. And, as the very nature of winter is now changing, from the drying of many tropical regions to the warming of polar ocean currents and the acceleration of spring events faster than some birds can shift to that season, we need to appreciate the scope of these impacts so we can, individually and collectively, take appropriate action to preserve and restore vital habitats and implement the practices that will reduce rates of climate change. Winter is too fascinating a season to lose!

ACKNOWLEDGMENTS

MY PRIMARY DEBT, of course, is to the legions of scientists who have worked through intense cold or extreme heat, unending nights, too short days, rolling seas, and long years to learn firsthand what birds are really doing during winter. Some of them I would have liked to join, especially those working in habitats that have since been degraded or lost. It is the diligence and insights of all these scientists that have provided me the material necessary to create this book. Two anonymous readers from that community made helpful suggestions on the manuscript, mainly about its structure. Any errors of fact or interpretation are my own.

In the Department of Ornithology at the American Museum, Joel Cracraft, Paul Sweet, Lydia Geratano, Peter Capainolo, and Mary LeCroy all literally and figuratively opened doors for me. In the museum's main library, Tom Baione, Mai Reitmeyer, and Annette Springer showed me through the several floors of stacks and helped me find other publications available only on the internet.

At the American Bird Conservancy, Michael Parr, Aditi Desai, and Grant Sizemore, and my former colleagues at the National Audubon Society, Matt Jeffery and Gary Langham, gave me information on current conservation activities and examples of their organizations' successes in protecting the habitats of wintering birds.

Margaret La Farge's line drawings animate these pages. I wish there could have been more!

At the Princeton University Press, my editor Robert Kirk was supportive from the very beginning and especially helpful when I needed advice on reconfiguring some of the chapters. Ellen Foos and Kristin Zodrow meticulously oversaw all the steps that metamorphose a larval typescript into perhaps a butterfly of a book. Laurel Anderton, the copy editor, spotted words and numbers in the text needing correction with the acuity of a kinglet making its way through the winter woods, finding prey no other bird could see. Thanks to them all.

CHAPTER 1

RESPONDING TO WINTER

A S LONG AS EARTH has circled the sun and tilted on its axis during the course of an annual rotation, some parts of the planet have received less heat and light for a portion of each year. This annual tilting of each pole toward and then away from the sun produces the seasons, the coldest and darkest being what we call winter.

In some earlier epochs of Earth's history, the entire planet was much warmer than it is today, with subtropical conditions extending as far as latitude 50° north, and the impacts of winter were probably minimal; at other times they were more intense.

Birds, on Earth 155 million or more years, have evolved and responded to the opportunities that these changing conditions have presented to survival and reproduction. The last five million years have been Earth's coolest, and therefore most seasonal. During the last 1.8 million years there have been at least 22 major periods of glaciation that made the entire planet colder and drier. Most of the birds species alive today lived through at least some of these oscillations, and by the end of the most recent glaciation, about 12,000 years ago, probably all of them had evolved their current form and annual life cycle.

This annual cycle, once a bird is an adult, can most easily be divided into the time spent in reproductive activities—finding a mate, nesting, and raising young until their independence—and the rest of the year, when each bird fends for itself. For the birds that must deal with winter—at least a third of the species living today—this season usually runs longer than the breeding season and is more challenging. At the higher latitudes of all continents, north and south, the features of winter include lower ambient temperatures, increased wind, and fewer hours of daylight.

These factors directly affect many of the food sources that were consumed by birds during the warmth of the breeding season. Diminished sunlight and increased cold substantially reduce or curtail production of plant material, in turn diminishing the abundance or availability of insects and other small animals that feed on plants, while snow may cover the ground and many water bodies may freeze, further limiting accessibility of food. For what food remains, there are fewer hours with enough light to forage, and longer, colder nights during which birds must fast, burning the energy they have gained through what they consumed the previous day. For small birds, those weighing less than 100 grams, insulation against cold is another challenge.

Warm-bloodedness has prevented birds from simply shutting down through an annual cold period, as can reptiles and amphibians, while the mobility that comes with flight has enabled them to search more widely for the food and shelter they need to survive, whether over permanent home ranges or more distant hospitable latitudes. Birds' responses to winter vary. Some stay in place, others migrate annually from and then back to the breeding site, while a smaller number are nomadic or have periodic irruptions when populations build beyond what the local winter food supply can sustain. Winter therefore affects not only the birds that remain to endure it and those that travel from higher latitudes to avoid it, but also the birds in regions where migrants settle, creating potential competition between residents and transients, often at a time when equatorial weather patterns also influence the availability of food.

Since migration is the response of most birds of high and middle latitudes, we will consider it first.

MIGRATION

ORIGINS OF MIGRATION

The fossil remains of birds show that some had the anatomy required for long-distance flight at least 100 million years ago. Climatic conditions have probably prompted migration in many current bird families for the past 15 million years or longer (Steadman 2005). Migration evolved among both birds of high latitudes and those of the tropics, where seasonal variations in rainfall and the resulting abundance or decline of food resources led to more local annual movements.

For birds originating at high latitudes, the eras of cold forced a retreat toward lower latitudes where exploitable habitat remained. Through the annual and long-term fluctuations in warmth, the birds were able to return poleward to breed. Their migrations to escape winter were relatively short and may not have taken place every year if birds could survive without moving. During glacial periods, the range of some birds contracted, but these birds had already evolved to exploit resources available during the cold season and were able to remain farther north than birds of tropical origin that may have been their summer neighbors.

In North America, most of the nonmigrant or short-distance landbird migrants are from families that originated in the Palearctic and crossed the land bridge that at various times of glacial maxima joined Siberia and Alaska. These include woodpeckers, corvids, tits, nuthatches, and finches, in which some species, or populations within species, are nonmigratory and others are short-distance migrants or have irregular movements in response to annual variations in food supply. For them, lifetime reproductive potential may be greater when they can remain on their breeding territory or have only a short migration to reach it; the risk is that they may not survive through the cold winter.

Long-distance migrants retreating to the tropics, in contrast, may increase their chance of survival by avoiding the cold, even if they reduce their reproductive potential the next spring by using time to establish a new territory and by missing some of the food sources the residents and earlier arrivals will already have consumed. This must, however, be an acceptable trade-off, since today, in all hemispheres, a large portion of the species breeding at higher latitudes migrate to the tropics. Some are from families that originated in the tropics, such as the uniquely Western Hemisphere hummingbirds and tyrant flycatchers. Other families that have many long-distance migrants—for example, wood warblers and blackbirds—are descended from birds that, like the ancestors of many families with short-distance migrants, came across Beringia from the Old World and then radiated into the families we know today in the Western Hemisphere. Parts of the radiation into current species may have taken place in the tropics. These families have been present long enough that some members could have evolved migratory behavior, lost it, and evolved it again several times in response to different climatic conditions. Thus, some of the migratory species living today may owe their habit to ancestors that originally retreated from cold weather at northern latitudes, while others are descended from birds that later expanded north out of the tropics (Winger et al. 2014). Since some of these families of Old World origin, such as sparrows, also include species that expanded far into the Southern Hemisphere and now migrate back toward tropical latitudes to escape the austral winter, both scenarios have been factors at different stages of their families' presence in the New World. Similarly, among Palearctic birds that go to Africa in winter, nearly all have breeding populations or close relatives there that may have been the ancestral source of the current migrants (Bohning-Gaese 2005).

Why would birds in the tropics leave? Competition for resources between species, and for breeding space within species, may, in the interglacial eras when regions farther from the equator warmed, have led some birds, especially young ones seeking their first territories, to edge their way beyond their birthplace. In the same way, on a shorter timescale, changes in habitat and climate over the last century have enabled several North American species to expand their breeding range significantly northward, with some retreating to their historical range each winter. The seasonal blooms of food sources at higher latitudes are short but intense, providing brief abundance of protein-rich food ideal for feeding nestlings; these resources, however, diminish rapidly, forcing a retreat back to warmer regions. During the warmer interglacial periods, such as today, the distance between exploitable breeding habitat and viable wintering areas was longer than in the glacial phases when cold reduced how far tropical birds could spread from their place of origin. The breeding ranges of many birds expanding from the tropics therefore went through periods of growth and contraction, while the winter range to which they returned each year may have been more stable.

Many tropical birds have annual seasonal movements that make it a short evolutionary step to long-distance migration. The Yellow-green Vireo (*Vireo flavoviridis*), for example, breeds from northern Mexico to Panama; in Panama, most have left by the end of August for the upper Amazon basin, returning in December. Their diet is small fruits, and their migration is linked to their food's annual cycle of abundance and decline (Morton 1977). Other intertropical migrants do not traverse significant latitudes; many, like the Resplendent Quetzal (*Pharomachrus mocinno*) of Mexico and Central America, move up and down mountains following the timetable of fruiting plants at different elevations. From this intratropical origin, it was an easy evolutionary step during interglacial periods for some birds within the tropics to expand their ranges north or south as changes in climate prompted the growth of food sources not previously available (Rappole and Tipton 1992).

The 1.8 million years of glacial and interglacial periods have given birds of both tropical and temperate origin time to evolve complex variations of these general patterns, refined by modern species in the last several thousand years when Earth's climate has been most like today's. The Western Hemisphere thrush genus *Catharus*, for example, contains 12 species; 7 are restricted to the Neotropics, where the genus originated, and 5 breed in North America: the Bicknell's (*C. bicknelli*), Gray-cheeked (*C. minimus*), Hermit (*C. guttatus*), and Swainson's (*C. ustulatus*) Thrushes and the Veery (*C. fuscescens*). All but the Hermit Thrush return entirely to various parts of the tropics in winter; the Hermit Thrush is a short-distance migrant wintering primarily from the central and southern United States to Mexico and Guatemala.

TYPES OF MIGRATION

Migration and other less regular large-scale movements to avoid winter take many forms. These reflect the origins of each species as well as its adaptations to different habitats over the course of a year.

PARTIAL MIGRATION   Many birds have both migrant and resident populations. Some individuals move to lower latitudes, while others remain in the breeding range year round. In the Northern Hemisphere, the entire high-latitude population may shift southward into the permanent range of the southern population and beyond, or some individuals from all parts of the range may move while others remain. Often, adult males stay closest to the breeding site, the better to regain their territory for the following year, while females and young move farther (Gauthreaux 1982). In parts of the range where birds are found all year, it is impossible without marking every bird to determine which are residents and which are migrants. The American Robin (*Turdus migratorius*), for example, leaves

Alaska and most of Canada for the winter and ranges as far south as Guatemala. Some years, more birds retreat, and some may go farther south in other years. Is this a stepwise migration, with all birds moving southward to some degree, or are some robins traveling various distances and others not moving at all? The term "partial migration" covers many variations in migratory behavior.

SHORT-DISTANCE MIGRATION   The entire population moves out of the breeding range, but usually the distance between the limits of the breeding and winter ranges is less than 2000 km. McCown's Longspur (*Calcarius mccownii*) and Chestnut-collared Longspur (*C. ornatus*) have overlapping breeding ranges in the northwestern Great Plains and southern Canadian Prairie Provinces, and both winter from Oklahoma and Texas south into northern Mexico. They have to move south in winter to reach latitudes where the grasslands and deserts are not covered with snow. Similarly, the Rusty Blackbird (*Euphagus carolinus*), which breeds in bogs and muskeg swamps from maritime Canada to Alaska, vacates that vast range for the central and southern United States from the Atlantic to about the 95th meridian, where swamps, wet woodlands, and pond edges are less likely to be frozen (Avery 1995).

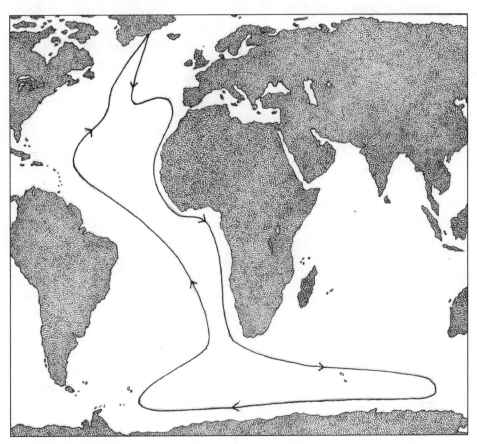

Recent tracking of Arctic Terns has revealed that, for some, the extent of winter travel is even greater than previously believed. Some terns from Greenland spend part of the austral summer far in the Indian Ocean. (redrawn with permission from Egevang et al. 2010)

LONG-DISTANCE MIGRATION   The entire population moves, usually more than 2000 km, often to a different continent. As with the short-distance migrants, their destinations are determined by the reliable availability of food to sustain them through winter. The most extreme example is the Arctic Tern (*Sterna paradisaea*), which nests around the globe almost entirely above latitude 59° N (some almost at 84° N) and winters at sea south of 58° S, primarily at the edge of the pack ice around Antarctica; some birds from Greenland colonies go south through the Atlantic and spend part of the winter in the Indian Ocean before returning to the South Atlantic. They travel 80,000 km each year and spend several months at both ends of their migration in perpetual daylight (Egevang et al. 2010). Unlike most terns, Arctic Terns feed substantially on crustaceans as well as fish; in their high-latitude nesting and wintering areas crustaceans are abundant, and the Arctic Tern avoids competition with several close relatives in the lower latitudes of all oceans (Hatch 2002).

ALTITUDINAL MIGRATION   Some birds, as already noted, make an altitudinal migration. Appalachian and some other mountain populations of the Dark-eyed Junco (*Junco hyemalis*) descend to lower elevations in winter, while more-northerly juncos move south; where two junco subspecies occur together during winter in the Great Smoky Mountains, most local breeders settle higher on the slopes than the birds from farther north (Rabenold and Rabenold 1985). In Bolivia, where from the Andes to the Amazon lowlands 34 flycatchers are found year round, 15 shift their elevational range 500 m or more, one as much as 2000 m. Other flycatchers there are partial migrants, in which a portion of the population shifts downward for winter, either withdrawing from the upper elevations of the breeding range or moving to elevations below the breeding range (Chesser 1997).

In addition to the distance traveled by each species, the annual movements must be considered in terms of which population goes where. For any migratory bird, competition for winter resources is reduced if the population is spread through the largest possible winter range. This can be achieved in different ways. Among Northern Hemisphere birds, all may move southward stepwise, more or less the same distance, rather than to the same destination, so that the more northerly breeding population remains farther north during winter even after all birds have shifted southward. Other birds, such as the Common Yellowthroat (*Geothlypis trichas*), have a leapfrog migration, in which the northernmost population goes farthest south; migrant yellowthroats, in fact, also "leap" over a resident population in California and the southeastern states. Birds that have a wide east-west range may move due south—if geography allows—in a parallel pattern, or they may cross each other.

In many species, birds from different parts of the breeding range mix during winter. This is especially likely to take place where the winter range is much smaller than the breeding range, as for the warblers that nest in the vast boreal forest of North America and winter in Central America and islands of the Caribbean.

In other species, however, there is a strong link between the breeding and winter ranges of certain populations, called "migratory connectivity," discussed at length in chapter 3. Long-billed Curlews (*Numenius americanus*) nesting in Oregon, Nevada, and Montana winter almost entirely in separate regions. Oregon birds go to the Central Valley of California;

a few Nevada birds go there as well, but more go to Baja California; and Montana birds winter on the Mexican Plateau, in the Texas Panhandle, and near the Laguna Madre on the Gulf of Mexico (Page et al. 2014).

While selection pressures may be significant in any part of the annual cycle, for most birds the survival challenges of winter are greater than in the breeding season, which is usually much shorter. Among long-distance migrants, winter selection has the greatest evolutionary impact on species with high migratory connectivity, when the breeding population of a species in one region, already somewhat genetically isolated, also winters together, thereby subjecting all birds to that season's local pressures as well. Breeding populations that disperse in winter and meet conspecifics from other regions will be less affected by distinctive selection pressures, because not all members of a regional breeding population will be exposed to the same selection factors during any winter.

Impacts of winter selection may be especially strong on birds that spend more of the year at their wintering sites than in their breeding range, such as Arctic-nesting shorebirds and geese. Even many birds breeding in temperate latitudes spend most of the year where they "winter," although winter, as we think of it with colder weather, may be at most only a fraction of their time there. Long-billed Curlews that nest in Oregon return to the Central Valley of California as soon as mid-June, remaining there until the following March or April (Page et al. 2014). The same is true for many birds going to tropical regions where there is no cold at all: Kentucky Warblers (*Geothlypis formosa*) and Louisiana Waterthrushes (*Parkesia motacilla*) that nest in the southeastern United States arrive in Veracruz, Mexico, by late July or early August and stay until late March (Rappole 1995:79).

Kentucky Warblers "winter" in Veracruz, Mexico, from late July until late March, making this their home for far more of the year than their breeding range in the United States.

Body size is one of the most widespread aspects of winter selection pressure. Within any species of bird (and many other animals), overall body size varies among populations based on latitude, usually with those of colder regions being larger, because larger bodies retain heat more efficiently. Nonmigratory Hairy Woodpeckers (*Dryobates villosus*) resident in Alaska are half again larger than those in Mexico. In migratory Canada Geese (*Branta canadensis*), however, the smallest races nest farthest north. They have a leapfrog migration, with the most northerly races wintering farthest south, on the Gulf Coast and in Mexico. For them, there is evidently a selective advantage to being smaller, so their body dissipates heat more easily during the many months of the year they spend in relatively warm latitudes.

OTHER SEASONAL MOVEMENTS

Not all birds that leave their breeding area in winter have a regular latitudinal or altitudinal migration. In some species, notably certain seabirds, populations disperse broadly from their colonies over a wide area of water or coast. Adult males usually remain closest to their nesting site and are the first to return at the end of winter, while adult females and young birds move farther; young birds sometimes eventually settle at other colonies. In passerines, winter dispersal may be only a few tens of kilometers within the breeding range as birds search for better short-term feeding sites in places that may not be suitable for breeding (Newton 2008:513–514).

Dispersal may be an appropriate response to some local winter conditions but not others. Among Herring Gulls (*Larus argentatus*), young birds from around the Great Lakes disperse during their first winter throughout eastern North America from James Bay to the Gulf Coast and Mexico, while first-year birds born on the Atlantic coast move south as far as the Gulf, and the birds of the St. Lawrence region are resident (Eaton 1934).

Some northerly birds can survive most winters where they breed but must move if their food source fails. Irruptions bringing these birds sometimes far south of their usual range happen every few years, for some species with more regularity than for others. Archetypal irruptive species include finches like crossbills (*Loxia* spp.) and redpolls (*Acanthis* spp.), and the Snowy Owl (*Bubo scandiaca*). Their cyclical foods precipitate these movements: tree seeds for the finches, and rodents, especially lemmings, for the owl.

Trees of the boreal forest produce different quantities of seeds each year, usually on a two-year cycle. A study correlating 30 years of Christmas Bird Count tallies, breeding bird surveys, and cone and seed crop data in North America found that the irruptions of most finches occur in years when poor crops follow a year with a large conifer seed crop. For Common Redpolls (*Acanthis flammeus*) and Black-capped Chickadees (*Poecile atricapillus*) in the West, however, irruptions followed directly after large seed crops, not after subsequent crop failures. In no cases were irruptions linked with milder winters that led to higher overwinter survivorship, nor did severe winters result in abandonment of normal winter range, or warm conditions in the breeding season generate greater reproductive success (Koenig and Knops 2001). When seeds are easily available, bird populations usually increase and remain; when the food source drops off or has been consumed, these expanded bird populations must move elsewhere.

"Winter finches"—the various species that in some years appear in numbers well south of their breeding range—shift their diet in late summer or autumn, when seed crops are maturing; their movements may begin long before winter if local resources are poor (Bock and Lepthien 1976). Crossbills, which feed on the seeds in pine and spruce cones, will wander widely when a cone crop diminishes or fails. As flocks search for food over long distances, they may even nest in midwinter if they find a reliable source. Both species of redpolls depend on birch seeds and may also move far from one breeding area to another. Common Redpolls banded one winter in New England appeared in Alaska the next winter, and one banded in Michigan was recovered a year later near Okhotsk in eastern Siberia, 10,200 km distant (Troy 1983). Another found in Belgium one winter was in China the next, 8350 km away (Newton 2008:541). Where any of these birds nested during the intervening summer is unknown.

Other small birds of the northern forest are also irruptive, often in synchrony, even though their diets do not overlap. General environmental conditions in these forests may lead to good or poor years for both plants and insects. An analysis of Christmas count data from 1959 through 1997 showed strong two-year synchrony, especially in eastern North America, among Red-breasted Nuthatch (*Sitta canadensis*), Black-capped Chickadee, and Pine Siskin (*Spinus pinus*)—but only during some decades. Invasions of some other irruptive species, however, from the insectivorous Golden-crowned Kinglet (*Regulus satrapa*) to the frugivorous Bohemian Waxwing (*Bombycilla garrulus*) and the bud- and seed-eating Purple Finch (*Haemorhous purpureus*), did not correlate strongly with the movements of any other species (Koenig 2001). Another study found little correlation between North American irruptions of several highly synchronized species in boreal forests and populations of the same species in similar forests at high elevations in the western United States (Bock and Lepthien 1976). It will take even longer-term studies, including more data on changes in food sources in different parts of the breeding range, to reveal the causes and patterns of these irruptions.

Snowy Owls preying on Arctic rodents—which themselves respond to food availability on the ground—go through cycles of abundance and die-off averaging three to five years. A study of Snowy Owl nesting productivity from 1994 to 2011 on Bylot Island in the Canadian High Arctic and Daring Lake, Northwest Territories, in the Low Arctic found that winter irruptions were correlated with the most successful breeding seasons, which create the greatest potential competition for food the following winter. Since young Snowy Owls do not themselves breed for several years, they have less incentive to remain near their breeding range. By going south they spend at least their first winter in milder conditions, perhaps with more easily available prey, and with less competition from adults (Robillard et al. 2016).

The Snowy Owl is the most conspicuous and regular of owls in cyclical irruptions, but it is not unique among northern rodent predators. When necessary, Boreal Owls (*Aegolius funereus*) and Northern Hawk Owls (*Surnia ulula*) move in winter, mainly within the boreal forest zone. Short-eared Owls (*Asio flammeus*) and Long-eared Owls (*A. otus*) wander south in irregular numbers each winter, returning north the following spring and settling wherever voles are plentiful. Larger raptors that prey on snowshoe hares (*Lepus americanus*) move in response to the hare's roughly 10-year cycle; Great Horned Owls (*Bubo virginianus*) and Northern Goshawks (*Accipiter gentilis*) may irrupt for as many as three winters running while hare populations are low (Newton 2008:579).

MIDWINTER AND FACULTATIVE MIGRATION

At middle and high latitudes, many short-distance migrants live where weather may change suddenly and dramatically, with freezing temperatures, snow, and ice transforming the landscape from one day to the next and eliminating food or preventing birds from finding it. Facultative migration, the decision a bird makes to move weeks after it has finished its normal autumn migration and reached a destination within the species' usual winter range, may come irregularly rather than annually, in response to unpredictable conditions during the severest part of an exceptional winter. Some individual birds capable of facultative migration may not experience the need for it during their lifetime.

These sudden transformations of the landscape may provoke substantial flights of birds that are otherwise scattered across a broad region. On December 28, 1964, 20,000 Eurasian Skylarks (*Alauda arvensis*) were counted passing westward over the Axe Estuary in Devon, England, and on January 30, 1972, a group of 10,200 Northern Lapwings (*Vanellus vanellus*) flew over Portsmouth, Hampshire, in 2.5 hours. There are no comparable figures from North America. In Europe, this mass evacuation phenomenon has become less frequent in recent decades as winters have grown milder (Newton 2008:469–470). But anywhere that water bodies may freeze rapidly in midwinter, ducks or geese first become concentrated in the diminishing area remaining open, and then they disappear overnight—with an influx farther south where water is open. Often, also flying at night, these birds will return a few days later when conditions become more favorable, thereby saving themselves from continued competition in the densely packed area where they found refuge with other waterfowl already present.

The Eurasian Golden-Plover (*Pluvialis apricaria*) is normally a short-distance migrant, with birds from Scandinavia wintering in northwestern Europe, where they feed primarily in pastures and agricultural fields, foraging for earthworms and insects. Plovers from Swedish Lapland tracked in 2011–2012 with light-level geolocators began their winter in the Netherlands, Belgium, and northwestern France. When temperatures in some parts of that region dropped below freezing between January 30 and February 1, the plovers there moved within a day to southern Spain and Portugal, and some then moved in the next day or two to Morocco. They remained at these locations until March, while other tracked plovers that were in England and western France, where the weather was milder, stayed there all winter (Machin et al. 2015).

Facultative migration has been well studied in a few passerines. In the area around Phoenix, Arizona, Yellow-rumped Warblers (*Setophaga coronata*) commonly winter in isolated riparian valleys with permanent water that are surrounded by the Sonoran Desert and agricultural lands. Here, they feed on insects on foliage, but occasional cold fronts that generate freezing temperatures, high winds, and hail may cause the leaves to fall from all trees. When this happens, there is a substantial decrease in warbler numbers around Phoenix—at one site where the warblers were counted all winter, from 285 birds/40 ha to 34 birds/40 ha in 11 days—with a corresponding influx in northern Sonora, Mexico, 325 km southeast, which had 11 birds/40 ha and 10 days later had 90 birds/40 ha (Terrill and Ohmart 1984).

Further evidence of facultative migration in Yellow-rumped Warblers, as well as proof that the birds migrate at night, as they do in their regular migration, comes from Florida, where midwinter counts of warblers killed at night by hitting a radio tower in Leon County correlated with cold fronts. During seven mild winters, the number of Yellow-rumped Warblers killed between December 15 and February 20 was as low as 5; in the severe winter of 1962–1963 it was 264 (Terrill and Crawford 1988). In that winter, this species was common in Jamaican lowland forests and mangroves into April, while in other winters it is conspicuously absent (Gochfeld 1985).

Experiments with two regular facultative migrants showed how this type of migration is triggered. Dark-eyed Juncos held in captivity during winter showed little or no restlessness at night when given large amounts of food. But when the quantity was restricted, their nocturnal activity paralleled migratory restlessness typical of autumn. This could be induced during December and January, but after late January, the migratory behavior could not be reactivated with food reductions comparable to what birds had received earlier in the winter. Similarly, Garden Warblers (*Sylvia borin*) deprived of food in midwinter will show the same nocturnal activity as during their autumn migration season, and they will cease this activity as soon as food is returned and they begin to regain weight. In winter, however, unlike in autumn, they will not seek to migrate again after they have returned to normal weight (Terrill 1990).

For some birds, facultative migration may continue until the end of winter. Altitudinal migrants, which are much closer to their breeding sites in winter than are latitudinal migrants, may go back and forth depending on weather conditions. Yellow-eyed Juncos (*Junco phaeonotus*) in the Chiricahua Mountains of southern Arizona nest as high as 2560 m and commonly descend to winter 8 or 10 km away where the elevation is 1000 m less. In late winter, they return to higher elevations to stake out or reclaim territories even though they will not begin nesting for another several weeks. After storms, however, some individuals depart again for their wintering area at lower elevations as soon as the next morning. Within a few days, they are back in their breeding territory (Horvath and Sullivan 1988).

MAJOR MIGRATION SYSTEMS

On every continent, substantial numbers of species migrate to avoid winter at high latitudes. Some 318 North American landbird species move to the Neotropics, from the Palearctic 185 go to Africa, and South Asia receives 338 species from the eastern Palearctic. To these numbers must be added the many migratory species that stay in temperate latitudes through the winter. In South America, at least 220 species of all kinds move northward for the austral winter, while in Australia about 40% of landbirds migrate, and the movements of other birds are governed by irregular patterns of rainfall, resulting in more nomadism than classic migration. In Africa's midbelt south of the Sahara, seasonal rains rather than significant shifts in temperature also govern most migration; 40% of West African landbird species, for example, may move within the tropics (Rappole 2013:260–269). Farther south in Africa, many birds move north during the austral winter. In each region, the migratory response of landbirds to winter reveals a different pattern set by climate, geography, and evolutionary history. Similarly, seabirds that touch land only to breed move the rest of the year through different parts of the global ocean to reach areas where seasonal resources are abundant; this is most striking in the many shearwaters and petrels that breed at high latitudes in the Southern Hemisphere and avoid winter there by migrating to the North Atlantic and North Pacific, where it is summer.

NEARCTIC/NEOTROPICAL

Among landbirds of North America, most of the migrants to the Neotropics breed in forests. When foliage that sustains their arthropod prey falls off the trees in deciduous forests and cold reduces similar prey in coniferous forests, these birds migrate to tropical and semitropical latitudes where arthropods remain reliably available—even as some of these birds broaden their diet to include fruit and nectar, which are scarce or absent farther north. Most of these forest birds concentrate in the latitudes where they may have originated: between southern Mexico and northern South America, as well as on Caribbean islands, seeking habitat that is at least superficially similar to the woody vegetation they use in spring and summer. Mexico and the Pacific slope and mountains of Central America are the winter home of many of the birds breeding in western North American grasslands, scrub, and coniferous forests.

The total winter range harboring all these birds is approximately one-seventh the landmass of their collective breeding range. For each species, the proportion will be different, but the overall effect is to crowd many more birds together in a region that has even more year-round residents. This has produced fine ecological and behavioral partitioning, which will be discussed in subsequent chapters.

PALEARCTIC/AFRICAN

Far fewer birds breeding in Eurasia winter in Africa, only about 185. Consider the differences between continents in two basic ecological niches: of comparable insect-eating foliage gleaners, in winter the Neotropics hold 46 migrant wood warblers (Parulini) while Africa has 29 migrant warblers (Sylviini), and among fly-catching birds, 23 tyrant flycatchers (Tyrannidae) return to the Neotropics and only 3 Old World flycatchers (Muscicapini) (Gauthreaux 1982:136). North of Africa's savanna belt, the Sahara

Desert and the Mediterranean Sea have been formidable barriers to expansion for birds originating in Africa. Unlike the essentially forest birds that spread north from the Western Hemisphere tropics, birds of African grassland and scrub were the successful colonizers of Europe and western Asia where they found similar habitat. Of these Eurasian colonizers returning to winter in Africa, 42 (23%) have conspecific populations breeding in tropical Africa and 97 (52%) have congeners (Rappole 1995:124, 129). Nearly all the Palearctic landbirds wintering in Africa live in savannas. Today, only three Palearctic breeders—the Pied Flycatcher (*Ficedula hypoleuca*), Collared Flycatcher (*F. albicollis*), and Wood Warbler (*Phylloscopus sibilatrix*)—winter primarily in forested parts of Africa.

The winter range of many migrants to Africa, however, is not static for all the months they are there, because the rains that bring a flush of insects shift southward between September, when the first birds arrive south of the Sahara, and March. Many wintering birds therefore move long distances every several weeks to reach another latitude where the rains are beginning. The birds eventually turn north to return to the Palearctic. This contrasts with the Neotropical pattern, where most passerine winterers remain in the same location for the entire season; here, too, wet and dry seasons shift during the course of the winter, but enough food of one kind or another remains, favoring a stable residency.

PALEARCTIC/SOUTH ASIAN

This is the least known of migration corridors. Its most distinctive feature is the absence of desert and major water barriers between breeding and wintering areas. Most of the long-distance migrants are forest birds that travel east of the Himalayas. The Southeast Asian forests are highly seasonal in rainfall, but we lack information on how that affects the movements or other adaptations of wintering birds.

AUSTRAL SOUTH AMERICAN

At least 220 South American birds migrate to avoid the austral winter (Chesser 2005). About one-third are tyrant flycatchers (76 species), the most numerous and diverse family of Neotropical origin. Ducks and geese (17 species), swallows (9 species), and sparrows and finches (22 species) are other major components (Chesser 1994). Most return from higher latitudes only to the temperate regions of the continent, south of the Amazon basin. They winter in habitats similar to where they bred, often occupied by conspecific nonmigrant populations. Very few have migrations as long as most North American birds that winter in the tropics. And, while the North American birds are compressed into a region much smaller than their collective breeding range, the bulge of South America makes for more extensive potential wintering areas for the austral migrants from the continent's tapering south end. Finally, as in the Palearctic–South Asian migration, South American birds encounter no significant barriers to their annual movements, thus facilitating range expansion of birds that originated in the tropics.

Only a few South American migrants regularly cross the equator. Not surprisingly, these are all very able fliers. The Argentinean race of the Brown-chested Martin (*Phaeoprogne tapera fusca*) is abundant in Panama during the austral winter, Blue-and-white Swallows (*Notiochelidon cyanoleuca patagonica*) from Patagonia are sometimes common in Panama and have also been found in Nicaragua, and the Argentinean subspecies of the Ashy-tailed Swift (*Chaetura andrei meridionalis*) has been recorded in Colombia and Panama in August (Eisenmann 1959). In Suriname, Argentinean races of the Gray-breasted Martin (*Progne chalybea domestica*) are common, and the Southern Martin (*P. modesta elegans*) is occasional between May and early September (Eisenmann and Haverschmidt 1970). Most of the populations of these two subspecies, however, winter in the Amazon basin, where in March and April and from late July to September they overlap with Purple Martins (*P. subis*) from North America (Eisenmann 1959).

Among the flycatchers, the best-studied group of austral migrants, more depart from northern Argentina than from any other region, and the greatest number of wintering flycatchers is in and around southwestern Amazonia. Most of the migrant flycatchers both nest and winter in edge, secondary, and scrub habitats, suggesting that it was residents in these habitats that spread south to exploit comparable seasonal environments there. Even the migrants that return to central Amazonia in winter seek out similar microhabitats such as forest edges, second growth, and lake margins rather than the interior of primary forest, which covers most of this vast region and is occupied year round by nonmigrant flycatchers. The White-throated Kingbird (*Tyrannus albogularis*), for example, breeds in Bolivia and southern Brazil in shrubby areas and edges of gallery woodland and winters, among other open places, on islands in Amazonian rivers (Chesser 1997).

The Slaty Elaenia is one of the few South American austral migrants to cross the equator each winter, traveling from Bolivia and Argentina to northern Venezuela.

Only 14 of the austral migrant flycatcher species have entirely disjunct winter and breeding ranges (Stotz et al. 1996:75–77). Among these, the Slaty Elaenia (*Elaenia strepera*) breeds in Andean forests in southern Bolivia and northern Argentina and winters 3000 km north in the hills and mountains of northern Venezuela (Marantz and Remsen 1991). Other Andean flycatchers remain in the Andes, shifting to similar habitat farther north, where, closer to the equator, their preferred zone may be at higher elevations. Thus, while in the Northern Hemisphere most altitudinal migrants move downslope in winter, the Cinnamon-bellied Ground-

Tyrant (*Muscisaxicola capistratus*), which breeds in southernmost Chile and Argentina mostly from sea level to 500 m, winters in the Andes as far north as southern Peru, primarily between 2000 and 4000 m (Ridgely and Tudor 1994). The South American migrants that travel farthest are the White-crested Elaenias (*E. albiceps*) of southern Chile. This species breeds from the Andes of southern Colombia to Tierra del Fuego. Elaenias from Navarino Island, at 54° S in the world's southernmost forest, were tracked with light-level geolocators and were found to winter in central Amazonia at 6–8° S. The average distance between their breeding and wintering grounds was 5932 km (Jiménez et al. 2016).

AUSTRAL AFRICAN

In southern and south-central Africa, the rainy season, October to April, is when all local birds breed, taking advantage of the flush of insects and other protein sources to feed their young. These summer months are also when Palearctic winterers, from storks to raptors to swallows, are present in large numbers, evidently not impinging on the needs of the breeding birds when these are greatest. During the dry season of the austral African winter, the Palearctic birds return to the Northern Hemisphere for their own breeding season, and many local birds also move northward. Of the 677 breeding landbirds of southern Africa, 53 leave the region entirely, and another 79 are partial migrants (Dowsett 1988a). Farther north, in tropical Zambia and Malawi, there are 676 breeding birds of all kinds; 61 of these depart entirely, and 39 are partial migrants (Dowsett 1988b). The greatest share of the birds moving closer to the equator in winter are aerial insectivores—swifts and swallows—and cuckoos, which consume large-bodied insects.

Among the partial migrants that do not entirely leave southern and central Africa, some move relatively short distances, or up and down slopes. In the Starred Robin (*Pogonocichla stellata*), adult males remain year round on their montane forest territory above 1300 m; in April, females and immature males descend to lower elevations, sometimes traveling as far as 120 km from the nearest breeding locale (Harrison et al. 1997). In south-central Africa,

The Torres Imperial Pigeon retreats in winter from Australia to New Guinea.

during February through August some Black-headed Orioles (*Oriolus larvatus*) move from the lowlands to mountains as high as 2000 m (Dowsett 1988b).

Madagascar has been isolated from Africa for at least 120 million years. Of its 201 current resident species, only 3 are definitely known to migrate to eastern Africa for the winter, the Malagasy Pond Heron (*Ardeola idae*), Madagascar Pratincole (*Glareola ocularis*), and Broad-billed Roller (*Eurystomus glaucurus*), while several insectivores not restricted to forest descend from the island's High Plateau (Langrand 1990:12–14).

AUSTRALIA AND NEW ZEALAND

On the Australian mainland, ranging from temperate in the south to tropical in the north, about 40% of the landbirds migrate. These include approximately 70 nonpasserines and 100 passerines. Most species have both resident and migrant populations (Chan 2001). A few cross water to winter in New Guinea and Indonesia. The Torresian Imperial Pigeon (*Ducula spilorrhoa*) departs northern Queensland between February and April for New Guinea, returning from there in August to October; a few individuals remain in Australia all through the year (Higgins and Davies 1996:1002–1003). The Sacred Kingfisher (*Todiramphus sanctus*) has a much longer migration, from as far as southern Australia to New Guinea, the Solomon Islands, Indonesia, and Borneo. In New Zealand, where winter is much cooler, however, Sacred Kingfishers do not make the long water

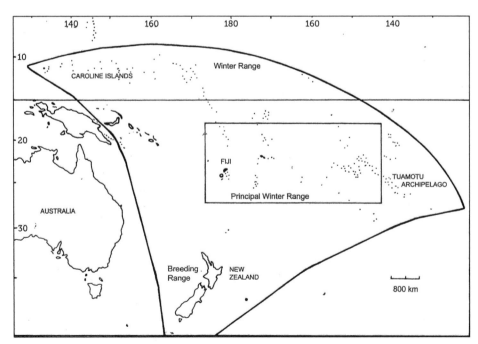

Long-tailed Cuckoos winter over a vast range of islands in the Pacific. Some individuals move westward from island to island during the course of the winter. (reproduced with permission from Gill and Hauber 2012)

crossing to Australia but simply move from higher elevations to the coast (Higgins 1999:1181–1186).

At least 20 Tasmanian species fly across the 160 km Bass Strait to Australia for the winter, but only 4 completely vacate the island (Chan 2001). As in Africa, consumers of large insects (4 cuckoos) and aerial insects (5 swallows and woodswallows) are prominent among the ecological specialists that retreat for winter. Four Tasmanian parrots are partial migrants. The nearly extinct Orange-bellied Parrot (*Neophema chrysogaster*) breeds only in southwestern Tasmania; most of the birds winter near the coast in Victoria and South Australia, where they feed in coastal salt marshes and nearby agricultural grasslands (Higgins 1999:560–563).

New Zealand is the winter home of many Arctic shorebirds, but most indigenous birds remain on the islands year round. Some move northward for winter. The South Island Pied Oystercatcher (*Haematopus finschi*) breeds inland on South Island and southern North Island; in January and early February most birds move to coastal estuaries in the northern part of North Island, some birds traveling as much as 930 km, returning to the same wintering sites year after year. Most remain there until late July and early August

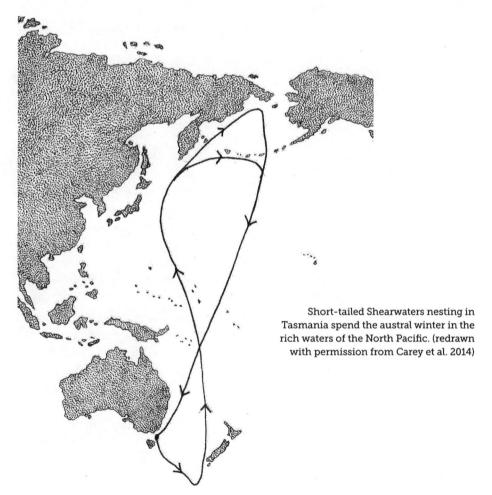

Short-tailed Shearwaters nesting in Tasmania spend the austral winter in the rich waters of the North Pacific. (redrawn with permission from Carey et al. 2014)

(Marchant and Higgins 1993:728). The Double-banded Plover (*Charadrius bicinctus*) breeds widely from coastlines to mountains on both islands; those nesting above 600 m, where there is often snow cover in winter, migrate. South Island birds go to southern and eastern Australia, where individuals return annually to the same coastal sites. North Island birds stay on the island, with those nesting inland moving to the coast, where they join sedentary resident plovers (Marchant and Higgins 1993:849–851).

One New Zealand bird, the Long-tailed Cuckoo (*Eudynamys taitensis*) has a unique migration. It breeds only in New Zealand, where, like many cuckoos all over the world, it lays its eggs in the nests of other birds, to be raised by foster parents. The Long-tailed Cuckoo therefore needs to spend relatively little of the year in New Zealand, since it is neither incubating eggs nor feeding young. Adults are in New Zealand only from October through December. They then depart north and northeastward across the Pacific, to winter on islands from Henderson Island in the Pitcairns westward to Palau in Micronesia, an arc of 11,000 km. Some adults start at the eastern end of the arc, moving westward from July until October, from one small island to others, aided by prevailing winds. In the principal winter range, the center of this arc, the islands have cuckoos all through the year, suggesting that young birds, which do not arrive there until March, remain until they are two years old before returning to breed in New Zealand (Gill and Hauber 2012).

SEABIRDS

Seabird responses to winter are guided by ocean currents, water masses, and climatic zones. In the Northern Hemisphere, many seabirds of high latitudes in both the Atlantic and Pacific, such as fulmars and alcids, remain at sea not far from their nesting colonies. None go as far south as the warmer, equatorial belt where the far fewer nutrients at the base of the marine food chain could not produce harvestable resources for the millions of birds of northern colonies. The North Atlantic and North Pacific can sustain these birds all through the year, as well as additional millions of austral migrant shearwaters and petrels wintering there during the Northern Hemisphere summer.

A study using global location sensors tracked Short-tailed Shearwaters (*Ardenna tenuirostris*) from the Furneaux Group of islands north of Tasmania, where 23 million shearwaters nest. The devices were affixed to 27 birds when they were in their burrows and were retrieved when they returned the following season. Nonbreeding shearwaters leave the islands early in March, followed by breeding adults in April and May, and then fledglings. The sensors revealed that from the Furneaux Group at latitude 40° S, the tagged birds went first to a stopover region south of the Antarctic Polar Front and north of the Ross Sea, between latitudes 60° and 70° S and longitudes 150° and 180° E. Their average stay was 26 days. From there, they traveled north rapidly across the warm equatorial belt, at a mean of 10,920 ± 732 km in only 13 ± 1.5 days (traveling speed 840 ± 121 km per day). Moving northwest, the shearwaters reached latitude 30° N off the coast of Japan. From there, some birds continued north off the Japanese coast while others went to the Bering Sea west of the Pribilof Islands. All birds spent the austral winter in these waters between

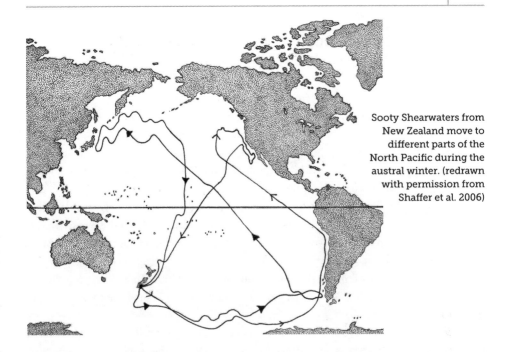

Sooty Shearwaters from New Zealand move to different parts of the North Pacific during the austral winter. (redrawn with permission from Shaffer et al. 2006)

latitudes 40° and 50° N, for a mean time of 148 ± 9 days. They began their return flight in mid-September to early October, flying southwest over the Central Pacific and arriving 18 ± 2.6 days later, with an average traveling speed of approximately 692 km per day. The average travel distance of this annual migration was 59,600 ± 15,700 km (Carey et al. 2014).

A similar study of Sooty Shearwaters (*Ardenna grisea*) nesting in New Zealand used miniature archival geolocator tags to reveal how this species exploits different parts of the Pacific during the austral winter. Sooty Shearwaters from New Zealand fly across the entire Pacific Ocean in a figure-eight pattern that totals 64,037 ± 9,779 km over 262 ± 23 days. The birds leave their nesting colonies in early April to travel eastward, some as far as the coast of Chile, to fuel up before a rapid northwesterly crossing of the equatorial belt. At about latitude 30° N, some shearwaters continue to the waters east of Japan and Kamchatka, others go to the Gulf of Alaska, and the balance to the waters off the Pacific Northwest and California; each of the tagged birds stayed in a single region. They return to New Zealand after traveling southwestward from the Central Pacific. The birds take advantage of trade winds on the long legs of their movements, traveling as rapidly as 910 ± 186 km per day (Shaffer et al. 2006).

Australasian Gannets (*Morus serrator*) of New Zealand use yet another part of the ocean during winter. Birds equipped with geolocator sensor tags left their colony on North Island in mid-March and early April and went west across the Tasman Sea to the eastern and southern coasts of Australia. They spent two to four months in Australian waters before returning briefly to their colony and then setting off again for the same destination or remaining off the east coast of North Island, finally returning to nest in September and October. Total travel distances ranged from 5800 to 13,000 km (Ismar et al. 2011).

CONTINUING EVOLUTIONARY RESPONSES OF MIGRANTS TO CHANGING WINTER CONDITIONS

These migratory systems are the result of long-term evolution, often over millions of years. But birds are also very responsive to changing conditions. For some birds, migration patterns and wintering locales have shifted in historical times, as is easily seen in North America, and today such shifts are visible even over decades as a result of changes in habitat and climate. These indicate how rapidly birds can adapt and evolve both physically and behaviorally to new winter conditions.

Acquisition of a new winter destination can come within a few generations through individual learning. The mid-twentieth-century winter expansion of Tundra Swans (*Cygnus columbianus*) to Britain from the Netherlands was probably accelerated because young birds travel with their parents. Similarly, an experiment that placed eggs of sedentary British Herring Gulls in nests of migratory Lesser Black-backed Gulls (*Larus fuscus*) found that some of the young Herring Gulls migrated to the winter destination of their foster parents (Lack 1968).

In North America, two hummingbird species have extended their winter range significantly over the last half century. Since the 1970s, the Rufous Hummingbird (*Selasphorus rufus*) has become common around the Gulf and Atlantic coasts north to South Carolina. The distance between its breeding range in the Pacific Northwest and this region is the same as to its traditional wintering area in western Mexico, so the new migration pattern involves only a change of orientation, which might have occurred as a *dis*orientation in a few birds as long as Rufous Hummingbirds have existed. But the conversion of so much of the Gulf region's original hardwood and pine forests to open lands and suburban gardens, and, since the mid-twentieth century, the proliferation of hummingbird feeders, has enabled these out-of-place hummingbirds to survive and their genes to spread. Because hummingbirds migrate singly—with adult male Rufous arriving on the Gulf Coast by late July, then immatures of both sexes, and finally adult females—learning from parents is unlikely (Hill et al. 1998).

Still more recent is the expansion of wintering Ruby-throated Hummingbirds (*Archilochus colubris*) over much of the same area from their usual winter range of central Mexico to Panama. Climate change, which has accelerated since the earlier expansion of the Rufous Hummingbird, as well as the availability of flowering gardens and feeders, has likely enhanced overwinter survival. A banding study in South Carolina found more males, especially immatures, than females, and the average return rate from one winter to the next was 19.4% (Cubie 2014).

The best-studied recent shift in winter destination is of the Blackcap (*Sylvia atricapilla*) in Europe. The traditional winter range of Blackcaps breeding from the British Isles east to Austria has been in Iberia and Africa. Beginning in the 1960s, Blackcaps have been common winterers in England and Ireland. Banding confirmed that these were not local nesters becoming nonmigrants but birds from central Europe. An experiment comparing migratory orientation of captive birds that wintered in England and ones breeding in

Germany, most of which go to Africa, found that in autumn the English-wintering birds oriented northwest, as though they were coming from central Europe, while the German Blackcaps oriented southwest toward Iberia, from where they crossed to Africa. The genetic basis for these orientations was confirmed in experiments with captive-born young of these two populations; each oriented as had its parents (Berthold et al. 1992).

While the original causes of the Blackcap's migratory shift cannot be determined, deliberate introductions of birds to a new region have demonstrated how rapidly even sedentary birds can develop migratory behavior in response to different winter conditions. In March 1890, a group of New Yorkers seeking to introduce to America all the birds mentioned by Shakespeare released 60 European Starlings (*Sturnus vulgaris*) in Central Park. Another 40 were released there in 1891 (Long 1981:360). From this stock of English nonmigratory birds, starlings have since spread all over the continent. In some regions, they migrate hundreds of kilometers each winter and shift wintering sites from one year to another, sometimes more than 300 km away, a feature that has surely contributed to their rapid dispersal over North America (Dolbeer 1982).

Caged House Finches (*Haemorhous mexicanus*), probably from the Los Angeles area, were released on Long Island, New York, in about 1940. Their descendants have now spread along the entire Atlantic coast from southern Canada to Florida and westward to where they have met the original, mostly sedentary, western population. By the early 1960s, some of the eastern birds were migrating; as the House Finch colonized new areas, the migration distances became longer and longer, although marking individual birds has shown that in certain years some previous migrants do not move at all. Like many partial migrants of eastern North America, females and young birds usually go farther south than adult males (Able and Belthoff 1998). A comparison of migratory House Finches in Ithaca, New York, and sedentary ones in Boise, Idaho, part of the original range, found that the wings of eastern birds were more pointed, although not longer, because of

Snow Geese historically wintered in the marshes along much of the Atlantic and Gulf coasts. Today, many stop short and spend the season in agricultural fields of the lower Midwest.

changes in the length of the primary feathers, a pattern found frequently between closely related migrant and resident populations of many birds (Egbert and Belthoff 2003).

A more complex set of anatomical, ecological, and behavioral factors may be influencing the Snow Geese (*Chen caerulescens*) that winter in the central United States. This population originally traveled from the High Arctic tundra to coastal brackish marshes around the Gulf of Mexico, where they feed by excavating roots and tubers. Since the 1950s, some Snow Geese have stopped short as much as 160 km inland to winter in rice fields on land that formerly supported tallgrass prairie not used by the geese; here they graze on green vegetation. In the 1970s, Snow Geese began settling for the winter approximately 1000 km north of the Gulf of Mexico, where they feed primarily by picking corn kernels off the ground on intensively cultivated land. Climate change and reduced snow cover may have opened a new belt of rich and easy food sources to the geese (Alisauskas 1998).

This expansion to two new winter ecosystems has brought physical changes to the Snow Goose. The two southern populations have thicker bills, longer skulls, and a longer culmen, which may be related to their feeding method. The northern grain eaters are the smallest of the three, and the rice field grazers have the greatest range of body size and bill shape. This variation may be becoming genetically encoded, but there could also be deliberate habitat selection by the geese, with those less adapted for excavating roots settling where they can graze or pick grain from the ground. The nutritional differences in the three winter food types may also be influential, since young Snow Geese continue growing until the May following the year of their birth. Finally, the smaller Ross's Goose (*C. rossii*), which formerly wintered primarily in the lower San Joaquin Valley of California, has increased in the midcontinent since the 1970s, perhaps for the same reasons; some now winter in the two agricultural habitats used by Snow Geese, but not in the Gulf Coast marshes. The two species hybridize—forming pairs in winter rather than on their separate breeding grounds—and this may also contribute to the variation in body features (Alisauskas 1998).

SUMMARY

The seasonal changes at middle and higher latitudes that produce winter conditions directly affect about one-third of all birds living today. Some are able to remain through the period of reduced temperature and daylight; more birds go greater or lesser distances to latitudes with climates that provide habitats where they can survive over the months when winter transforms their breeding range. Migration has likely evolved in two forms. Birds of families that originated at higher latitudes may have withdrawn toward the equator during the eras when colder conditions made their initial range uninhabitable over the course of the year, or over far longer periods. Birds from families of tropical origin may have expanded during interglacial periods when warming opened new areas they could exploit at least during the months when their food sources are abundant.

Today, migration to winter destinations takes several forms. In some species, part of the population may move while other individuals within the breeding range remain through the winter. Among short-distance migrants, the entire population migrates, but usually less than 2000 km. Long-distance migrants go much farther, sometimes traveling over oceans, deserts, and other habitat barriers to entirely different ecosystems. In mountainous regions, a short descent to lower elevations has the same result, if climate conditions at different altitudes are significantly different. Still other birds are nomadic over the winter, depending on how long food supplies last in their breeding area. Finally, some species of high latitudes are periodically irruptive, linked to the multiyear cycles of their usual food source.

Migration in response to winter takes place in all hemispheres and involves birds of all kinds, both landbirds and those that spend the nonbreeding season far at sea. While the most studied have been those of North America and Europe, the austral winter also prompts many birds of South America, Africa, Australia, and New Zealand to move northward to warmer regions.

These seasonal movements are the result of long-term evolution, but birds can respond rapidly to changes in habitat and climate that make new areas attractive in winter. Range shifts can take place in decades or less, enabling us to witness directly how finely attuned birds are to local winter conditions and opportunities. Today, scientists have the benefit of rapidly advancing tools that enable them to mark and follow the movements of individual birds over the course of the winter, revealing far more than was previously possible about how and where they spend that season.

NEW TOOLS FOR STUDYING MIGRATION AND OTHER MOVEMENTS

Much of the research described in this book results from technology that enables scientists to follow individual birds remotely, so that their daily and seasonal movements can be tracked, providing much more precise and extensive information on how and where birds spend their time. In addition to radiotelemetry, which has been used for many years, in which a tiny radio transmitter is affixed to a bird that can then be tracked locally with an antenna receiving the signals, several rapidly advancing technologies now let scientists follow birds as they move around the globe over the course of a year or longer (Rappole 2013:285–291).

■ SATELLITE TRACKING uses transmitters mounted on birds and receivers mounted on satellites. The devices placed on birds are now much smaller than the earliest generations were in the 1990s, which were applied to large birds like storks migrating to Africa. Today, even small transmitters have sufficient battery life and transmission power

Lightweight tracking devices like this light-level geolocator that can be attached to small birds without harm enable scientists to follow their seasonal movements precisely and have revealed much information that helps develop conservation plans.

to track birds for a year; they can be used without harm on sandpipers, passerines, and other small birds. Their accuracy is the finest of all the tracking technologies, to within approximately 1 km.

— GLOBAL LOCATION SENSOR LOGGERS, "light-level geolocators," can be attached to birds to record visible light intensities every 60 seconds and maximum light per 10-minute interval. When the data loggers are recovered from the bird, they can be downloaded and read by a computer program. While much cheaper than satellite tracking, the geolocators cannot pinpoint location as accurately, currently only to within 100 km or more, while latitude cannot be determined within 15 days of an equinox, and proximity to the equator also reduces accuracy.

— FINALLY, STABLE ISOTOPE ANALYSIS uses the isotope ratios of chemicals in bird feathers, most often carbon, nitrogen, and deuterium, collected from birds in different parts of their breeding range and compares these averages with the ratios in feathers from birds of the same species taken on migration and in winter; this reveals which populations are moving where. A problem with this technique is that the ratios may vary by year, by age of the bird, by the elevation at which it was found, and by diet at the time it acquired the feathers being tested. Thus, the variety of molt sequences among long-distance migrants must be understood so that accurate comparisons can be made between feathers collected at different localities and seasons, even within the same year.

These technologies have begun to reveal the daily and seasonal movements of birds in individual detail as was never previously known, adding vastly more to our understanding of behavior, ecology, and distribution and enabling conservationists to develop much better measures for species and for critical sites. A study of Harlequin Ducks (*Histrionicus histrionicus*), for example, using satellite transmitters showed not only that different populations wintered in regions as much as 2400 km apart, but that the bird was far more numerous than previously believed. The Harlequin Ducks of eastern North America, which all breed in Quebec and Labrador, were best known from the counts of wintering birds on the New England coast. Because these number fewer than 1,000, the Canadian government concluded that the population was endangered. Radio-tagging Harlequins nesting in different parts of Quebec and Labrador, however, revealed that while the birds from southern Quebec to eastern Labrador indeed winter on the New England coast, birds from Hudson Bay winter on the coast of southwestern Greenland, where they mix with Harlequins that nest locally. The Hudson Bay birds perform the only northward winter migration known for the Northern Hemisphere. The origin of this directional split between the two eastern populations may reflect the effects of the Laurentide Ice Sheet during the last glaciation; the current breeding range may have been uninhabitable, with Harlequins at that time nesting farther south in eastern North America as well as in ice-free western Greenland. These populations may then have spread to their current range, returning in winter to their breeding range of some 12,000 years ago (Brodeur et al. 2002).

CHAPTER 2

PREPARING FOR WINTER

WHEN THE ESSENTIAL ACTIVITIES of the breeding season are complete, most birds, except those that will lead their young on migration to a wintering site, such as swans, geese, and cranes, have no responsibility except for their own survival and welfare until the following year. The birds of temperate and high latitudes, where winter will in due course arrive, begin preparing. Three major activities may occupy them: molting, food storage among birds that will not migrate, and the actual departure of those that do.

MOLT

Most birds replace their entire plumage at least once a year. Some larger birds, including albatrosses and raptors, may take more than a year, sometimes two or three, to complete their molt, especially of the wing and tail feathers, which require the longest time to grow. For all birds, feather growth is energetically costly and so does not usually occur when birds are undertaking other demanding activities, such as raising young or migrating. Of these three key activities, only molt can be flexible in time. As a result, molt schedules vary widely, reflecting the ecology of each species and the seasonal and geographic availability of the food resources necessary for feather growth.

Most year-round resident birds maintain the same or a very similar plumage throughout the year. Having no migration or other competing significant energy demands, they begin a prolonged molt of all feathers after the nesting season; they can replace their feathers slowly while maintaining the same plumage. A few nonmigrants of high latitudes that depend on camouflage to protect them from predators, like ground-dwelling ptarmigan (*Lagopus* spp.), change their plumage over the seasons so they are inconspicuous both in low vegetation in the warmer months and against snow during winter. From October to April, all three North American ptarmigan species are entirely or almost entirely white. In April and May, males and females of each species replace different amounts of their white winter feathers with brown ones.

Many short-distance migrants also replace all their feathers once annually, after breeding, and complete their molt in less time than nonmigrants. European Starlings (*Sturnus vulgaris*), resident or partial migrant, molt between late May and September; the replacement process is easy to observe, because most of the new feathers have brown edges or whitish tips that will wear away over the winter to reveal the glossy plumage of breeding birds. Adult Dark-eyed Juncos (*Junco hyemalis*) may take about 100 days in late summer and autumn to replace all their feathers; between February and April, they may again replace just a few feathers, mainly on the head (Nolan et al. 2002).

Other short-distance migrants have two molts of body feathers, sometimes into a significantly different plumage. In male American Goldfinches (*Spinus tristis*) the progress of body feather molt is very visible as the bright yellow feathers are patchily replaced with browner ones from late summer into October or later, depending on latitude; similarly, the yellow begins to return on some birds in January, but for most it begins substantially in March. The American Goldfinch is the only cardueline finch that changes its appearance so dramatically (Middleton 1993). The unusually late timing of the spring molt be may linked to the goldfinch's breeding schedule, which also begins later than that of many songbirds; it feeds its young thistle seeds, which are not available until summer.

Males of many ducks, most notably among the dabbling ducks, have an inconspicuous, female-like "eclipse" plumage from June into the autumn, when they molt into their more familiar plumage. Males court and pair with females in winter; the initiation of courtship does not begin until males have returned to their brighter plumage, October for Mallards (*Anas platyrhynchos*) and as late as December for some Northern Shovelers (*Spatula clypeata*).

The greatest variety in molt schedules is among long-distance migrants. For them, timing is a more critical factor. Depending on their diet, there may be more or less food available where they have bred, as well as more or less time before they need to depart. Thus, while most adult birds initiate their complete ("prebasic") molt after breeding is concluded, some or all of the molt may take place when they have reached their winter home. Birds born that year have more complicated molt sequences, because they may replace their very rapidly grown initial set of feathers with more durable ones either before or after their first migration.

The urgency of reaching a distant wintering site may influence molt schedule. Hooded Warblers (*Setophaga citrina*) compete for a limited number of territories in their winter range, so birds that arrive first have an advantage, especially if they have occupied a site during a previous winter. Because they have their prebasic molt before departing at the end of summer, they may shorten their breeding season by not attempting a second brood even if there is enough time to renest; Hooded Warblers that raise a second brood begin their molt two to three weeks later than individuals that were not feeding young late in the season, and they probably migrate later as well, reducing their chance of securing a winter territory. The territorial prospects for their young, arriving later still, are even poorer (Stutchbury 1994).

As the following descriptions of molt sequences for adult long-distance migrants show, replacing feathers is a significant winter activity for many birds. It can influence their movements over the course of the winter as well as local shifts in habitat choices and diet.

- *Breeding—complete molt—migration to wintering site:*
  In regions where food resources are abundant at the end of summer, such as eastern North America, most long-distance migrants have a complete molt before migrating. Fifty of the 55 northeastern passerines that winter in the Neotropics molt before migrating. The nomadic Common Redpoll (*Acanthis flammea*) also molts between July and September, when food is more reliable than over the vast areas it may search in winter (Knox and Lowther 2000). In Europe, some species with a wide latitudinal range have several strategies. Southern European Barn Swallows (*Hirundo rustica*) are resident or short-distance migrants; they molt during June–August, after breeding, while northern European swallows begin molting in September and October, when they have reached sub-Saharan Africa (Cramp 1988:276).

- *Breeding—partial molt—migration to wintering site—finish molt on wintering site:*
  Long-distance migrants that leave their breeding range by late summer or early autumn and have lengthy once-annual molts may require more time than is available either when they finish caring for their young or where they winter. They begin molting in summer and resume after migration. Adult Ospreys (*Pandion haliaetus*) of middle and high latitudes molt some body and flight feathers between July and September, before they depart; molt resumes when they reach their wintering area—for North American birds, mainly Central America, the Caribbean, and South America—where it may continue until March, when Ospreys begin their northward migration (Poole et al. 2002). Similarly, in Purple Martins (*Progne subis*), another bird that also pursues prey in flight and has a long migration, adults begin body and flight feather molt in July and interrupt it while migrating through Central America. When they reach their wintering grounds, primarily in Amazonian Brazil and eastern Bolivia, molt resumes and is probably completed by late February (Brown 1997).

- *Breeding—migration to wintering site—complete molt:*
  Some 40% of Old World passerines wintering in Africa have complete molts while
  there. Timing depends much on local rains and resulting insect availability. In
  northern Ghana, nine warblers (Sylviini) and the Woodchat Shrike (*Lanius senator*)
  molt very rapidly in November, at the beginning of the dry season when insects are
  still abundant; some of these birds remain there through the winter, while others
  move farther south. In other parts of Africa, such as Malawi and Uganda, where
  rain is reliable for several months, warblers molt more slowly, from November to
  March or April (Bensch et al. 1991).

- *Breeding—short migration—molt—longer migration to wintering site:*
  Some birds have a two-part migration in which they stop to molt and then continue to
  their winter destination. In North America, the lowland West is dry and unproductive
  at the end of the summer; several birds there, including Bullock's Oriole (*Icterus
  bullockii*), western Warbling Vireo (*Vireo gilvus*), Lazuli Bunting (*Passerina amoena*), and
  Western Tanager (*Piranga ludoviciana*), move after breeding to New Mexico, eastern
  Arizona, and northern Mexico, where in late summer the Mexican monsoon typically
  generates a large flush of insects that can fuel the growth of new feathers. After
  molting, these birds continue farther south. Painted Buntings (*Passerina ciris*) in the
  central part of their range also move to the Mexican monsoon region, while eastern
  birds molt before migrating (Rohwer et al. 2005). Similarly, in Africa, Great Reed
  Warblers (*Acrocephalus arundinaceus*) start arriving in northern Ghana in late September,
  at the end of the rainy season, settle into marshes, and molt rapidly. During the final
  stages of molt, as the marshes are drying, the birds begin accumulating fat and then
  depart in December for wetter areas farther south (Hedenstrom et al. 1993).

- *Breeding—complete molt—migration to wintering site—partial molt:*
  The male Scarlet Tanager (*Piranga olivacea*) has an especially conspicuous molt, easy
  to see in midsummer when the red body feathers are gradually replaced by green
  ones. Flight feathers are replaced then also, and the complete molt is finished by
  September, before tanagers depart for South America. Between January and March,
  male Scarlet Tanagers again molt their body feathers to red, as well as their tail
  feathers, which, like the wings, are black in all plumages. Female Scarlet Tanagers
  have a similar molt sequence but do not change their colors significantly (Mowbray
  1999). The closely related Summer Tanager (*P. rubra*) has the same molt sequence,
  but males are red in all plumages. Summer Tanagers winter from Central America
  south to Bolivia and Brazil, overlapping with Scarlet Tanagers on the eastern side of
  the Andes (Robinson 1996). Summer Tanagers are territorial in winter while Scarlets
  are not, but conspicuous plumage is not always associated with winter territoriality.

- *Breeding—partial molt—migration to wintering site—partial molt—partial molt:*
  Many Arctic shorebirds replace some body feathers soon after breeding and move
  rapidly to migratory staging areas where they can add the substantial layer of fat

that will fuel their long southward migration. Only when they reach their winter destination will they molt the balance of body feathers plus all flight feathers. White-rumped Sandpipers (*Calidris fuscicollis*) wintering in Argentina and Chile take 60 to 90 days, between November and February, to replace their flight feathers (Harrington 1999). Many North American shorebirds that winter in Argentina complete their prebasic molt on arrival in the austral spring and then, during the late austral summer and autumn, acquire their alternate plumage by again molting body feathers (Myers and Myers 1979).

- *Breeding—migration to wintering site—complete molt—partial molt:*
  The Bristle-thighed Curlew (*Numenius tahitiensis*) has a distinctive variation on the shorebird pattern. It is unique in wintering exclusively on islands in the tropical and subtropical Pacific. On Laysan Island in the northwestern Hawaiian Islands, it arrives from western Alaska for the winter as early as July and begins a complete prebasic molt that takes about 92 days, considerably faster than that of other shorebirds wintering in tropical and southern latitudes. Between early September and late November, some curlews shed so many wing feathers almost simultaneously that they become flightless for about two weeks until the new feathers reach full size. This is a viable strategy only for birds on remote islands with no predators; it allows them to replace feathers more rapidly. By February, they are again molting body and tail feathers, coming into the alternate plumage they will wear until they return. Unlike most shorebirds, Bristle-thighed Curlews gain mass gradually through the autumn and winter. Their rapid prebasic molt may allow them more time for fat gain prior to their 4000-km flight from Laysan back to Alaska, since food sources on the island—which lacks extensive mudflats and intertidal habitat—do not enable rapid buildup prior to departure (Marks 1993).

- *Breeding—complete molt—migration to wintering site—complete molt:*
  A very few birds have two complete molts each year. In the Western Hemisphere, Franklin's Gull (*Leucophaeus pipixcan*) and the Bobolink (*Dolichonyx oryzivorus*) both winter in South America, the gull along the Pacific coast, the Bobolink in the southern grasslands. The Bobolink has a complete molt in North America between August and early October, and then another during the austral summer. Only in the males is there a significant change in plumage—to a female-like basic plumage in autumn; the male's new feathers grown later in winter have buffy edges that abrade by the time it arrives in North America, revealing the distinctive bold black-and-white pattern underneath (Martin and Gavin 1995). Bobolinks wintering in Bolivia arrive there in November, begin their molt sooner, and take 13 days longer than the Bobolinks that go to Argentina, another 1200 km southeast, where they arrive between mid-December and mid-January. For both, molt is completed between mid-February and March 1 (Renfrew et al. 2011).

The Eurasian Willow Warbler (*Phylloscopus trochilus*) is the only member of its widespread genus of many very similar species to have two complete molts. It breeds in scrub, second growth, and semiopen habitat almost to the edges of the Arctic Ocean, from Ireland to eastern Siberia, and winters in all habitats in sub-Saharan Africa except forest (Brooks 1992:641). It is not obvious what environmental or evolutionary factors led to these distinctive and energetically demanding molt patterns in such very different birds in two hemispheres, since other birds sharing their winter habitats do not have two complete molts.

Postbreeding food resources, distance to wintering areas, environmental conditions on arrival or at intermediary stopover sites, and pairing schedules all influence the timing and extent of molt, generating the range of sequences just described. Among *Calidris* sandpipers, the prebasic molt schedule seems most closely correlated with distance between breeding and wintering areas. The farther a species migrates, the later it begins its molt. Northerly wintering Dunlins (*C. alpina*) and Purple Sandpipers (*C. maritima*) molt entirely on the breeding grounds. Western Sandpipers (*C. mauri*), which winter in temperate and tropical latitudes north of the equator, start their prebasic molt where they nest and complete it on their wintering grounds. Semipalmated (*C. pusilla*), Baird's (*C. bairdii*), and Pectoral (*C. melanotus*) Sandpipers, and others wintering in the Southern Hemisphere, molt entirely there (Marks 1993).

In addition, the farther south an Arctic-nesting shorebird winters, the longer the molt takes. At northern latitudes, food may decline during the winter, so molt is best completed where birds have bred, if food is still easily available, or as soon as possible after arrival at the winter destination. The pattern holds for populations of species that have a vast latitudinal winter range: Ruddy Turnstones (*Arenaria interpres*) wintering in South Africa take 45 more days to molt than those wintering in Scotland (Summers et al. 1989).

Brighter colors and bolder patterns are used by many birds, males especially, as part of territorial display and mate attraction. In winter, few migrants except ducks are seeking mates, but many passerine migrants of both sexes establish individual territories. Molt schedule, however, is not obviously linked to winter territoriality. Among 102 Neotropical migrants, two-thirds of the 48 nonterritorial species molt before reaching their wintering grounds; four-fifths of 54 winter-territorial species also molt before arrival. Delaying molt to arrive as soon as possible so as to secure a better territory therefore seems not to be a factor in the molt schedule for most of these birds (Rohwer et al. 2005).

Nor are bright colors and bold patterns in male winter plumage of Neotropical migrants necessarily associated with territoriality. While males of most winter-territorial birds are brightly colored or boldly patterned, most males in nonterritorial species are also. Even among species that molt into a much more conspicuous plumage in late winter, like many wood warblers, there is no consistent correlation with winter territoriality when the birds are in either plumage (Froehlich et al. 2005).

Most Eurasian passerines wintering in Africa are sexually monomorphic; many are more mobile through the winter, following the rains. Some bright, sexually dimorphic migrants like the Golden Oriole (*Oriolus oriolus*) form flocks, while males of three territorial wheatear (*Oenanthe*) species molt into cryptic plumages. Diet may be the determining factor in territoriality, but not plumage: fruit eaters like Golden Orioles cannot defend a territory large enough to hold fruiting trees all through the winter; nor could one defend an entire tree when it bears fruit. Insectivores like wheatears, however, can space themselves in territories that will reliably provide food for the entire season.

A further distinction between Neotropical and African passerine winterers is the timing of molt. Most Neotropical migrants replace their flight feathers before departing their breeding grounds, or interrupt their autumn migration to do so (as in the western birds moving to the Mexican monsoon region), or molt upon arrival at their wintering site. Many of the African winterers replace flight feathers only later in the winter, shortly before migrating north. Some of these birds also have two complete body molts, at the end of summer and again during the winter, early or late depending on the rainy season timetable at their wintering latitude (Froehlich et al. 2005).

These African-wintering birds with two complete body molts, as well as the Bobolink and the Willow Warbler with their two molts of both body and flight feathers, live all winter in open habitats that may be more abrasive. They are also exposed to more ultraviolet radiation and airborne dust than most Neotropical passerine migrants, which typically live in forest or second growth. Comparative studies of molt schedules and patterns in sedentary African relatives of the Eurasian migrants would help demonstrate the importance of these factors, because the resident birds are subject to these conditions more of the year. In the Neotropics, wintering birds already in a conspicuous plumage have little body molt before migrating north; those with a dull winter plumage have more—while some, like the Prothonotary Warbler (*Protonotaria citrea*), become brighter as the olive tips of the feathers they acquired the prior summer wear off (Petit 1999).

FOOD STORAGE

Long before winter, some nonmigrants of high and middle latitudes begin storing food for future use. Various tits, nuthatches, woodpeckers, and corvids—crows, jays, nutcrackers, and others—begin caching food in August, while it is abundant, and continue into November. During these months, birds can easily provision themselves for the day as well as devote time and energy to future needs. Most of what they put away is seeds, nuts, and insects and other arthropods that will not spoil when left concealed, often for months. Some hawks and owls, both sedentary and migratory, also cache food, but not until winter, because their animal prey is more perishable. Raptors, with prey in larger packages than the seeds and insects stored by passerines, cache fewer items.

Passerines working through summer and autumn may store tens of thousands of items. Some also store smaller amounts during winter, usually to consume the same day or within two or three days. Retrieving cached food requires an aspect of memory not deployed in many other avian activities. Black-capped Chickadees (*Poecile atricapillus*) remember cache locations for at least four weeks; they also remember which sites have already been emptied and avoid revisiting them (Sherry 1984). A Clark's Nutcracker (*Nucifraga columbiana*) remembered its caches as much as 285 days later (Balda and Kamil 1992).

The hippocampus is the part of the brain responsible for this kind of memory, and a comparison of all major Northern Hemisphere passerine families found that it was largest, relative to body weight and brain size, in the three families in which food storing is most widespread—corvids, tits, and nuthatches (Sherry 1989). Comparisons of similar-sized corvids and tits between species that store and those that do not—Black-billed Magpies (*Pica pica*) versus Jackdaws (*Corvus monedula*), and Marsh Tits (*Parus palustris*) versus Blue Tits (*Cyanistes caeruleus*)—found that hippocampal volumes did not differ as nestlings but were larger in the storing species as adults. Hand-reared Marsh Tits began storing food at age 44 days, when they were able to handle seeds correctly and could reliably distinguish between suitable and inappropriate items. This ability depended on age, not experience, as was shown by depriving young tits of storage opportunities until they were as much as 115 days old and seeing that they then took up storage rapidly when suitable food was available. Thus, unlike acquisition of song and some other behaviors in certain birds, there was no "window" or sensitive period for learning. With active storing came an increase in the size of the hippocampus, until it reached a fixed level, while Marsh Tits that never stored continued to lose cells in the hippocampus. The hippocampus growth could be stimulated in young Marsh Tits by other memory-based activities, but not in Blue Tits, which do not store (Krebs et al. 1996).

For small birds, food storage is an efficient and sometimes necessary way to meet the severe energy demands of a long winter that has far fewer hours of daylight than other seasons for finding food—and many more hours that must be passed without any replenishment while cold makes maintaining an adequate body temperature more challenging (Pravosudov and Grubb 1997). A 10-gram chickadee can store on its body only half a day's fat to meet the energy requirements of metabolism over 24 hours. A 75-gram Blue Jay (*Cyanocitta cristata*) can store about twice its daily requirements. Typically, more of what the corvids cache is for use farther in the future.

Food storage takes two forms—scatter-hoarding, where food is put in many separate locations, and larder-hoarding, where it is kept together. Scatter-hoarders conceal their items to make them difficult for conspecifics or others to find, while larder-hoarders defend their sites aggressively, since they have more to lose from any competitor's discovery of the site. Within the woodpecker genus *Melanerpes*, the Acorn (*M. formicivorus*) and Red-headed (*M. erythrocephalus*) Woodpeckers are larder-horders, the Acorn Woodpecker in family groups and the Red-headed singly, each favoring certain kinds of trees with bark appropriate to its storage method, while the Red-bellied Woodpecker (*M. carolinus*) is a scatter-hoarder. Among raptors, the Northern Pygmy Owl (*Glaucidium passerinum*) is a larder-hoarder, and the Eurasian Sparrowhawk (*Accipiter nisus*) and American Kestrel (*Falco sparverius*) are scatter-hoarders. All passerines are scatter-hoarders.

Passerines do most of their caching in the middle of the day. At dawn, their first priority must be refueling. At the end of the afternoon, they must consume enough to carry them through the following night. That leaves midday for less immediate needs, but in winter when food supplies are unpredictable, birds cache more in the afternoon to retrieve the next morning. Over the course of the autumn and winter, some of what has been cached is inevitably lost to spoilage and to pilferage by conspecifics or other competing species, or made inaccessible by snow or breakage of the storage site. During the winter, birds recache some of their stored items to another place they feel is more secure.

Among the best-studied hoarders are tits (Paridae), which can be followed easily in the wild, observed closely when they come to feeders, and monitored still more thoroughly in aviary experiments focused on impacts of day length, food type and quantity, and other variables. These studies have revealed a wide range of caching strategies that reduce competition both among conspecifics and between species that spend the fall and winter together in mixed-species flocks. Tits carry single items or small collections of food in their bill, usually storing each load in a separate location not more than 100 m from where they found it. They never go back to the same storage site to add more. They place items in crevices in bark, under lichens or moss, in foliage and hollow stems, or buried in soil. To secure seeds or nuts, tits may hammer them into bark grooves or cavities. Some food may be retrieved the same day, but farther north, birds are more likely to resort to their caches weeks or months later, depending on them during the coldest, darkest portion of the winter (Vander Wall 1990:309–314).

In North America, where all tits and chickadees are hoarders (Sherry 1989), no more than two species live in the same region, and in much of the continent only one species is present. In Eurasia, several parids may inhabit the same woodland year round. This has led to further refinements in caching techniques that reduce competition (Vander Wall 1990:309–314). Great Tits (*Parus major*) and Blue Tits, which avoid purely coniferous woods and in some winters leave their nesting area, rarely, if ever, store food. Among the nonmigratory conifer specialists, Crested Tits (*Lophophanes cristatus*) alone secrete a glue-like saliva around their stored items to further secure them. Crested and Coal Tits (*P. ater*) may then conceal their cache with lichen or debris, while other tits do not.

Each species favors certain storage sites as well as distinctive food items. In Norway, Crested Tits cache spruce seeds and insects equally along dead main branches, the bare inner portion of live branches, and the needle-covered ends of live branches. Coal Tits in the same forest specialize in seeds of hemp-nettle (*Galeopsis*), which they put in spruce bud

Of the several European tits that live in the same forests and each store food, the Crested Tit is unique in using a glue-like saliva to secure the items it caches.

capsules at the tips of branches. In the same trees, the Willow Tit (*P. montanus*) stores most food on the trunks and bare inner branches.

These storage preferences match the part of the tree on which each species most often forages, leading to different height preferences for caching. Coal Tits usually conceal food higher in trees than do Crested and Willow Tits. Marsh Tits and North American Boreal Chickadees (*Poecile hudsonicus*) bring food down from the canopy and up from the ground to store in the portion of the tree least likely later to be covered with snow. In Britain, where there is usually less snow than in Scandinavia, Marsh Tits cache food closer to the ground.

In winter, most tits form small territorial flocks that may include other species. The match between preferred foraging area within a tree and caching sites helps reduce interspecific competition and pilferage. Individual tits develop their own specialties for concealment and are alert for competition from other birds. Some Marsh Tits hide more items in moss, others in bark crevices, while still others are generalists. Marsh Tits with a territory that overlapped that of a pair of Eurasian Nuthatches (*Sitta europaea*), themselves hoarders, put sunflower seeds in the ground rather than in bark crevices where the nuthatches were more likely to find them.

Dominance—by age and sex—affects foraging and caching behavior. In flocks of Black-capped Chickadees, where adults outrank birds born earlier that year, dominants do more caching, first flying away from others in the flock, while subordinates that succeed in observing them do more pilfering. Experiments in an indoor aviary showed that 9% of caches were taken by other birds that saw the caching; in natural situations, the likelihood of such observation must be lower (Hitchcock and Sherry 1995). The loss to the dominant birds is not significant enough to cost them their rank, since they are the more experienced foragers and can drive younger birds away from a source of food. In a pair of European Nuthatches in Sweden, the male stored more food on the periphery of their range, and the female hid more of her caches under lichen than did the male. The male sometimes watched her hide her food and then removed it to cache elsewhere (Moreno et al. 1981). Pairs of New Zealand Robins (*Petroica australis*), the only known food storers in the vast family of Muscicapidae (Kallander and Smith 1990), share a territory during winter, and both birds frequently pilfer and recache prey stored by the other, but females remove more than males (Burns and van Horik 2007).

Many corvids are food hoarders. It is easy to see Blue Jays leaving bird feeders with their cheeks and esophagus bulging with food they will hide in their territory. Both artificial feeders and nut trees like oaks attract jays from beyond their own territory; each returns there to cache what it carries away from the common source. Blue Jays store food singly or in small collections, putting the items in loose soil they then cover, or wedging them in the bark of tree trunks or in dense coniferous foliage. Unretrieved seeds in the ground may germinate the next spring; a clump of saplings of a seed- or nut-bearing tree may well be evidence of an old cache. Gray Jays (*Perisoreus canadensis*) secrete a sticky mucus that adheres to any surface and hardens around the food; they make no effort to conceal their cache. Crows cache food all through the year, but most of it is perishable and is consumed quickly (Vander Wall 1990:297–307).

Clark's Nutcrackers spend the summer and autumn storing pine seeds that may constitute their entire diet the following winter.

The most sophisticated food storers are the nutcrackers. The North American Clark's Nutcracker specializes in pine seeds. Caches made in the summer and autumn may constitute their entire diet during winter. Nutcrackers store the seeds in a variety of places that become accessible at different times over the course of the winter; at high elevations, they put some seeds in the soil of windswept ledges where little snow accumulates and on south-facing slopes, where snow melts faster (Vander Wall 1990:303). A multiyear study on the eastern slope of the Sierra Nevada in California found that nutcrackers begin storing seeds in late summer from the pines at subalpine elevations; they later move downslope to harvest seeds from other pine species. They remain at the lower elevations through the winter and begin nesting in February. The young are fed on pine seeds from trees and from stores made months earlier. When the young fledge in late spring, families migrate to the higher slopes and feed on the seeds cached the previous summer. By August, the young are independent and the storing cycle begins again (Tomback 1977).

Among raptors, both residents and migrants that remain in colder latitudes cache some of their animal prey, but not far ahead as do the passerines that begin in summer. Many falcons store food, but winter caching has rarely been recorded among hawks and eagles, which in the breeding season sometimes leave unconsumed prey at the edge of the nest for the young to eat later (Vander Wall 1990:283). The most detailed observations have been made on American Kestrels, which hunt and store in open country and are easy to monitor. In the pasturelands of Humboldt County, California, they are scatter-hoarders, putting most of their food in grass clumps, placing the animal dorsal side up for better camouflage. Fence posts and shrubs are occasionally used. Kestrels store only vertebrate prey; in one winter this was 88% shrews and small rodents, and in the following year, which was warmer and wetter, 68% of the cached items were frogs and snakes. These preponderances reflected the birds' total captures. The prey was decapitated and sometimes partially consumed before storage. Birds seemed to determine their next caching site before going to it with prey—they flew directly to it from their perch. Kestrels usually continued hunting after caching; while they stored food throughout the day, they returned most often to retrieve their caches late in the day, landing next to the site or hovering briefly above it (Callopy 1977).

Kestrel studies elsewhere found the same pattern of storage and retrieval the same day or within two or three days (Vander Wall 1990:287). As in the passerines, short-term storage of food consumed before roosting may help kestrels get through the night. They may also be more agile if they weigh less during the peak hunting hours, just as small birds may be more successful in evading predators if they do not consume all the food they find each day until it ends.

In owls, food storage is more widespread in high northern latitudes. In winter, most owls cache their food in larders rather than scattering it. They frequently decapitate, eviscerate, or eat some of each item before storing it. Owls sometimes put their prey in cavities or nest boxes where they roost in winter, otherwise on branches near a regular perch. Many owls also cache food during the breeding season, when it may often be kept at or near the nest, so their continuing habit of larder-hoarding at a roost site through the year is not surprising, especially among nonmigratory owls (Vander Wall 1990:288–292).

An exception is the Northern Hawk Owl (*Surnia ulula*), a nomadic diurnal species that moves long distances in response to prey availability and in some winters comes well south

Northern Hawk Owls are nomadic in winter and so scatter the prey they hoard rather than accumulate many items in one place, as do more sedentary owls.

of its usual range in the boreal forest. Birds wintering in Minnesota cached each prey item separately, covering it on the ground with snow or placing it at various heights in trees (Schaefer et al. 2007). Depending on local food availability, Northern Hawk Owls may continue moving through the winter. For them, long-term larder-hoarding has fewer benefits than for sedentary owls.

Nonmigratory Northern Pygmy Owls (*Glaucidium passerinum*) may put as many as 18 to 26 items in a single larder. They hoard between October and April, with the largest caches from November to February; caches are most frequently made when the daily maximum temperature is below 0° C (Solheim 1984). Their hoarding behavior is influenced by competition with the larger Boreal Owl (*Aegolius funereus*), which inhabits some of the same forests and preys on the same voles. Pygmy owls prefer to cache their prey in holes with small entrances that larger owls cannot penetrate; where the Boreal Owl is present, pygmy owls do not use holes with larger entrances, and even in the smaller cavities they put fewer items than they do where Boreal Owls are absent (Suhonen et al. 2007).

The cold of northern forest winters helps preserve caches of animal matter, enabling owls to continue using them when energy demands increase. Many species begin the breeding cycle long before spring, with courtship feeding and egg development. But these caches must be thawed before they can be consumed. Great Horned (*Bubo virginianus*), Boreal, and Northern Saw-whet (*Aegolius acadicus*) Owls, and probably other species, thaw frozen prey by assuming an incubating posture, warming the carcass with heat from their lower abdomen (Vander Wall 1990:290). Even if in owls caching may have originated in the breeding season with unconsumed food left at the nest, the thawing behavior must have evolved separately, because male owls do not incubate eggs.

In order to eat it, owls like this Northern Saw-whet sometimes need to thaw the prey they have stored.

DEPARTURE

Most migrants begin their preparations for departure to wintering sites long before there are obvious changes in the environment where they have bred. After whatever molt may take place in the breeding range, migrant birds undergo other physiological changes focused on gaining body mass to fuel their flights. They use the time while food is still easily available to add the necessary subcutaneous fat reserves. Long-distance migrants especially increase their feeding. They gain mass rapidly, often by as much as 50%. Birds that have long flights over regions such as deserts and oceans that lack places where they can rest and refuel put on the most weight, which will be dissipated before they reach their destination.

The now probably extinct Eskimo Curlew (*Numenius borealis*) may have doubled its weight before leaving the coasts of Labrador and the Maritime Provinces, and occasionally New England, on its nonstop oceanic flight to South America (Gill et al. 1998). It and the similarly fattened American Golden-Plover (*Pluvialis dominica*), following the same route, were considered the most delectable of shorebirds in the nineteenth century and were therefore the preferred targets of both sport and market hunters, who shot them by the cartload, both as the birds were going south and in the spring when, also fattened, they returned to the Gulf Coast and migrated north through the prairie grasslands.

The triggering mechanisms prompting the increased intake are likely a combination of external cues like diminishing amounts of daylight together with internal rhythms under genetic control, varying by species. The hormone corticosterone, for example, plays a role in migratory fattening; when it was blocked in Dark-eyed Juncos and Yellow-rumped Warblers (*Setophaga coronata*) that were increasing their food intake in preparation for migrating, the birds continued to eat but were not able to put on fat. The protein hormone leptin may help signal that sufficient fat has been acquired and a bird is fit to go. The chemistry critical to this phase of the annual cycle is complex and much of it is still uncertain; to date, most of the research has been done on passerines (Holberton and Dufty 2005).

Passerines are among the birds that show the greatest shift in diet before autumn migration. Those that feed themselves and their young protein-based animal matter often shift to fruits that provide fructose and lipids. This is easy to observe in late summer when many migrants gather at fruiting trees and shrubs. Some birds of woodland interiors leave their breeding territory for edges and open areas that have more fruiting plants. There is likely some degree of coevolution between avian frugivory and the plants that have ripe fruit at the end of summer and into autumn; the birds swallow the fruits whole and disperse the seeds when they defecate, which may be far from where they fed. The fruits of some plants, such as pokeweed (*Phytolacca americana*), contain an emetic that assists the process; thrushes and the Gray Catbird (*Dumetella carolinensis*) are especially fond of pokeweed—its toxins, so harmful to mammals, seem not to affect them.

Actual departure dates are linked to the ecology of each species and the latitude at which it breeds. In the Arctic, shorebirds that feed on insects and berries, for example, leave sooner than waterfowl, which consume aquatic vegetation that remains available longer. At 68° N in Alaska, most birds leave in August when the mean temperature is above freezing and any frost is brief; some birds there move south earlier, just after the peak of summer's heat when the weather is still warm. American Robins (*Turdus migratorius*) are the last passerine to depart (Irving 1960:314).

In birds with a wide latitudinal range, usually the populations at higher latitudes begin their autumn migration before those at lower latitudes. There are some exceptions, especially among species where the southernmost birds begin and complete their nesting cycle much sooner than those breeding farther north. Purple Martins leave Georgia in July and August, while New York martins depart primarily in late August and early September (Rappole 2013:104). Since Purple Martins all winter in Brazil and Bolivia, where they search widely for aerial insects and form roosts containing hundreds of thousands of birds, a rush to secure the best feeding sites cannot be the reason for the earlier departure of more southerly nesters.

Long-distance migrants generally leave any given latitude sooner than short-distance or partial migrants, and their timing is usually more consistent. The timing cue is probably more closely linked to changes in day length than to local weather, which more often prompts some short-distance migrants to vacate their nesting area and can vary from year to year. Similarly, in winter when long-distance migrants are in the tropics, their cue for beginning spring migration is independent of conditions at their final destination.

Age and sex also play a role in departure sequence. Among long-distance migrants, adults often leave the nesting area before juveniles, especially in species where adults do not molt until they reach their winter destination. But in species where adults do molt before migration, juveniles may depart sooner. Where one sex has more responsibility for care of the young, the less-involved parent usually leaves first. Failed breeders may migrate even earlier (Newton 2008:348, 434–437). These staggered departures reduce competition in the breeding range and may give the early migrants an advantage in reaching declining food sources on their migration route or in establishing a winter territory.

Each of these patterns has its exceptions. Male and female Common Loons (*Gavia immer*) both stay with their young for 10–12 weeks after hatching, but adult males leave the lakes on which they nested before females, flying to an ocean coast where they will remain all winter, even though males typically winter farther north, thereby traveling less far (Paruk et al. 2015). Conversely, among Long-billed Curlews (*Numenius americanus*), which also have biparental care, females leave Montana and Oregon as much as 22 days ahead of their mates but do not winter in significantly different areas (Page et al. 2014).

In waterfowl, swan and geese families migrate together to their winter destination. Young of the year stay with their parents all winter, often returning with them to their natal area the next spring, but do not actually nest until the following year. Pair formation for adult swans and geese takes place during winter; a bird is therefore likely to mate with another that follows a similar route both south and north, reducing any navigational conflicts the following autumn when experienced pairs lead their naive young south. In contrast, female ducks are solely responsible for incubation and raising the young; they

Unlike other ducks, the young of Harlequin Ducks stay
with their mother through at least part of the winter.

leave them prior to fledging, and some species have a complete molt before migrating. Harlequin Ducks (*Histrionicus histrionicus*) are distinctive because they molt on the wintering grounds and young accompany their mothers there. Among Harlequins wintering on the coast of British Columbia and Washington State, young in some families separated soon after arrival, while others were with their mothers as much as several months (Regehr et al. 2001).

A 44-year banding study in Wisconsin that trapped more than 23,000 raptors found that among migratory hawks and falcons breeding at northern temperate latitudes, young of the year usually depart before adults. Except for American Kestrels, none of the North American species nest before they are two years old, so they have little incentive to stay where they were born to familiarize themselves with the area in anticipation of the following breeding season. In many species, young females migrate before young males. Adults stay on their territory longer, especially the sex that normally defends it. In most hawks, this is the male, but in Northern Goshawks (*Accipiter gentilis*) and possibly in Cooper's Hawks (*A. cooperii*) it is the female; in these two, adult females migrate last (Mueller et al. 2000).

For Arctic raptors, departure is closely linked with food supply. Peregrine Falcons (*Falco peregrinus*) and Rough-legged Hawks (*Buteo lagopus*) have a much shorter nesting season than more southerly birds of prey. Peregrines feed on small birds that have even shorter incubation and nesting periods and are ready to depart sooner, resulting in a rapid diminution of prey at summer's end. Adult Peregrines of both sexes leave as soon as their young fledge, reducing competition for food when the young begin fending for themselves. There is no gender difference in the departure times of the young birds. Rough-legged Hawks in the Arctic are lemming specialists; both sexes and all ages seem to leave during the same time (Mueller et al. 2000). Farther north still, Gyrfalcons (*F. rusticolus*) nesting on the coast of Greenland at 77° N have an abundant food supply in the millions of seabirds living nearby until sea ice forms in October or November; adults and young both remain there until about September 22, while Gyrfalcons nesting at 65° N in interior Alaska, where prey is thinly distributed, depart around August 27 (Burnham and Newton 2011).

Passerine departure dates and sequences vary widely by latitude, by species, and, as we have seen, within species. Of 18 passerine species near Fairbanks, Alaska, only adult Alder

Flycatchers (*Empidonax alnorum*) leave before their young, an average of 13 days sooner. Adult Alder Flycatchers, in fact, stay only 48 days in this part of Alaska. They may depart early because they do not molt until they reach their winter home, on the eastern side of the Andes from Colombia to Bolivia. In the same part of Alaska, the closely related Hammond's Flycacher (*E. hammondii*) spends an average of 86 days, and adults and young depart at the same time. Hammond's Flycatchers molt before migrating and have a much shorter migration, to Mexico and Central America. Eleven other local breeders have immatures preceding adults by several days. These include both short-distance migrants like American Tree Sparrows (*Spizella arborea*) and long-distance ones such as Blackpoll Warblers (*Setophaga striata*) that will fly across Canada to the Atlantic coast and then over water to Brazil. In the remaining six species, most adults and young leave within the same period. This group also includes short-distance migrants such as the Savannah Sparrow (*Passerculus sandwichensis*) and Neotropical ones like the Northern Waterthrush (*Parkesia noveboracensis*) (Benson and Winker 2001).

Farther south, in Idaho, 22 of 31 Neotropical and short-distance migrants show significant differences in timing of southbound adults and young. Among the Neotropical migrants, in species in which adults molt before migrating, the adults remain longer than their offspring in their breeding range. Adults of species that molt after migration leave before young birds. Some of the Idaho breeders are among the species that move to the Mexican monsoon region to molt; for them also, adults depart sooner. All the short-distance migrants molt before moving; they remain or pass through Idaho later than the Neotropical migrants, which typically feed on small insects that decline rapidly at the end of summer and therefore must move south sooner than short-distance migrants eating seeds or fruit (Carlisle et al. 2005).

A different group of Neotropical winterers with more easterly ranges, banded as southbound migrants at stations in Ontario, Pennsylvania, and Alabama, showed a variety of patterns. Most adult Red-eyed Vireos (*Vireo olivaceus*) preceded young birds by 21 days at the northern banding stations and by 17 in Alabama (Woodrey and Chandler 1997). Red-eyed Vireos have a distinctive molt that may influence the difference in timing. Adults are unique among vireos in beginning their molt of body feathers and secondaries in late July and early August but replacing their primaries on the wintering grounds between January and March; young birds have acquired all their first basic plumage when they are approximately 57 days old, which in the northern part of their range is probably late July (Cimprich et al. 2000). While perhaps sufficiently well feathered to leave as early as their parents, the young birds may benefit by having their natal woods to themselves, to feed and bulk up with less competition. In Magnolia Warblers (*Setophaga magnolia*), which have an overlapping northern breeding range with the vireo, immatures preceded adults by an average of as much as 16 days at the northern banding stations and 19 days in Alabama. Adult Magnolia Warblers have a complete molt before migrating. The three other species in the study, Swainson's Thrush (*Catharus ustulatus*), American Redstart (*S. ruticilla*), and Common Yellowthroat (*Geothlypis trichas*), showed no consistent differences, although there was some variation among the three banding sites and between years (Woodrey and Chandler 1997).

SUMMARY

Birds begin preparing for winter long before that season arrives. Many birds molt at the end of summer into the plumage they will wear through the winter and, for some, into the following breeding season. Among migrants, some replace only certain of their feathers before departing and then complete their molt on arrival at their winter destination. Still others delay their entire molt until they have reached their winter range, where some have a second molt, usually only of body feathers, that returns them to the plumage of the breeding season. The timing of molt, which also may vary between adults and birds born that year, is linked to the relative availability of time before departure and of food at each end of the migration. While many birds, both permanent resident and migrant, defend territories wherever they winter, there is no evident correlation between either molt schedule or distinctive plumage and winter territoriality.

As early as late summer, when food is still abundant, many nonmigrant passerines, from jays to tits and nuthatches, as well as some woodpeckers, begin storing nonperishable seeds, nuts, and arthropods for retrieval during the winter. Some individuals cache tens of thousands of seeds. Jays that bury acorns in the ground, where some never retrieved may germinate, are incidental dispersing agents for the trees on which they depend in winter. Most of these birds are scatter-hoarders, secreting one or a few items each in many concealed sites used only once until they return as much as several months later. The need to remember all these locations through the winter has led to a proportionally larger hippocampus, the part of the brain responsible for this kind of memory, than in related species that do not cache. Birds of prey, especially some owls and falcons, also store food; because their prey is larger and more perishable, they do not begin until winter itself when cold temperatures will preserve it. Most storing raptors are larder-hoarders, putting their catches in a single place they actively defend.

Finally, the migrants prepare for their departure, which may be triggered by changes in day length, by internal rhythms, or both. Most long-distance migrants depart on a highly fixed schedule, often while food is still abundant, while short-distance migrants may remain longer, responding to local conditions. Long-distance migrants, especially those traversing extensive areas like oceans or deserts where they cannot rest or feed, put on substantial weight to fuel their long flights. Many passerines shift their diet to fruits that provide fructose and lipids. Age and sex influence departure and are in turn linked to molt schedules. Adults of species that molt only when they reach their wintering site usually depart sooner than young, while adults of species that molt before departure remain longer. When one sex is more involved in raising young than the other, it may also remain longer, and in diurnal raptors the sex that defends the breeding territory stays there longest.

CHAPTER 3

WINTER RANGES AND HABITAT SELECTION IN MIGRATORY BIRDS

FOR MOST SHORT-DISTANCE OR partial migrants, the wintering site is not substantially different from the place they departed. Waterfowl winter on water bodies—fresh, estuarine, or coastal—that do not freeze; they move to another if the initial site does freeze. Forest raptors stay in forest. Raptors of open country that take prey from the ground move to open country less likely to be covered with snow. Passerines that do not travel far similarly remain loyal to their breeding habitat, be it woodland, scrub, or grassland—even as these change when trees become leafless, other vegetation dies back, and weather conditions alter so that some birds must shift to a different food source. American Robins (*Turdus migratorius*), wintering where the ground freezes and worms are no longer accessible, instead feed substantially on fruit from trees, vines, and shrubs. Most predators and seed eaters, however, continue to have a similar diet year round, though the actual species they consume may be different.

Long-distance migrants traveling to tropical latitudes enter a very different world. Most are, as we have seen, returning "home" to their evolutionary place of origin. For individual birds that have made the trip before, as well as for young ones arriving for the first time, there is an instinctual if not experiential familiarity with the region where they will spend the longest portion of their year. The idea that these birds were marginal to the tropical environment, competing with the permanent residents and less well adapted to exploit the region's resources, has given way—with the recognition that most of the migrants originated in the tropics—to the concept that each migrant species is fully integrated into the ecosystem it inhabits. Some may in fact be more successful in patchy and ephemeral habitats than are the many permanent residents that never leave the forest interior. These habitats are becoming more common as forest and other old growth is increasingly disturbed or destroyed by human activity, but the extent to which wintering birds can survive in degraded habitats depends on the resemblance these have to natural areas the migrants have exploited for millennia.

CONTRASTS IN BREEDING AND WINTER RANGES

Most long-distance migrants encounter several features in their winter range that are significantly different from the region where they bred. In addition to the presence of many more other bird species, for Northern Hemisphere migrants the entire available landmass is less than their breeding range, and the local climatic conditions are distinctive. The spatial limitations of winter range are most acute for northern migrants to the Neotropics. A glance at a map will show that the entire region used by most Nearctic passerine migrants—from Mexico and the Caribbean islands to northern South America—is much smaller than the North American forests from which most of the wintering birds came. The range of each species within this area is usually smaller still. But even in the region of highest wintering migrant diversity, northern Central America, the seasonal birds are less than one-quarter of the local avifauna (Stotz et al. 1996:65). Thus, the number of potentially competing birds—within each more tightly packed species, as well as among the different migrant species and between migrants and the much more numerous resident species—means that each of the wintering birds has developed a viable niche and successful strategy for survival.

Some shorebirds wintering in southern South America are concentrated even more densely. The Pectoral Sandpiper (*Calidris melanotos*) breeds from central Siberia to the western side of Hudson Bay in Canada and winters in parts of Argentina, Uruguay, and Paraguay. American Golden-Plovers (*Pluvialis dominica*) in Argentinean grasslands occur at densities of 10 birds per hectare, while in northern Alaska where they breed there are rarely more than 0.2 birds per hectare. Most of the North American shorebirds wintering in the pampas are territorial—but their territories are usually from one-tenth to one-hundredth of what they defend during the breeding season (Myers 1980a). In contrast to the species that have a concentrated winter range, North American Sanderlings (*C. alba*), which breed in the coastal region of Arctic Canada and Alaska, winter on sand beaches from 50° N on the Pacific coast and 35° N on the Atlantic down to Tierra del Fuego, at 50° S. For them, finding enough suitable habitat requires stretching thinly over 100 degrees of latitude in a very narrow band where the sea meets a beach.

For Palearctic birds that winter entirely in Africa, those with the largest breeding ranges have the most extensive winter ranges. But for 69% of those species, the total winter range is nevertheless smaller than their breeding range. For 57 landbirds, the winter range is two-thirds the size of the breeding range, and for many of these species, only portions of this range are used at any one time, because the birds move long distances within it over the course of the winter to exploit food sources that come and go with the rains. Thus, in any given month, the bulk of any species' population may occupy a fraction of its total African winter range (Newton 1995).

Just as some species with vast breeding ranges are compressed into a much smaller wintering area that provides a reliable source of food, the scarcity of a key ecological feature such as roosting sites that can safely accommodate large flocks may similarly concentrate populations from several regions. For terns that in winter feed at sea but roost on land at isolated coastal sites and islands, these may be shared by birds of different populations. The first known major wintering site of the threatened Roseate Tern (*Sterna dougallii*) in South America was discovered only in 1995. It is a sandy point in Bahia, Brazil, at the mouth of the Rio Real. By day, three local tern species gather there at Mangue Seco. After dark, an estimated 10,000 Roseate and Common (*S. hirundo*) Terns come in to roost; they depart before dawn. Three years of netting operations captured many terns that had been banded at widely scattered breeding sites. The Roseates came from colonies on the coast from New York to Maine as well as from colonies in Puerto Rico and the US Virgin Islands. Common Terns came mostly from New York and Connecticut as well as

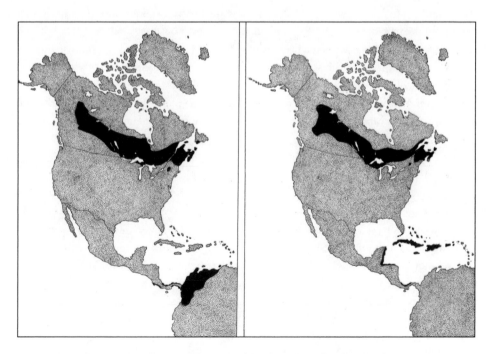

While Bay-breasted (left) and Cape May (right) Warblers have practically identical breeding ranges, they are almost entirely separate in winter.

from the Azores, at a similar latitude in the eastern Atlantic—the first known case of a transatlantic migration for this species (Hays et al. 1999).

Warblers, with 47 species leaving Canada and the United States to winter from Mexico and the Caribbean south, demonstrate the fine partitioning of range, habitats, and ecology that enables them all to make it through so many months in a region that, in total, is less than half the size of their collective breeding range. Consider some for which the winter range is a far smaller fraction: The Cape May Warbler (*Setophaga tigrina*) breeds from northern New England and the Maritime Provinces west and north to Mackenzie and Alberta; it winters primarily on the northern Caribbean islands, with a smaller number on the Caribbean coast of Veracruz, Mexico, and Central America. The Bay-breasted Warbler (*S. castanea*) has almost the same range within the boreal forest and winters primarily in lowland Costa Rica and Panama and northernmost Colombia and Venezuela. The Chestnut-sided Warbler (*S. pensylvanica*) overlaps much of the breeding range of these two and extends south into the Appalachians, but it lives in early successional habitats, not evergreen forest. In winter, however, it shares lowland forests with the Bay-breasted Warbler in Costa Rica and Panama and extends north on the Atlantic side of Central America to southern Mexico.

Different feeding methods, even among superficially similar foliage gleaners like the Bay-breasted and Chestnut-sided Warblers, reduce the competition where they winter together. The slightly larger Bay-breasted is a methodical searcher that peers carefully for insects on leaves, while the Chestnut-sided moves rapidly, shaking the leaves to frighten and dislodge prey. They probably each specialize in different insects on the same foliage—immobile ones for the Bay-breasted, flying ones for the Chestnut-sided. In contrast, the Black-and-white Warbler (*S. varia*) has a foraging method unique among warblers; it climbs on the trunks and major branches of trees looking for insects on the bark. With no competition from other warblers, both its breeding and winter ranges are more extensive. In winter, it is found from coastal North Carolina through Florida and the Caribbean islands and from northern Mexico to northern South America. Unlike the Cape May and Bay-breasted, which in winter sometimes flock at fruiting plants and are common in much of their compact range, the Black-and-white is solitary, territorial, and thinly distributed.

WEATHER AND CLIMATE

The arriving migrants find distinctive conditions in various parts of the tropics. In Panama, the rainy season continues into December. There, many insectivores settle first in wetter habitats and microclimates such as mature forest and mangrove swamps. These will remain moister through the dry season, which begins in January and continues until after the birds depart in spring. Foliage gleaners that may have begun their stay in young forests move to mature forests in early January and may then shift back in March when the first fruit appears (Morton 1980). In South America, migrants wintering north of the equator have a dry season for most of their stay, while south of the equator, where far fewer Nearctic migrants winter, the rainy season usually begins in October, soon after they arrive, and continues until May or June.

In northern tropical Africa, the rainy season is from May to September, ending before or just as most migrants are arriving—which is also when some local African breeders move farther south (Fry 1992). Food dwindles through the winter and is not replenished by new rains until after the Eurasian birds have left. Water is rarely available even as dew, and most small migrants must get all their moisture from insects and the fruits produced in the dry season. Nevertheless, many birds remain in this belt until they return north. Seed eaters can stay there through the dry season and would find little food farther south in the rainy season, when most available seeds germinate. Many insectivores, as we have seen, leave to follow the rains that begin in September south of the equator; insect and plant foods are renewed throughout their stay. The rains also precipitate the emergence of swarms of winged termites and winged ants, major food sources for wintering aerial insectivores like the Common Swift (*Apus apus*) and Bank Swallow (*Riparia riparia*) (Moreau 1972:53–54, 66–69).

Daylight and temperature are also significantly different in breeding and wintering areas for long-distance migrants. Only birds that migrate from one hemisphere's high latitudes to the other's, like the Arctic Tern (*Sterna paradisaea*), the many Arctic-nesting shorebirds that winter in southern South America, Africa, Australia, and New Zealand, and the austral shearwaters and petrels that go into the northern Atlantic and Pacific for the Northern Hemisphere summer, have the same long days all through the year. At the equator, the day is 12 hours 7 minutes (Moreau 1972:55). Shorebirds that depend on the ebbing tide to expose feeding areas also lose valuable hours in the short equatorial day; many of them compensate by feeding at night, as do those wintering at northern latitudes where the winter day is shorter still. But most birds that locate their food by sight cannot forage after dark—and for those that live in lowland tropical forests where the sun may be filtered by 50 m of foliage, it is too dark to feed near the ground until well after dawn and before dusk.

To escape the intense heat of the African desert, wintering birds like this Northern Wheatear may spend many hours resting in the shade.

For many birds wintering in the tropics, the heat of midday further shortens the hours available for foraging. In much of Africa from the tropics to South Africa, daily temperatures exceed 27° C and often 32° C, dropping most nights to 14° C or above. These are "standard" temperatures, but close to the soil, temperatures are often 5° C warmer during the day. In Sudan, open-country ground feeders, like Northern Wheatears (*Oenanthe oenanthe*)—which may have spent the summer well above the Arctic Circle—can experience air temperatures of 40° C, while the surface of the unshaded sand on which they walk is over 80° C. When not feeding, wheatears seek out a perch even slightly elevated over the sand; in Nigeria, they may spend as much as six hours of daylight resting in the shade (Moreau 1972:55–59, 253).

For Palearctic birds wintering in Africa, the daily maintenance needs may be 17% less than in the breeding season because of shorter periods of activity, and another 16% less because of the greater warmth. Not having to care for young birds further reduces energetic needs, by perhaps 10% (Moreau 1972:258). There are no comparable estimates for Nearctic or austral landbirds that winter in tropical latitudes, but some of the same considerations should apply—note, however, that many of these birds winter in forest or second growth that is shaded and cooler than arid parts of Africa. These factors mean that most birds make fewer demands on their winter home than where they breed. This may help explain how so many more birds can be accommodated at these latitudes for so many months of the year.

HABITAT CHOICES

Whether birds remain where they have bred or depart, whether they wander opportunistically as do crossbills and redpolls, move at regular intervals over the winter as do so many Eurasian birds in Africa, or are faithful year after year to a single wintering site, all birds must make a choice about the habitat they use during that season. Reliable availability of food is of course the first consideration. For some birds, safety from predators by day is another factor. Roosting sites may also be chosen for safety or for protection from the elements and may be different or at some distance from the areas used for feeding.

Gyrfalcons (*Falco rusticolus*) from Greenland demonstrate flexibility of habitat choices in a seemingly limited and challenging environment. Falcons equipped with satellite transmitters were tracked from their nesting area at 75–77° N. None consistently wintered north of 67.9° N, where midwinter twilight is almost 6 hours, but one female was at 70.6° N on the winter solstice, December 21, when twilight was 3.5 hours. Some Gyrfalcons established home ranges farther south on the Greenland coast or had a succession of as many as four use areas each occupied for one to three months, staying on the sea ice as it extended from shore and then retracted over the winter. The core of their home ranges ran from 160 sq km to 26,157 sq km, which must reflect territories of varying quality, with their prey, wintering seabirds, close by or more scattered. Still other Gyrfalcons moved continuously. One juvenile female traveled over 4548 km during approximately 200 days, spending more than half that time at sea between Greenland and Iceland, presumably resting on icebergs and ice floes, feeding on the abundant murres and ducks wintering on the open sea. But because the seabirds are not evenly distributed, a falcon at sea must travel farther each day to find them than do the birds with prime coastal territories (Burnham and Newton 2011).

Most migrants select a habitat as similar in structure if not in species as widely differing latitudes allow. Thus, birds that breed in early successional habitat usually winter in the same, and forest birds choose forest. About 60% of Nearctic migrants that winter entirely in the Neotropics prefer some type of primary forest. Pine and pine-oak forests of Mexico and northern Central America hold the greatest diversity of migrants. Lowland evergreen forest, especially in Central America, supports many of the other wintering birds, and in northern South America, montane forest hosts more than lowlands (Stotz et al. 1996:69). Among migrants to the Neotropics, a few species shift to very different habitats. Forest birds wintering in scrub include Least Flycatcher (*Empidonax minimus*), Eastern Wood-Pewee (*Contopus virens*), Blue-gray Gnatcatcher (*Polioptila caerulea*), and Rose-breasted Grosbeak (*Pheucticus ludovicianus*), and scrub breeders shifting to forest include Chestnut-sided Warbler, Blue-winged Warbler (*Vermivora pinus*), and White-eyed Vireo (*Vireo griseus*) (Askins et al. 1983:41).

The warbler species that breed exclusively west of the Rocky Mountains winter almost entirely between northern Mexico and Guatemala, on the Pacific slope, where they each settle into habitats very similar to where they bred. The desert scrub, riparian oak woodlands, and forests, at increasing elevations, of pine-oak, pine, and pine-fir, are almost contiguous with similar habitats in the United States and Canada. This proximity and similarity have enabled each species to specialize in a particular habitat type it can exploit all through the year (Hutto 1985:470). These species may well have originated in the region where they now winter and expanded northward when their habitats were continuous. While they may be extremely densely packed in this narrow strip in winter, they have had little or no evolutionary experience to expand their range into the moister, more tropical habitats used by their eastern counterparts.

These eastern warblers, in species and populations more numerous than the exclusively western ones, each have their geographic and foraging niches that, as we have already seen, enable them to partition their collective winter range more finely and reduce competition. A short tour, focusing on habitat, through some of the key wintering sites will supplement the examples already given and those that follow in the context of subsequent chapters. Most of the research on winter habitat preference has compared what remains of pristine landscapes with each locale's range of agricultural and abandoned lands in various stages of regeneration, to determine the extent to which migrants can adapt to and exploit these less natural areas.

In the US Virgin Islands, 13 warbler species are found through the winter, when in some areas they are 99% of the individual winter residents found on surveys. St. John retains more of the island's original habitat, with 88% of the land undeveloped, while in 1987 adjacent St. Thomas was 62% urban and suburban. On both islands, nearly all individuals of the 13 warbler species were found in moist forest and the border with adjacent dry woodland, with both density and diversity higher in the moist forest. The extensive forests on St. John held many more birds at greater densities than the remnant patches on St. Thomas (Askins et al. 1992).

The Yucatán Peninsula is both a corridor for migrants crossing the Gulf of Mexico at the shortest distance to the United States and a winter home for many of the migrants that travel no farther. On the northern peninsula, 21 common wintering species, including flycatchers, catbirds, thrushes, and finches as well as 13 warblers, were surveyed in a cross section of widespread habitats, revealing different degrees of habitat breadth and adaptability. Twelve of the 21 species were commonest in early successional scrub and farmland resembling it, 8 were more frequent in later second growth and mature forest, and 9 occurred regularly in the entire range from pasture and cultivated fields to mature tropical forest. The species most restricted to forest were Wood Thrush (*Hylocichla mustelina*), Black-throated Green (*Setophaga virens*) and Kentucky (*Geothlypis formosa*) Warblers, American Redstart (*S. ruticilla*), and Ovenbird (*Seiurus auricapillus*) (Lynch 1992).

Another survey elsewhere in the northern Yucatán, in the Sian Ka'an Biosphere Reserve, Quinana Roo, where forest is more widespread, found that most migrant landbirds were more common in forest than in early second growth. Over the course of the winter, as arthropod prey diminished in the second-growth habitats, migrants declined there. The forest-dependent migrants that did use second growth concentrated in small patches of trees. Black-throated Green Warblers there favored acacias and other leguminous trees with small leaves, perhaps because these most resemble the needle foliage of coniferous forests where they breed. White-eyed Vireos, the only significant fruit eater among the migrants, was most often near *Bursera* trees. Worm-eating Warblers (*Helmitheros vermivorum*), which forage for arthropods in curled-up dead leaves attached to understory plants—"aerial leaf litter"—were uncommon in the dry forest and second growth, because few plants there produce the large leaves this warbler searches (Greenberg 1992).

Some Yellow Warblers (*Setophaga petechia*) wintering on the Yucatán Peninsula favor coastal mangrove forests, which they share with resident Yellow Warblers. This may have been the species' original habitat, with the migrant birds adapting as they expanded northward to wetland and streamside edges with willows, and the most northerly birds living in Arctic willow scrub and dwarf birch. In mangroves, they forage lower beneath the canopy than in other habitats, and they take more food while perched, rather than when fluttering to pick something off a leaf, as they do in other seasons. In these small shifts, they come closer to the feeding style of the resident Yellow Warblers in the mangroves (Wiedenfeld 1992).

On the southern part of the Yucatán Peninsula, in Belize, migrants are common in mature moist forest but are even more abundant in early successional stages adjacent the forest. Some 57% of the wintering species in a study in southern Belize favored edges and forest borders. But here, too, Black-throated Green Warblers as well as Black-and-white and Magnolia (*Setophaga magnolia*) Warblers were restricted to the canopy of trees, while Wood Thrushes, Kentucky Warblers, and Ovenbirds were most frequently in moist forest. The birds favoring successional sites were species that live in similar habitat during the breeding season—Gray Catbirds (*Dumetella carolinensis*), Common Yellowthroats (*Geothlypis trichas*), and Yellow-breasted Chats (*Icteria virens*) (Kricher and Davis 1992).

Farther south, on the Atlantic slope of Costa Rica, two studies, one in Braulio Carrillo National Park and the other at La Selva Biological Station, found mostly similar results. In a three-year survey of habitats below 1000 m, Acadian (*Empidonax virescens*) and Yellow-bellied (*E. flaviventris*) Flycatchers, Wood Thrush, Kentucky Warbler, and Louisiana Waterthrush (*Parkesia motacilla*) were entirely in primary forest. In more-open habitats, from second growth to grassland, the wintering birds stayed in forest remnants (Eastern Wood-Pewee and Baltimore Oriole [*Icterus galbula*]) or foraged in grass—but only when it was near woody vegetation (Yellow Warbler) or a marsh (Common Yellowthroat) (Powell et al. 1992).

In other parts of the same protected areas, a five-year study that surveyed wintering birds in new and older second growth as well as primary forest found Wood Thrushes and Kentucky Warblers in both forest and second growth, which had abundant fruit. Several other species, including Chestnut-sided Warbler and Ovenbird, which surveys elsewhere, already cited, found in forest, were also in young second growth (Blake and Loiselle 1992). This study extended into April, while the other Costa Rican one ended in February, and the presence of forest birds in the second growth may reflect the dietary shift of so many migrants to fruit before their departure. Not only does young second growth have more plants that bear fruit in March and April, but more of these fruits are small enough to fit into the mouths of wintering birds, while most of the resident frugivores are bigger and consume larger fruits (Martin 1985).

In Panama, where migrants arrive in the wet season, many shift in January from their initial use of second growth and other drier habitats to moister, mature forest when the dry season begins. Among the warblers, many Bay-breasteds begin their winter in shrub habitats and shift to forest; Chestnut-sideds choose late second growth and mature forest throughout the winter; Kentucky Warblers stay in moist forest; Northern Waterthrushes (*Parkesia noveboracensis*) are in mangrove swamps, forested streams, and other damp places; and Tennessee Warblers (*Leiothlypis peregrina*) seek out nectar sources in all habitats. Baltimore and Orchard (*Icterus spurius*) Orioles also congregate at nectar-rich flowering trees and shrubs in all habitats. Omnivorous species that consume both insects and fruit, such as Great Crested Flycatchers (*Myiarchus crinitus*) and Summer Tanagers (*Piranga rubra*), are also widely distributed (Morton 1980).

Far fewer Nearctic passerines winter south of the equator. At 11° S latitude, migrants in Manu National Park and adjacent Andean slopes in southeastern Peru are finely partitioned by habitat and elevation. In the lowlands at 350 m above sea level, the most-used habitat is the stands of *Tessaria integrifolia*, a treelike composite that is the first woody plant to colonize newly created beaches on the inner side of river meander loops. Within two years, it may be joined by *Gynerium sagittatum*, a cane. *Tessaria* is subject to outbreaks of lepidopteran larvae and to swarms of flies that are an abundant, but ephemeral, food source. As the rainy season progresses, the stands are inundated, so permanent residents cannot live exclusively in this habitat, while wintering birds can simply move on. The winterers make up 17% of the birds in this and adjacent early successional habitats, but fewer than 1.5% in the primary forest on higher ground. The Alder Flycatcher (*Empidonax alnorum*) is common in *Tessaria*, and the Eastern Wood-Pewee in the taller vegetation of the transitional forest farther behind the beaches. During the dry months, June to October, when *Tessaria* is always on dry ground, it supports austral migrants; of the 19 austral migrants to Manu (including 11 flycatchers), most use *Tessaria* (Robinson et al. 1988). The tall forest away from the river edge is used by only a few Nearctic migrants, notably Black-billed (*Coccyzus erythrophthalmus*) and Yellow-billed (*C. americanus*) Cuckoos and Scarlet (*Piranga olivacea*) and Summer Tanagers, and these are thinly distributed (Munn 1985).

The Andean foothills immediately to the west support some of the same species as well as additional migrants. Both tanagers are commoner here. Eastern Wood-Pewees range up to 1400 m in elevation in forest openings, and Olive-sided Flycatchers (*Contopus cooperi*) are found at tree gaps between 500 and 2000 m. Here, treefalls and landslides on the steep slopes create the edges and disturbed landscape that are structurally similar to the succession of colonizing vegetation on the lowland riverbanks. Cerulean (*Setophaga caerulea*) and Canada (*Cardellina canadensis*) Warblers use the forested foothills and lower slopes, while Blackburnian Warblers (*S. fusca*) occur up to at least 2900 m and Rose-breasted Grosbeaks between 2500 and 2600 m (Robinson et al. 1988).

Much of the research cited above has measured the extent to which migrants and wintering birds will exploit disturbed or regenerating land compared to their use of the natural forest landscape. As we have seen, this varies by region and is affected by the extent of any remaining patches of woody plants as well as those that bear fruit. As any agricultural land loses the original vegetation that was spared when the forest was cleared, its value to both wintering and resident birds diminishes toward zero.

A study focused specifically on croplands from the Caribbean islands through Mexico and Central America to Venezuela found some forms of cultivation much more usable by wintering birds than others. In Puerto Rico and the Virgin Islands, the Northern Parula (*Setophaga americana*) is the commonest wintering warbler; it is found equally in cropland and in native forest. Black-throated Blue Warblers (*S. caerulescens*) in Puerto Rico and Jamaica are frequent in shade coffee plantations but rare in other agricultural habitats. American Redstarts on those islands and in Belize wintered in stands of both sun and shade coffee as well as in plantations of cacao, citrus, pine, and mango. Magnolia Warblers in Belize have adapted to cacao, citrus, and mango plantations. The Hooded Warbler (*S. citrina*) in Belize was found regularly in pine and cacao plantations and was scarce or absent in other agricultural habitats. Ground feeders seem less adaptable, with the Ovenbird present in only small numbers in cultivated land on the islands and in Belize and Costa Rica, and the Wood Thrush and Kentucky Warbler not at all. Finally, in Belize nectar feeders like Baltimore and Orchard Orioles were more abundant in citrus and cacao plantations than in native forest, and rice fields attracted Common Yellowthroats and Indigo Buntings (*Passerina caerulea*) (Robbins et al. 1992).

Many Arctic shorebirds traditionally winter in South American wetlands, but much of that habitat has been converted to agriculture. In interior northern Argentina and southern coastal Brazil where 12 Nearctic plovers and sandpipers winter, rice fields have become the default habitat in the absence of any remaining significant wetlands. Grassland specialists, the American Golden-Plover and the Buff-breasted Sandpiper (*Tryngites subruficollis*), are common in the rice fields when these are still dry. As the fields are first flooded, muddy

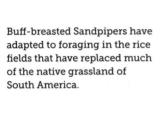

Buff-breasted Sandpipers have adapted to foraging in the rice fields that have replaced much of the native grassland of South America.

depressions especially attract White-rumped Sandpipers (*Calidris fuscicollis*). Most of the 12 species are commonest in the second stage of rice cultivation, when the fields have been lightly flooded and the rice is less than 20 cm tall. As the water level rises, only the longer-legged Pectoral Sandpiper and Lesser Yellowlegs (*Tringa flavipes*) remain common in the now denser vegetation; other wintering species are scarce or absent (Dias et al. 2014).

In Africa, most terrestrial Palearctic migrants use savanna, as we have seen, with very few wintering in forest. Desertification of the Sahel from overgrazing, intensified crop cultivation, and climate change has reduced usable habitat in that zone. Farther south, clearings in what were once forested areas have created new opportunities for a few ground feeders that are mobile and opportunistic. In Liberia, which was once extensively forested, conversion to agriculture has benefited wintering Yellow Wagtails (*Motacilla flava*) and Tree (*Anthus trivialis*) and Red-throated (*A. cervinus*) Pipits, which can find even small plots of open land surrounded by forest (Leisler 1990).

DIFFERENTIAL MIGRATION AND HABITAT PARTITIONING

The prevalence of territoriality, aggression, and dominance hierarchies among wintering birds demonstrates the intensity of the competition for the key resources essential to survival—food and shelter—and the importance of securing an advantageous place when settling in for the winter. For many birds, direct competition is reduced or avoided when each sex and age group differs in habitat choice, feeding specialty, anatomy, or winter range. For some, the winter partitioning of foraging sites and food sources is subtle and may replicate habits that evolved to meet the needs of other seasons; for others, the natural selection pressures of winter, which have led to gender-based habitat separation and dietary distinctions, may be more responsible for anatomical and behavioral differences than are aspects of the breeding season.

The winter movements of three owl species that breed in Scandinavian forests demonstrate the variety of potential responses to the same environmental conditions by different sexes and age classes. Ural Owls (*Strix uralensis*) are generalist feeders that nest in tree cavities, a scarce resource; each pair remains together on its territory, thereby securing the nest site for the following year. Long-eared Owls (*Asio otus*) prey on voles and use the abundant twig nests made by other birds; they are migratory. Boreal Owls (*Aegolius funereus*) also feed on voles but nest in tree cavities; to ensure that a nest site is retained but to reduce winter competition for food, males remain on their territory through the season while females and young birds migrate. In Ural Owls, the familiarity with the territory gained over the winter increases long-term survival, while the size difference between larger females and smaller males reduces competition for food. For Boreal Owls, male survival is enhanced by the absence of females and young; females do not return in spring to their former territory, so they lose no useful local experience by migrating (Lundberg 1979).

SHORT-DISTANCE MIGRANTS: AGE AND SEX PARTITIONING

Permanent residents can reduce competition when each sex and age class selects a different habitat for the winter. Migrants can do the same, adding another degree of separation if some sex or age classes travel different distances from their breeding range. Differential migration occurs more often in short-distance migrants, which often have a larger winter range than the long-distance migrants that concentrate at specific latitudes. Three major factors may play a role in where each class winters: body size, dominance, and distance from breeding territory. These are not mutually exclusive and are difficult to separate experimentally. The significance of each varies among species.

The body-size hypothesis assumes that larger individuals of any species are more likely to survive cold temperatures than smaller individuals because they lose body heat more slowly, can store more fat, and can fast longer. That enables them to remain at higher latitudes or elevations, closer to their breeding site. The dominance hypothesis asserts that dominant classes—based on sex, age, or size, which are often interrelated—will winter in the most favorable areas, forcing subordinate classes to less optimal places that may require additional travel, with its attendant risks. Finally, the arrival-time hypothesis posits that the class seeking to return earliest to its breeding range to establish a territory for the next season—usually adult males—winters at the least distance. But in most species at least some individuals of different sizes and sexes do winter at the same places, sometimes without notable differences in survival, while survival rates are sometimes higher at greater distance, that is, usually at more benign latitudes. And birds seeking to be first at their breeding sites can leave these more distant wintering areas earlier or travel more rapidly (Cristol et al. 1999). Thus, while differential migration and habitat segregation by age and sex are widespread, the significance of each factor behind the pattern in various species is challenging to separate.

Behavior as well as ecology influences the likelihood and nature of differential migration. Among waterfowl, swans and geese remain paired throughout the year and lead their young to wintering sites and so are not likely to separate by sex or age class on arrival. Dabbling ducks begin pairing in autumn or winter for the following spring, so an even sex ratio at any wintering site is beneficial. Where there are more males than females, this is usually due to a lower survival rate of female ducks before they reach their winter range; female dabbling ducks have sole responsibility for incubation and care of young and so are at greater risk than males, which have no role in reproduction after inseminating the females.

Among diving ducks like the Canvasback (*Aythya valisineria*), however, mating takes place when birds return to their breeding sites; in winter the elements of all three hypotheses for differential migration are evident, as is habitat partitioning. Males are larger than females, tend to winter farther north, dominate at preferred sites, and return first to their breeding range. In Chesapeake Bay, males are in large rafts in the estuaries and main tributaries, while females are in small flocks in creeks and ponds. When both sexes are in large rafts, females are on the periphery, where they are more vulnerable to predators. They are generally in poorer condition, and in harsh winters their survival is lower than that of males (Nichols and Haramis 1980).

On the Pacific coast, first-year male Surf Scoters (*Melanitta perspicillata*) tend to winter south of adults. Where they occur with adults and females of all ages on the coast of southern British Columbia, habitat preferences are significant: more first-year males are in areas with low exposure to wind and waves, while females do not seek more sheltered sites. Males of all ages feed in waters with sandy substrates, where clams are their chief prey; females dive for mussels in rocky or muddy substrates. The first-year males cluster in flocks, especially within larger rafts, but females do not preferentially associate with their own sex (Iverson et al. 2004).

In birds of prey, females are larger, but in many migratory species the females winter farther from their breeding range. In the American Kestrel (*Falco sparverius*), the pattern of dominance and habitat use is complex. Adult males of migratory populations winter north of most females and juveniles and return to their breeding range approximately seven days before females. They are also the last to leave their nesting area and, where they overlap with females and juveniles in winter, may arrive more than a week later. The females and young birds have already established their territories on the better feeding sites—those with more open ground—leaving the adult males to settle in areas with more trees, less open ground, and less visibility (Smallwood and Bird 2002). In south-central Florida, territories of adult females average 77% open substrate, while territories of males have 55%. Both sexes there feed exclusively on arthropods found in low vegetation, but males, in their inferior territories, need to spend more of the day hunting than do females, even though females have a 9% greater body mass and presumably require more food (Smallwood 1988).

Migratory European Robins (*Erithacus rubecula*) from Scandinavia and northern Europe overlap resident ones in winter at mid-European latitudes, with males commoner farther north. In southern Portugal, where no robins breed, robins are abundant in winter in woodlands and shrubby areas. Nearly 18% of the robins there are male, but they are 29% of the birds occupying territories in woodland and only 13% in shrubland, where females, young, and small individuals of both sexes are prevalent. Since the woodland birds, both male and female, were found to be in better condition, it is likely that dominance rather than any innate habitat preference creates the habitat separation (Catry et al. 2004).

Similarly, in the robin's closest North American ecological equivalent, the Hermit Thrush (*Catharus guttatus*), males tend to have superior territories—those with more arthropod prey available through the winter. In South Carolina, fruit dominates the diet of both sexes in the first part of the winter, but as the season progresses, larger birds, mostly males, eat more arthropods, which provide protein. The birds of inferior territories have more fat—for many species, a feature of birds in territories with fluctuating rather than reliable food sources (Diggs et al. 2011).

The Townsend's Solitaire (*Myadestes townsendi*) consumes only fruit in winter, primarily juniper berries. In Arizona, males establish territories large enough to have an excess of juniper berries; the dimensions change from winter to winter, sometimes by a factor of five, depending on that year's crop. In good years, both adult and subadult males have territories, while in poor years the subadults may be excluded. Females and some males are floaters, moving inconspicuously and quietly, while the territorial males spend much of the day perched high in a tree, scanning for trespassers. These dominant birds expend much less energy than the floaters, which are chased and must fly from territory to territory and therefore have less access to replenishment (Salomonson and Balda 1977).

Differential migration has been most thoroughly studied in eastern populations of the Dark-eyed Junco (*Junco hyemalis*). In northern New England and west through the Great Lakes region and beyond, 80% of wintering juncos are male. From South Carolina to the Gulf Coast and Texas, 70% of the birds are female. As in so many birds, adult juncos of each sex are larger than hatching-year birds of the same sex, but what may be unusual compared with other species that have differential migration is that in juncos these first-year birds tend to winter north of adults. Thus, adult females are farthest south and young males farthest north. In males, the age separation is less distinct, with many adults wintering at the same latitude as the bulk of hatching-year males, and another peak farther south. First-year male juncos may winter closer to their future breeding range in order to arrive there early and establish a superior territory, but adult males are dominant, even if returning later, and so do not need to get there first. Since

From the plumage variations in different sexes and ages of Dark-eyed Juncos, one can see which have overlapping ranges in winter.

overwinter survival improves for birds going farther south, adults of both sexes have little incentive to remain at higher latitudes. Adults, males especially, are therefore less faithful in subsequent years to their first wintering site. And, unlike the species that have individual winter territories, in junco flocks there is no sexual partitioning of the habitats used (Ketterson and Nolan 1983).

LONG-DISTANCE MIGRANTS: HABITAT PARTITIONING

When migrants depart for lower latitudes in autumn, they reduce the competition for resources among the resident species they leave behind, but they may increase it for themselves and their winter neighbors, because the warmer latitudes support more species at all seasons. Competition is further increased for many long-distance migrants because their winter range in the tropics or subtropics is significantly smaller than their breeding range. Individuals of many species may thus be more tightly packed, as well as sharing space with both local residents and potentially competing wintering birds of species they do not encounter during spring and summer. Winter niche partitioning within and among these different sectors is therefore more complex than at any other time of year, even before considering sex and age divisions of available habitat.

On Caribbean islands where the resident avifauna is small, wintering birds substantially increase the pressure on local resources. Jamaica has 35 breeding passerines, including 4 that are absent when North American winterers inhabit the island. Eighteen migrant warbler species regularly winter in Jamaica, joining the 2 resident warblers. Another 12 warblers pass through Jamaica on migration, continuing farther south to other islands or to the mainland, presumably because more distant destinations offer less competition. A few of the warblers that winter in Jamaica have elevational or habitat restrictions: the Northern Parula (*Setophaga americana*), Prairie Warbler (*S. discolor*), and American Redstart are more common in the lowlands, but most other species are found at all elevations and in all types of forest. The five ground feeders each use different habitats or feeding methods: Palm Warbler (*S. palmarum*) in grasslands, Northern Waterthrush in standing water, Louisana Waterthrush at rocky streams and rivers, Ovenbird in forest leaf litter, picking insects from the surface, and Swainson's Warbler (*Limnothlypis swainsonii*) in the same litter, tossing leaves and probing.

That these specializations are fostered by competition is shown by Louisiana Waterthrushes at elevations above Northerns; there, they also forage in standing water. Farther south, in Trinidad, where the Northern Waterthrush is the only wintering ground warbler in the forest, it feeds in leaf litter and along rocky streams as well as in standing water. Among the several species in Jamaica foraging in trees, there is also fine habitat partitioning. Black-and-white Warblers find insects on the bark of the trunks and major branches, redstarts catch them in the air, Cape May Warblers consume more nectar and fruit, while some foliage gleaners focus on leaves of different types or on twigs, reach their prey by hovering, or restrict themselves to different heights in the vegetation. Many of these specializations probably evolved in the breeding range, where some of the species also share the same forests, but in Jamaica even more of them occur together and the specializations may be more critical (Lack and Lack 1972).

The forests of southeastern Mexico provide another example of how three especially closely related warblers reduce competition. Black-throated Green, Townsend's (*S. townsendi*), and Hermit (*S. occidentalis*) Warblers are all common in pine-oak woodlands, with the Black-throated Green also in subtropical forests on the Caribbean coast. The Hermit Warbler strongly favors pines, the Townsend's somewhat favors oaks, and the Black-throated Green is most generalized as to vegetation type but is usually in the upper and outer branches, hopping more and hovering and sallying for insects, while the other two—each larger and heavier—forage on inner branches, which requires less agility. Their wintering specialization may reflect their evolutionary history. The Townsend's and Hermit Warblers may have evolved from successive westward expansions of the ancestor to the Black-throated Green Warbler, which in winter already occupied the more productive lowlands and middle elevations in Mexico and Central America. That left the less rewarding oak woodlands for the Townsend's and then the higher-elevation, still less productive pine forests for the Hermit, the most recent of the three species (Greenberg et al. 2001).

LONG-DISTANCE MIGRANTS: LATITUDINAL SEPARATION

Few families of birds have as many species of similar size and basic ecology wintering together as do the warblers. For most birds, potential winter competition is most acute not with closely related species but with conspecifics. Among long-distance migrants, differential migration with latitudinal separation by sex or age occurs, especially among species that are particularly able fliers. Extensive recoveries of Peregrine Falcons (*Falco peregrinus*) banded in Greenland have found that females from there winter between the Gulf of Mexico, at 28° N, and northernmost South America, 2° N, while males are found from 2° N to 26° S. The average male's wintering distance from Greenland is 4000 km greater than that of females, but they return no later the following spring (Newton 2008:443). Hatching-year Common Terns banded at a colony on Great Gull Island, New York, have been found on the South American coast as far as 27° S and adults to 37° S, some 1400 km more distant (Hays et al. 1999). Young Red Knots (*Calidris canutus*) born on islands north of mainland Siberia spend their first winter and the subsequent austral summer in southeastern Australia and move in their second winter to New Zealand, to which they will return in subsequent years after breeding (Newton 2008:444).

The factors prompting differential migration by sex—size, dominance, and distance from breeding site—can be further examined in shorebirds that have wide ranges in winter where their coastal habitats are a thin linear strip running through broad latitudes. Fourteen shorebird species that breed in western Alaska and Siberia winter in Australia; they concentrate in two separate regions, on the northwest and southeast coasts, some 1500–3000 km apart. Monthly mean maximal temperatures are 7–12° C warmer in the northwest. Thus, in contrast to the Northern Hemisphere examples of differential migration, where shorter distance correlates with colder winters and the birds expected to be farther north are either larger bodied or the first to return in spring to their breeding range, the transequatorial migrants that travel less far are in warmer areas when farther north. How does this affect the winter distribution of shorebirds, in which most species have larger females, as reflected by wing length, and in some cases also by bill length?

Among the migrants to Australia, the birds of both sexes that have shorter wings but longer bills are commoner in the warmer northwest. A larger-bodied shorebird is at a disadvantage there because it retains more heat, but the longer bill may help dissipate heat, as well as probe more deeply into sand or mud to reach prey that has burrowed farther to remain cool. The distributional differences among most of the shorebirds are less between the sexes; rather, they are based primarily on size differences within each sex. Of species with larger females, only the Curlew Sandpiper (*Calidris ferruginea*) has more females in the northwest. Sharp-tailed Sandpipers (*C. acuminata*), the only species in which females are smaller, also have more in the northwest. For all these birds, the need to survive cold Arctic springs constrains how much their body forms can evolve to enhance winter survival at much warmer latitudes; given these limitations, it seems that thermoregulation, not dominance or distance from the breeding range, is the key factor in differential migration among shorebirds wintering in Australia (Nebel et al. 2013).

The Western Sandpiper (*C. mauri*) breeds in western Alaska and northeastern Siberia. The winter range stretches on the Pacific coast from southern Canada to Peru and on the Atlantic coast from North Carolina to Suriname; the largest concentrations are on the Pacific coasts of west-central Mexico and Panama. The bills of females average 12% longer than those of males; in winter, females feed more by probing, males by pecking. Females are more numerous from Mexico south; those with the longest bills winter farthest south, to Peru (Mathot et al. 2007). More hatching-year birds of both sexes, however, are at the northern and southern extremities of the winter range. Most of the hatching-year birds from Mexico north return to their breeding range in their first spring, while those farther south are more likely to remain without migrating through their second winter. The timing and frequency of molt may be the determining factor. Adult Western Sandpipers molt after they complete their autumn migration, but the feathers that hatching-year birds grew soon after birth are retained through their first winter and subsequent summer. For those wintering farthest south, these flight feathers may not be adequate for two additional long migrations before being replaced; the hatching-year birds at the southern end of the winter range may therefore enhance their long-term survival by waiting another year until they have fresh flight feathers (Nebel et al. 2002).

Other small sandpipers with even more extensive winter ranges show different patterns. The Least Sandpiper (*C. minutilla*) winters from Oregon and North Carolina to northern Chile and central Brazil. Females are more numerous in the southern portion of the winter range, but there is no difference in age distribution. Both bill length and wing length are greater in individuals of both sexes wintering farther south; these may reflect the advantages of size for longer migrations and for probing more deeply into warmer substrates (Nebel 2006). Finally, Western Hemisphere Sanderlings have the most extensive winter range of all the sandpipers, from British Columbia and Massachusetts to Tierra del Fuego. Adults and young have the same overall range, and among adults there is no difference in latitude by sex, but hatching-year males tend to winter farther south than hatching-year females (Myers 1981). One might predict that the Sanderlings wintering farthest south have the longest wings and those in equatorial latitudes have the longest bills; this awaits further research.

Among long-distance landbird migrants to the Neotropics, relatively few are known to separate by sex or age over different latitudes. The more widespread partitioning by sex of habitats at the same latitude is discussed in chapter 4. A study based on examination of museum specimens of 45 species that winter in Mexico found eight examples of latitudinal segregation by sex over the birds' entire winter range from North America to Panama. Males tend to winter farther north than females in the Yellow-bellied Sapsucker (*Sphyrapicus varius*), Swainson's Thrush (*Catharus ustulatus*), Orange-crowned Warbler (*Leiothlypis celata*), Yellow Warbler, Black-and-white Warbler, Ovenbird, and Wilson's Warbler (*Cardellina pusilla*). In the Indigo Bunting, more females winter north of males (Komar et al. 2005). In observations of Summer Tanagers in Costa Rica, adult-plumage males were 87.5% of birds seen, in central Colombia 84.8%, but in northern Ecuador only 11% (Pearson 1980).

SITE FIDELITY

Winter survival requires skills migratory young birds cannot acquire before they reach their winter destination, and ones adults may not deploy in other seasons. Mastering these skills is often easier when birds remain in the same place for much or all of the winter. For some, this may be precisely the same territory, and for others the same general area, depending on the distribution of food and shelter. The experience they gain their first winter— finding food, locating roosting sites, avoiding predators, and so forth—is an advantage to them in future winters. The more that traditional research practices like banding and new technologies that track birds remotely are deployed to monitor individual movements, the more scientists are learning about the many birds that show strong wintering site fidelity, both those that are solitary in winter and species that form flocks.

The time spent on the initial wintering site also helps birds navigate to return there in subsequent years. The period during winter when the birds acquire this skill varies among species and with age. Young Pacific Golden-Plovers (*Pluvialis fulva*) that migrate to islands in the Central Pacific were found in experiments to fixate on their wintering site and the navigational cues necessary to reach it after their first migration in September and October; the sensitive period ends by February (Sauer 1963). When adult Northern Waterthrushes wintering in Venezuela were removed from their territories in January, all returned within 12 days, but none of the first-year birds returned; it may be that first-year birds fixate on their winter territory about the time of their spring departure (Schwartz 1963).

White-crowned (*Zonotrichia leucophrys*) and Golden-crowned (*Z. atricapilla*) Sparrows that were removed from their wintering site in central California did not return that winter, but many adults did the following year. The later in the winter, between November and early March, that adults were removed from their original site, the more likely they were to return to it the next winter; none returned to their introduced site. Among first-year birds, the earlier they were translocated, the more likely they would return to the new site, similarly indicating that longer winter residency increases site fixation (Ralph and Mewaldt 1975).

Further evidence that wintering site fixation is learned comes from observations of birds that in 1974 were displaced on their southward migration by two early autumn hurricanes in the Caribbean Sea that deposited many birds in El Salvador, beyond their normal range. Hermit Thrushes (*Catharus guttatus*), Nashville Warblers (*Leiothlypis ruficapilla*), one Hooded Warbler, and Lincoln's Sparrows (*Zonotrichia lincolnii*), all previously unrecorded in the country or recorded only once almost 50 years earlier, appeared at various elevations and were found again over the next several winters. One Nashville Warbler and the Hooded Warbler were banded; the Nashville returned to the same site the following winter and the Hooded was found there three winters later. It is likely these birds learned the necessary orientation to return to a genetically unfamiliar site on their first northward flight rather than when blown off course by a hurricane (Thurber 1980).

The degree of site fidelity depends on the reliability of food and other necessary resources. For shorebirds, this may vary by year and by region. On the Brazilian Amazonian coast, one of the most important wintering areas in South America for Semipalmated Sandpipers (*Calidris pusilla*), a marking and recapturing study found that some individuals returned to the same area as many as six years after their original capture (Rodrigues et al. 2007). The Pacific coast of Mexico is a major wintering area for the closely related Western Sandpiper. At one small estuary there, birds that arrived in September to spend the entire winter and other, later transients that stayed more briefly both returned in successive years. Birds past their second year, however, were more likely to shift from residency to transience, perhaps because two prior winters had given them the experience to evaluate local conditions and the likelihood of sufficient resources remaining for the rest of the winter (Fernández et al. 2001).

Sanderlings disperse more thinly in winter on coastal beaches than these other sandpipers; they have also been found to be loyal to their wintering sites. Most birds marked in a study at Bodega Bay, California, remained there that winter and returned the next, and most of the additional marked birds that were removed and released elsewhere on the coast returned within 20 days. A few did not return until the following winter. In contrast to the sparrows discussed above, adults were more likely to return if transplanted in October or November than in January, while first-year birds showed the opposite pattern (Myers et al. 1988).

Long-term marking studies are showing how widespread wintering site fidelity is among passerine migrants. The Northern Shrike (*Lanius borealis*) is an unusual passerine, having a winter diet of small birds and mammals. Depending on local conditions, it may not descend, or descend far, every year from its breeding range in Alaska and Canada, but recoveries of banded birds from Alaska to Quebec and Massachusetts include some individuals found again at the same wintering site as many as four years after having been banded there (Rimmer and Darmstadt 1996). They may have been elsewhere in the intervening winters, indicating a potentially intermittent site fidelity and a memory of the site that persists even when not deployed for more than a year.

Commoner and more consistent migrants yield more data showing annual fidelity. In Orange County, Florida, banding a variety of short-distance migrants using different habitats, from wetlands to thickets to forest, showed how many came back to the same places where mist nets were hung each winter. Some House Wrens (*Troglodytes aedon*), Gray Catbirds, Orange-crowned Warblers (*Leiothlypis celata*), and Ovenbirds returned to the same site at least four successive winters after first being marked. Ruby-crowned Kinglets (*Regulus calendula*) and Swamp Sparrows (*Zonotrichia georgiana*) also returned regularly, but for fewer years (Somershoe et al. 2009).

Recoveries of banded long-distance migrants to various Neotropical destinations show how differences in local habitat may affect site fidelity within a species. In a Jamaican forest inhabited by Black-throated Blue Warblers and American Redstarts, the Black-throated Blues returned as many as two years after being banded there, and redstarts up to three years later. Among the male redstarts, which take two years to acquire adult plumage, most juveniles returned only one year after banding. The rates of return for adults were approximately 50% per year for each species, which matches their estimated survival rate (Holmes and Sherry 1992). Elsewhere in Jamaica, in mangrove forest, which may be a more desirable habitat because it retains moisture through the dry season at the end of winter, at least one male redstart returned for eight years (Marra and Holmes 2001).

On Hispaniola, the island immediately east of Jamaica, in a shade coffee plantation in the Dominican Republic, these two species both had a return rate of approximately 33%. While males predominated in the plantation (70–77% in Black-throated Blue Warblers; 71–79% in American Redstarts), females had a slightly higher return rate (Wunderle and Latta 2000). In Puerto Rico, the next island east, where as many as 10 wintering warbler species (but not Black-throated Blue) inhabit a dry forest, 74% of the redstarts are female. The redstarts are not territorial, but some were recaptured as many as seven years after they were first banded. Of the several other species in that forest, only the Northern Parula equaled that record, while Black-and-white and Prothonotary (*Protonotaria citrea*) Warblers returned at least four winters after banding (Faaborg and Arendt 1984).

There are far fewer recoveries of austral migrants on their winter territories, but at a banding station in northeastern Venezuela, Slaty Elaenias (*Elaenia strepera*) and Small-billed Elaenias (*E. parvirostris*), both migrant flycatchers from southern South America, were found in subsequent winters at their original banding site, to which American Redstarts also returned (McNeil 1982). In west-central Argentina, Scale-throated Earthcreepers (*Upucerthia dumetaria*), Patagonian Mockingbirds (*Mimus patagonicus*), and Rufous-collared Sparrows (*Zonotrichia capensis*), all wintering species coming north from Patagonia, demonstrated that multiyear site fidelity also occurs in these three other passerine families (Jahn et al. 2009).

These few examples of wintering site fidelity could be augmented by many others of birds for which season-long resource needs can be met at a single location. Birds that depend on

White Storks wander opportunistically in Africa, stopping wherever wetlands meet their needs. They may spend successive winters in entirely different parts of the continent.

more ephemeral or widely scattered resources that vary annually have other strategies—and until recently were almost impossible to follow, because marked birds do not usually reappear at the original study site and may never be encountered again. New technologies are now opening this frontier. White Storks (*Ciconia ciconia*) tracked by satellite from a colony in Rodosze, northeastern Poland, were found wintering in many parts of sub-Saharan Africa. Each bird moved widely over the season from wetland to wetland—one started four winters in Chad, went on in two of those

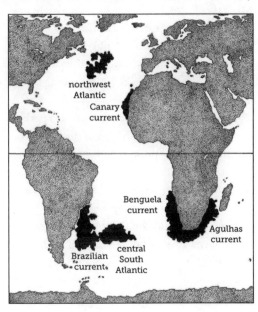

Cory's Shearwaters from the Canary Islands have several different wintering areas. Some individuals return regularly to the same one, while others use different ones in successive winters. (redrawn with permission from Dias et al. 2011)

years to Kenya, and spent much of a fifth winter in Sudan; another stork spent successive winters in Tanzania, Sudan, South Africa, and Botswana; and a third was in Sudan one year and Botswana the next (Berthold et al. 2002). At a wintering site in France where food was abundant and stable year after year, however, about 60% of the storks were found two winters after first being seen there (Archaux et al. 2008).

Some species combine fidelity and flexibility in their winter destinations. Different Cory's Shearwaters (*Calonectris diomedea*) that breed at Selvagem Grande Island in the Canaries winter in six widely dispersed areas: the northwest Atlantic, the Canary Current off northwest Africa, the Brazilian Current off southern Brazil and Argentina, the central South Atlantic due east of there, and the Benguela and Agulhas Currents, respectively off the west and east coast of southern Africa. Fourteen shearwaters tracked with geolocators over three winters showed that each of these oceanic wintering areas was used by some birds from the same colony for an entire winter. Nine of the birds returned to the same area in successive years, while five changed destinations between years, with two shifting from the South Atlantic to the North Atlantic, two others from the western to the eastern portion of the South Atlantic, and one from the Benguela to the Agulhas Current. Neither age nor sex of the birds nor any discernible ocean conditions seemed to correlate with fidelity or flexibility, with some destinations as much as 7000 km apart. But Cory's Shearwaters do not begin breeding until they have spent most of nine years at sea, and by then every bird may know each of the six traditional wintering areas. One tracked immature bird, four to five years old, visited all six regions in a single year, traveling 108,000 km (Dias et al. 2011).

SEQUENTIAL WINTERING SITES

Migratory birds wintering in regions where regular annual weather patterns produce significant changes in food availability over the season may exploit a succession of sites distant from one another. Unlike the White Storks that wander over Africa in response to year-by-year variations in wetland conditions and locations, the species that each winter move every several weeks or few months from one region to another are performing a fixed multistep migration.

This phenomenon is best known and perhaps most widespread in sub-Saharan Africa, where the rains shift from the north to the south side of the equator during the months Eurasian birds are present. In Kenya, Whitethroats (*Sylvia communis*) arrive in mid-October and are abundant through December; in January, most of them leave. Many Whitethroats have then put on more weight, as though in preparation for long flights farther south, where they will remain until March. During wetter years in Kenya, more stay through the winter (Cramp and Brooks 1992:464). Red-backed Shrikes (*Lanius collurio*) from southern Scandinavia first spend 34–62 days in the Sahel/savanna zone of southern Sudan (10° N)—almost as long as they are in Scandinavia—and in November go to Botswana and Angola (approximately 20° S), where they remain into March (Tøttrup et al. 2012). In central Nigeria, Yellow Wagtails at a roost in sugarcane numbered 16,000 in November and gradually diminished to 3,000 in March, in parallel with the decline of the insects they found in fields. The bulk of these birds moved farther south, while the dominant adult male wagtails that had secured territories along semipermanent streams remained (Wood 1979). Montagu's Harriers (*Circus pygargus*) from northern Europe, tracked by satellite over four winters in the Sahel, were found to move south-southwest over the six months they spent there, using an average of four territories, each approximately 200 sq km, the same size as their breeding territories, and each an average of 200 km distant from the previous one. The birds' movements every 45 days or so matched the peaks of their usual grasshopper prey in different parts of the Sahel; the birds tracked over more than one winter returned each year to the same sequence of territories (Trierweiler et al. 2013).

In the Atlantic Ocean, regular seasonal changes in major currents or the movements of key prey species may be the cause of long-distance shifts, especially for seabirds that breed at high latitudes and are somewhere over the ocean as much as three-quarters of the year, moving from one rich feeding area to another. Sabine's Gulls (*Xema sabini*) that breed on the east coast of Greenland migrate first to the Bay of Biscay and the Iberian Sea, arriving in August and September and remaining an average of 45 days. By the end of October, they depart for the Benguela Current off the west coast of southern Africa, mostly at latitudes 20–30° S, where they stay as many as 180 days. Between late March and early May, the gulls move to latitudes 9–27° N off the coasts of Senegal, Mauritania, and Morocco, staying an average of 18 days before returning over the Atlantic to Greenland. Their total annual migration, not counting local movements, may be as much as 39,000 km (Stenhouse et al. 2012).

In the Neotropics, where many wintering North American birds establish individual territories in which some remain for their entire stay, rainfall and other conditions lead some birds to move after exploiting one site for a few months. In Panama, mangrove forests on both the Pacific and Caribbean coasts are the winter home of many warblers. But some move from one coast to the other when the dry season begins in January. Northern Waterthrushes are common in mangrove forests on the Pacific coast in early winter and on the Caribbean coast 75 km away in late winter. American Redstarts, in contrast, reverse the pattern, while Prothonotary Warblers are abundant on both coasts from September into January, after which they evidently move farther south. A two-winter study that sampled the food resources used by these birds found that insects and other prey consumed by waterthrushes were more abundant on the Pacific coast during the wet season, into December, and more available on the Caribbean coast during the dry season. There was no correlation for the two other warblers. None of the birds the study found at either site were individually marked and later recovered on the other coast, so it is not known whether the midwinter population shifts represent the same birds or larger-scale movements (Lefebvre and Poulin 1996).

A more recent study of Veeries (*Catharus fuscescens*) that had the benefit of light-level archival geolocators found that individually tracked birds from Delaware arrived in the Brazilian states of Mato Grosso, Pará, and Amazonas, all south of the Amazon, between November 2 and December 2. Each of the marked birds departed for a second wintering site between January 7 and March 7. These were in the interior and the northern and southern periphery of the Amazon basin and in the headwaters of the Río Orinoco, Venezuela; the birds began their northward migration in mid-April. All the late-wintering sites were more than 100 km from where each bird spent the first part of the winter. It is likely that in winter Veeries feed in leaf litter on the ground, as they do the rest of the year, and that they are obliged to leave parts of the Amazon basin as these areas flood, moving to higher ground or places where flood waters have already receded. Since the various tributaries of the Amazon each flood their surrounding forest at different times, progressively eastward, Veeries, like some of the resident species, leave each site at different times between January and March (Heckscher et al. 2011).

New technology has also shown that at least a few Neotropical migrants have even longer sequential migrations. About 44% of the Purple Martins (*Progne subis*) that spend the first part of the winter in central Amazonia, for an average of 66 days, then move a mean distance of 776 km south or east for another 77 days. Of the martins that move, 18% leave the second site after some days or weeks for a third destination an average of another 763 km away, where they remain some 58 days. The cause of the sequential moves is not clear but may be overcrowding at the initial roosts. Purple Martins from the southern part of their breeding range depart the United States and arrive in Amazonia weeks before more northerly breeders; as these later birds arrive, the roosts may not be able to accommodate all the birds (Stutchbury et al. 2016).

The Bobolink (*Dolichonyx oryzivorus*) migrates even farther into South America, with all birds spending weeks in at least two separate regions and many birds moving to a third for a portion of the winter. From all over North America, Bobolinks migrate through the Caribbean and arrive in the grasslands of northern Venezuela and eastern Colombia in early October. They remain there approximately one month and then fly 2000–2500 km to grasslands in Bolivia. Some Bobolinks stay there the rest of the winter, while in December many move to Argentina, another 1000–2000 km. From both these latter regions, birds begin their return trip to North America in early April. The Bobolink is the only North American migrant to the Neotropics that in winter feeds exclusively on seeds. The timing of its three-step winter itinerary matches the sequence of peak seed production in each of its stopover sites (Renfrew et al. 2013).

Light-level geolocators have also shown that some South American austral migrants have sequential wintering sites. Forked-tailed Flycatchers (*Tyrannus savana*) nesting in Buenos Aires Province, Argentina, depart there between late January and late February, moving northwest to spend April and May in western Amazonia, between northwestern Brazil, Peru, and southern Colombia. Five of six tracked birds then moved east to spend the rest of the nonbreeding season in central Venezuela and northern Brazil. These movements may also be related to rainfall patterns. Rainfall peaks in northwestern Amazonia in April, when the flycatchers arrive, and begins to decrease in June, when the birds have moved northeast, where rainfall peaks in June and July. For aerial insectivores like the Fork-tailed Flycatcher, prey abundance may be linked with greater rainfall (Jahn et al. 2013).

MIGRATORY CONNECTIVITY

Individual wintering site fidelity will, in some species, reflect a larger scale population-wide site fidelity—when the population in one part of a species' breeding range also migrates to a single part of the winter range. This phenomenon was first given the name "migratory connectivity" at an ornithological conference in 2000 that brought together scientists using several traditional and recent rapidly advancing methods to identify, mark, and follow the movements of individual birds (Boulet and Norris 2006). These studies are confirming the variety of migration patterns described in chapter 1 that were first categorized in 1955 (Salomonsen) based substantially on recovery of banded birds.

But why connectivity should exist in some species but not in others that are ecologically similar is hard to explain, since in many highly migratory species, birds from the same colony, and even pairs, may winter in widely different areas. Among aerial insectivores, Common Swifts from a colony in Germany tracked with light-level geolocators were found to winter in West Africa between 15° N and 14° S, with individual birds faithful year after year to the same wintering locale in this vast expanse (Wellbrock et al. 2017). In contrast, Purple Martins from eastern Ontario to western Alberta wintering in Amazonian Brazil roost in the same trees (Fraser et al. 2017).

Between these two extremes of wide dispersal from a single colony and high winter concentration of birds from distant parts of the breeding range, extensive banding programs of another aerial insectivore have shown that most British Barn Swallows (*Hirundo rustica*) winter in South Africa, where the range has been shifting westward to Cape Province since the 1960s. Barn Swallows from Germany, however, winter primarily in the Democratic Republic of the Congo, 1600 km north of the British birds, although some German swallows nesting in the region midway between Berlin and Hanover were all found wintering in southern Ghana, indicating that the geographic partitioning among European populations may be very fine. The Congo region is much warmer and more equable than South Africa, where Barn Swallows sometimes die when the temperature descends toward 10° C. Resident swallow species there survive, so despite the sensitivity of Barn Swallows to low temperatures, there must be some compensating benefit to them in South Africa (Moreau 1972:122–128).

One benefit of connectivity is that a genetically similar wintering population can adapt and expand more easily to appropriate adjacent habitat, should this become available or should the traditional sites become unusable. Conversely, when connectivity is weak, as in the Common Swift, migration and adaptation to entirely new, sometimes more distant, wintering areas may be more likely in a genetically mixed population if the initial wintering area is lost (Webster and Marra 2005). Over the course of any species' evolution, both these situations may have occurred in varying frequency, reinforcing or weakening a trend toward connectivity.

Understanding the extent and patterns of connectivity is especially important for the conservation and management of migratory birds that are losing breeding or winter habitat, or both, to development. A seven-year study that used light-level geolocators to track a rapidly declining species, the Wood Thrush, from seven different breeding sites across its North American range found that 91% of the birds from the Northeast south to North Carolina winter in a narrow belt from eastern Honduras to Costa Rica; these birds form 56% of the Wood Thrush's total population. Wood Thrushes that breed in the Southeast, from South Carolina to Florida, and in the Midwest, from Indiana south to Alabama and west, winter primarily (65%) between the Yucatán Peninsula of Mexico and western Honduras. All Wood Thrushes wintering elsewhere in Mexico also come from the Midwest. Since this species is particularly sensitive to forest loss and fragmentation in both its breeding and winter ranges, a full understanding of the annual movements and distribution of each region's population is essential to any conservation planning (Stanley et al. 2015).

The Golden-winged Warbler (*Vermivora chrysoptera*) is another declining species that breeds in much of eastern and central North America and has a high degree of connectivity between its breeding and winter ranges. Geolocators were used to track birds nesting in Minnesota, Tennessee, and Pennsylvania and found that they wintered, respectively, from southern Mexico to central Nicaragua, along the border of northern Colombia and Venezuela, and in north-central Venezuela. The autumn migration routes of each population were also distinctive, with Pennsylvania birds traveling 4000 to 5000 km farther each year than birds from Tennessee (Kramer et al. 2017). Another study that attached light-level geolocators to Golden-winged Warblers wintering in the northern highlands of Nicaragua at the El Jaguar Reserve found that these birds bred in eastern Minnesota, northern Wisconsin, southwestern Ontario, and Michigan's Upper Peninsula. This small reserve, only 168 ha of remnant cloud forest and shade coffee plantations in a region with little surviving forest, is therefore critical to the Golden-winged Warblers of the western Great Lakes region (Larkin et al. 2017).

For some North American birds, observational history has been long enough to give some clues as to the impacts that changes in either the breeding or winter range can have on populations with high connectivity. We have already noted, in chapter 1, the different winter ranges of three populations of Long-billed Curlew (*Numenius americanus*). On November 10, 1831, John James Audubon and his Charleston patron, the Reverend John Bachman, left the city for Cole's Island, where by evening at the nearby "Bird Banks" they found a roost containing several thousand Long-billed Curlews (Audubon 1967:36–37); two of the birds became the models for one of Audubon's finest paintings, with the city of Charleston in the background. Today, however, sightings are rare anywhere on the Atlantic coast. The curlew continued to be common around Charleston during autumn for a few decades after Audubon saw his thousands, but it was declining by 1879 and gone by 1899. The population that wintered on the Atlantic coast was probably from a portion of the now-empty former breeding range that ran through the grasslands of southern Michigan northwest to Manitoba and down to Arizona and coastal Texas (Dugger and Dugger 2002).

The Hudsonian Godwit (*Limosa haemastica*) is an Arctic nester with highly disjunct breeding areas that may have been separated by past glacial events. Today, one population nests on the south side of Hudson Bay, another in the northern Northwest Territories and northeastern Alaska, and a third in south-central and western Alaska. They winter, respectively, in Tierra del Fuego, in northern Argentina, and on Isla de Chiloé and the adjacent Chilean coast. The gaps between these three widely separated areas may once also have supported wintering godwits from populations no longer extant. While the birds of the three wintering regions have similar northbound migration routes before they split off for their final destinations, their departure times from South America at the end of winter vary by as much as a month, so they do not encounter one another (Senner 2012). Since their nesting schedules also differ by several weeks, they may not overlap in autumn either.

Broad-scale studies of other species with a wide North American breeding range have confirmed the east-west patterns known also from many Old World migrants—with birds shifting primarily on longitudinal lines to more southerly latitudes. Studies of American Redstarts (Norris et al. 2006a) and Yellow Warblers (Boulet et al. 2006), using several technologies to corroborate findings of the others, found that eastern populations of the two warblers did indeed winter in the more easterly part of the winter range (Caribbean islands for the redstart, southern Central America and northern South America for the Yellow Warbler), and the western breeding populations wintered farther west (Mexico and Central America for the redstart, Mexico to Panama for the Yellow Warbler). Subspecific western forms of the Yellow Warbler, from Alaska to Arizona, however, were all found in the same parts of Mexico, indicating that this small region can support the wintering birds from a very large part of North America. While many of the northeastern Yellow Warblers winter farther south than western birds and have a leapfrog migration, redstarts have a chain migration, where more-northerly breeders winter north of the region holding more-southerly breeders (Norris et al. 2006b).

Additional evidence of connectivity in birds with several easily recognized races can be determined from museum specimens collected on the winter range as well as from individual marking and tracking. The Swainson's Thrush (*Catharus ustulatus*) has six major subspecies across its vast breeding range from the Maritime Provinces of Canada west to Alaska and south into the Rockies and coastal California. The species winters mainly in lowland and midelevation forest, from southern Mexico to Argentina. The races are distinguished by their russet or olive back and the amount of spotting on the breast. Winter ranges of some races overlap while others are more distinct, but collectively they demonstrate the leapfrog pattern in which the most northerly breeders winter farthest south. The "olive-backed" races from eastern Canada and south into the Appalachians west to interior Alaska winter east of the Andes from Colombia to northwestern Argentina. The three western "russet-backed" forms from southeastern coastal Alaska south to southern California winter in Mexico and Central America, more often on the Pacific slope (Evans Mack and Yong 2000). For some of the California birds, at least, the connectivity has a fine grain. Light-level geolocators placed on 12 thrushes nesting in Marin County found that 11 wintered in the Jalisco region of western Mexico; the 12th bird went to the Sierra Madre Oriental or the Sierra Madre del Sur (Cormier et al. 2013).

SUMMARY

While most short-distance migrants, as well as permanent residents, winter in habitat similar to their breeding site, long-distance migrants are often in very different environments for much of the winter. Both North American migrants to the Neotropics and Eurasian birds wintering in Africa occupy a far smaller land area than in spring and summer, so that each winterer's own population may be more densely packed. Among closely related wintering species, competition is reduced when they have winter ranges with little overlap and when their foraging methods differ, even subtly. Within shared habitats, they may partition the environment very finely by feeding at different elevations or in different types of vegetation. They also share their wintering site with the many more species that live there year round. Some wintering birds reduce competition by specializing in habitats that would not support residents through their breeding season and are therefore little used.

Arriving landbird migrants find distinctive conditions in different parts of the tropics, based primarily on the timing of local rainy and dry seasons, which govern especially the abundance of nectar, fruit, and arthropods. In some regions where wintering birds concentrate, they are present for a season of scarcity. They may stay throughout or move on in a regular pattern of sequential destinations. In tropical latitudes, day length is also shorter than what birds have enjoyed in spring and summer, and local temperatures are much warmer.

For both short- and long-distance migrants, in some species competition is reduced or avoided when each sex or age class differs in habitat choice, feeding specialty, anatomy, or winter range. Among short-distance migrants, males and sometimes all adult birds typically winter closer to their breeding site than do females and birds born that year. Larger or more experienced birds may be able to survive in harsher winter conditions, and the incentive, for males especially, to return soonest to claim a good breeding site may be another factor. Some long-distance migrants also have latitudinal separation by sex and age, while others partition habitats, with adults, particularly males, in the superior ones.

Whatever their winter destination, most migrants usually return to it in subsequent years—except for birds that graduate to different, usually closer, sites as adults in their second winter. Imprinting of the location, both general and specific, seems to occur while the birds are there, sometimes soon after arrival, while for other species only at the end of the season.

Some birds are more mobile during the winter. Erratic changing winter conditions force some to wander irregularly in search of their preferred habitat or food source. Other birds have a regular sequence of moves as changing local conditions, such as an advancing dry season or, elsewhere, flooding, deplete or eliminate their food.

In addition to individual wintering site fidelity, in some species all the birds of a portion of the breeding range also migrate to and winter in the same region. Modern tracking technology is revealing that migratory connectivity is widespread, while also showing that within families some species may disperse more broadly across their winter range, mixing with individuals they would not encounter in the breeding season.

CHAPTER 4

SPATIAL AND SOCIAL ORGANIZATION

FOR MOST BIRDS, RESPONSIBILITY for their offspring ends sometime during the summer, and so does the need to maintain a pair bond—if it has lasted even that long. Until the next breeding season, each bird's activities are directed toward its own survival. Winter is thus the season that shows the widest range of spatial and social organizations among birds. Whether birds remain where they have nested or travel any distance, their options include maintaining a territory, moving about singly in a small area without a territory, flocking in a stable range, or wandering long distances singly or in flocks. For some birds, survival may be enhanced by maintaining or creating a pair bond during the winter, but birds do this when the benefits accrue to themselves, either immediately or in terms of potential future reproductive success.

The strategy each bird adopts during winter reflects the best option for securing food and ensuring safety, based on the species' ecology. Birds that were aggressively territorial while nesting may form flocks in winter, colonial nesters may become solitary, and some birds may interact with species they ignored or never encountered during the breeding season. Social and spatial responses may vary within the winter range of any species, as well as by sex and age, or over the course of the season.

Birds maintain territories when a space of defendable dimensions will reliably supply all the individual's needs. The dimensions of the territory depend, of course, on how much area is required to meet those needs—less for consumers of insects and seeds that are common and fairly evenly distributed, more for hunters of larger prey that may be more thinly dispersed. Specialists in highly mobile prey that move long distances, like schools of fish and swarms of aerial insects, and consumers of nectar or fruit that is patchy in time and space, are less likely to have a territory because the areas they need to search are so much larger than they can defend, and their food, when they find it, is more abundant than any single bird could harvest. These birds are often in flocks, which may or may not maintain cohesion over time. Some birds move together, like flocks of crossbills searching for ripe cone crops; others seem to gather from different points at an ephemeral source like a tree briefly laden with nectar-rich flowers.

TERRITORIALITY

Territoriality thus best suits birds living year round or seasonally in a fairly stable environment, where familiarity with local resources may take time to acquire and is a benefit to retain. Many nonmigratory birds that can find the same types of food throughout the year are territorial in winter. Similarly, migrants that have strong site fidelity to their winter home return not just to the same area but often to the territory they have established in previous winters. And some birds that move varying distances each winter, like Snowy Owls (*Bubo scandiaca*), may ultimately settle for the season on a territory they will defend against others if they find it has enough to sustain them. But territorial birds are also opportunists and may have a succession of territories, each claimed for only a few days or weeks, if nothing holds the birds any longer. Other birds that may retain a territory all through the winter, or all year, can leave it several times a day if food is abundant and easily available somewhere else close by—as at a tree in fruit or a feeder.

NONMIGRANTS

In much of the United States, the Northern Mockingbird (*Mimus polyglottos*) is one of the most widespread and easily visible of winter-territorial birds. In the southern part of their range, mockingbird pairs that breed together often defend the same winter territory together, while farther north each bird usually has a separate territory, although those of pairs may be adjacent. Males sing from September to November when establishing an autumn territory; some females, especially if they establish their own territory, also sing, but less loudly (Derrickson and Breitwisch 1992). Much or all of the mockingbird's winter diet, more so in the north, is fruit.

Mockingbirds seem able to assess in autumn the capacity of a potential territory to supply enough fruit through the winter, when with each day there will be less available. A study in New York that plotted mockingbird winter territories against the number of fruiting bushes found that territories ranged from 0.35 ha to 1.65 ha, but all had equal quantities of food, and none had been exhausted by winter's end (Safina and Utter 1989). In the Piedmont of North Carolina, individual territories shrank during an unusually cold winter—by over half between October and December—and then grew slightly in January. For each bird, it may have been more energetically efficient to defend only the smallest area necessary containing enough fruit. By spring, each bird's territory had expanded to include the vacated winter space (Logan 1987).

Titmice and chickadees, also easy to observe, have more complex winter-territorial ecology and behavior. Most are nonmigratory and establish their territory in late summer or autumn. In Eurasia, where several tit species may live in the same habitat, winter social organization ranges from territorial pairs, especially in tits restricted to coniferous woods, to small territorial groups including unrelated birds, to loose roving flocks of the deciduous forest specialists, notably Great (*Parus major*) and Blue (*Cyanistes caeruleus*) Tits (Matthysen 1990).

In North America, Oak Titmice (*Baeolophus inornatus*) maintain territories as pairs, probably of birds that bred together earlier that year, if both survive, and covering the same area. The more widespread and most frequently studied North American species— Tufted Titmice (*B. bicolor*) and Black-capped (*Poecile atricapillus*) and Carolina (*P. carolinensis*) Chickadees—have a variety of social and spatial arrangements. Each winter territory has at its core a pair, which likely nested in the territory. In late summer or early autumn, they may be joined by a few others—typically three to eight, but this varies among populations—when widowed adults and roving bands of juveniles join adult pairs. More rarely, juveniles form a group without any adults. Young Tufted Titmice sometimes stay in the territory of their parents until the following spring. In chickadees, the young birds joining pairs are not offspring; these disperse at least several territories away from where they were born. The chickadee groups that contain migrants are less cohesive than those in which all birds are local (Matthysen 1990).

On Marion Island at 46° S in the sub-Antarctic portion of the Indian Ocean, where winter temperatures average −2° to −5° C and there are occasional ice squalls, Black-faced Sheathbill (*Chionis minor*) behavior demonstrates the advantages of maintaining a winter territory in a challenging environment. Sheathbills superficially resemble large white pigeons as they walk around the penguin colonies where they nest, but they are in a distinctive family most closely related to shorebirds. On Marion Island, some sheathbills nest in colonies of Rockhopper (*Eudyptes chrysocome*) and Macaroni (*E. chrysolophus*) Penguins, others in colonies of King Penguins (*Aptenodytes patagonicus*). The two smaller penguins leave the island in April to winter at sea, while the much larger Kings require 14 to 16 months from egg laying to chick independence and are on Marion all through the year (Burger 1984).

Sheathbills that breed in colonies of the smaller penguins vacate those areas in winter to forage singly or in flocks on the shore or coastal plain. Some move to the still-active colonies of King Penguins, where the local sheathbills maintain territories year round and where the juveniles stay on their parents' territory and are chased less often than unrelated birds. In these colonies, the sheathbills feed primarily on penguin carcasses and on food they steal from King Penguins feeding their chicks. Territorial birds dominate at fresh

On Marion Island in the sub-Antarctic portion of the Indian Ocean, Black-faced Sheathbills with territories in the colonies of King Penguins are more likely to survive the winter than sheathbills spending that season along the shore.

carcasses and are more successful at robbing penguins than are intruders. Because they are not chased, the territorial birds can spend more time watching for the right moment to attack a King Penguin feeding its chick. Young of the territorial sheathbills also have access to better food and spend more time eating than do intruders; their survival rate is higher than that of juveniles wintering on the shore. By late winter, well before the Rockhopper and Macaroni Penguins return in the last week of October, sheathbills from their colonies spend several hours there each day reestablishing their territories. But spending a significant portion of each winter day in a part of the island that has little or no food until the penguins return may reduce their fitness compared with the sheathbills that never leave their territories in the King Penguin colonies (Burger 1984).

SHORT-DISTANCE AND PARTIAL MIGRANTS

Many migratory birds are also territorial in winter. For partial migrants, species in which some individuals do not migrate very far and others not at all, this represents little change of behavior. European Robins (*Erithacus rubecula*) in England are permanent residents; males and females each establish their own territory in autumn, sing frequently, and defend it aggressively, while in spring when the birds form pairs, only the males sing and defend the shared territory. Migrant robins wintering in parts of Italy where none breed also form individual territories, with both males and females singing and behaving aggressively. In both resident and migrant robins, females experience an increase in testosterone in autumn that diminishes by the time they join males in spring to form a joint territory (Schwabl 1992).

In North America, the Hermit Thrush (*Catharus guttatus*) is closest in size and ecology to the European Robin. In southeastern Louisiana, where Hermit Thrushes winter but do not breed, each male and female, adult and immature, has its own territory, averaging 0.56 ha, abutting or slightly overlapping those of its neighbors. A few birds sing a soft whisper song soon after arriving in October, but during the rest of the winter they use only various call notes and displays. At least some birds return to the same territory in subsequent winters. There are also some individuals moving around the interstices, looking for a vacant territory; when a vacancy appears, it is quickly occupied, usually within a week (Brown et al. 2000).

Because White Wagtails (*Motacilla alba*) feed on insects on open ground, their winter spatial organization is more complex than that of these two thrushes, which can spend the entire season alone in a small territory. In Israel, wagtails are common in winter in cities, towns, villages, farms, and along roads, wherever they can find food. Wagtails that have not acquired a territory feed in flocks on fields. But in small territories or extensive fields, the exposed places where they feed would be dangerous at night, so wagtails roost communally in dense vegetation where they are safer from predators. They leave the roost before dawn and return after dusk. The wagtails that have found defensible patches of food are territorial; in places where food is abundant, as in a rubbish dump, the territory may be only 10 m wide, while in cities it can be 100 m wide. Some return to the same minute territory in successive winters. Many of the territories support a female as well as a male, but these are not likely to be long-term pairings, since each sex arrives in and departs Israel on different dates. The females gain access to a reliable food source and the male has the benefit of a codefender (Zahavi 1971).

Barrow's Goldeneyes (*Bucephala islandica*) maintain winter territories when they are paired but not when single. In British Columbia, goldeneyes nesting on lakes in the

European Robins, including nonmigratory pairs, each have separate territories during winter.

interior winter along rocky shorelines; males arrive first and are joined several days later by their mate from the previous summer. It is likely that the initial pair formation occurs in winter, because in one study a marked pair departed together and appeared the next day on its breeding territory (Savard 1985). Females stay with their young longer than males and would know how to find their mate only if they had previously wintered at the same site. The reunited pair defends its stretch of coastline against other goldeneyes as well as other waterfowl with a similar diet (Savard 1988).

LONG-DISTANCE MIGRANTS

American Redstarts (*Setophaga ruticilla*) and Black-throated Blue Warblers (*S. caerulescens*) spend five to six months in Jamaica, where they are common and widespread in coastal forest scrub, mangroves, and wet limestone forests as well as in agricultural and urban areas. This is a wider range of habitats than the deciduous forests both species occupy during the breeding season. Redstarts are more frequent in the Jamaican lowlands, and Black-throated Blues at middle and higher elevations. A study focused on natural habitats found that in both species, males and females of all ages lived in every type of forest and each bird maintained an individual territory, using displays and various "chip" notes as well as chases to defend the territory. Neither species sang. Some males and females had overlapping territories, but never two birds of the same sex. The redstarts occupied the forests at the same density as in their breeding habitat; the Black-throated Blue Warblers were somewhat more dense. This may have been because the Black-throated Blues also consumed nectar and fruit and therefore needed less space, while the redstarts ate only insects. Some 80% of the redstarts banded in October and November were still on their territory in March and April, as were 66% of the Black-throated Blues, but whenever a marked bird disappeared from its territory it was replaced in a few days by another from elsewhere (Holmes et al. 1989).

In a shade coffee farm elsewhere in Jamaica, Black-throated Blue Warblers were also common but were attracted primarily to the *Inga* trees that grew over the coffee; *Inga* produces nectar-rich flowers and fruits that attract more insects than will feed on the leaves of coffee. Warblers were more aggressive against intruders in the *Inga* trees than in the coffee or the nearby forest understory, and males dominated the *Inga* trees while females were more often in the coffee (Smith et al. 2012).

Inga, a fast-growing "pioneer" tree that is one of the first to sprout in cleared areas, is also important for warblers wintering in Chiapas, Mexico, where in cattle pastures isolated patches of trees are the only place supporting migrants. Male Yellow Warblers (*Setophaga petechia*) dominate these little islands. Once they have established their territories, the Yellow Warblers focus their aggression on at least 29 species of smaller resident and migrant birds rather than on neighboring conspecifics. The males drive other birds out of the tops of the trees; female Yellow Warblers feed in the shrubby vegetation below. Magnolia Warblers (*S. magnolia*) often have territories overlapping those of the Yellows in these patches but are rarely allowed in the canopy, which has more arthropod prey than

Male Yellow Warblers with territories in small patches of trees surrounded by
pasture are dominant over at least 29 other species of resident and wintering birds,
including Magnolia Warblers.

the understory. Some 35% of observed chases by Yellow Warblers were of Magnolias,
occasionally resulting in a sprained or broken wing for the Magnolia. The only birds that
chased Yellow Warblers were Orchard Orioles (*Icterus spurius*), which came to the *Inga* trees
in flocks when they flowered in March and April (Greenberg and Salgado Ortiz 1994).

The ground itself is territory for some warblers. The very first study of territoriality
in wintering warblers began in 1958 on Northern Waterthrushes (*Parkesia noveboracensis*)
in the small botanical garden of the Universidad Central de Venezuela in Caracas.
Waterthrushes begin arriving there in mid-September and remain until late April and early
May. Territories range in size between 400 and 5000 sq m, averaging 2000 sq m, which
is about one-fifth the size of a breeding territory. Water is not required on the territory as
long as the ground itself has been watered, but every afternoon shortly before sunset each
waterthrush bathes in whatever water source or accumulation may be nearby. Birds will
chase their neighbors when one comes into the territory and occasionally also a nearby
American Redstart, but more often redstarts chase the waterthrushes (Schwartz 1963).

Social tolerance of conspecifics or other birds varies by species, habitat, and season. In
central Panama, the territorial warblers include both of the waterthrushes and Mourning
(*Geothlypis philadelphia*), Kentucky (*G. formosa*), Tennessee (*Leiothlypis peregrina*), Yellow, Bay-
breasted (*Setophaga castanea*), and Chestnut-sided (*S. pensylvanica*) Warblers, but the last four
of these will also occasionally join mixed-species flocks where they tolerate conspecifics
as well. Three other warblers that are more regular members of mixed-species flocks—
Golden-winged (*Vermivora chrysoptera*), Blue-winged (*V. pinus*), and Black-and-white (*S.
varia*)—treat the flock as their territory and will defend it against conspecifics. Finally,
Tennessee and Prothonotary (*Protonotaria citrea*) Warblers sometimes form flocks of their
own, and in the dry season Bay-breasteds and Chestnut-sideds will leave their territories
to cluster at trees with fruit (Morton 1980).

Of these species, Tennessee Warblers are most flexible. When part of single-species or mixed-species flocks, they probe for larvae on leaves and probably benefit from being in a group that has more eyes out for predators. Their vocalizations are soft and unchallenging. But when feeding on *Combretum* nectar, Tennessees are aggressive, giving harsh "chips." Each bird can defend only a small area of this flowering vine, but successful birds come away from the flowers with a coating of red pollen that may remain on their face and throat for a few days; this gives them dominance as they move to the next *Combretum* vine (Morton 1980).

In the flycatcher genus *Empidonax*, almost entirely insectivorous, winter territoriality is the norm among species that have been studied. Yellow-bellied (*E. flaviventris*) and Least (*E. minimus*) Flycatchers wintering in the Tuxtla Mountains of southern Veracruz, Mexico, each defend a territory in the same habitat; both sexes use song, other vocalizations, and various displays to advertise their ownership (Rappole and Warner 1980). In northwestern Costa Rica, Willow Flycatchers (*E. traillii*) also each have a territory, and both sexes defend it with song, calls, and displays. They favor shrubby areas adjacent to wetlands that have a stable supply of insects through both wet and dry seasons, and they retain the same territory from September until they depart in late April or early May. Their territories, each for a single bird, were 0.46–0.48 ha at one site and 0.77–0.82 ha at another, and 68% of the birds returned to their wetland edge territory the next winter (Koronkiewicz et al. 2006).

Along the bends of the Río Manu in southeastern Peru, in the *Tessaria* thickets where small insects are more abundant than in any of the adjacent rainforest habitats, Alder Flycatchers (*E. alnorum*) occupy single or joint territories. Of nine territories that were monitored on one beach, six had two birds, two had one, and one had three. The territories ranged in size from 0.04 to 0.25 ha. The group territories of the Alder Flycatchers each had a dominant bird that patrolled the territory borders regularly, sang, and displayed; throughout the day and for all activities, the one or two subordinate birds stayed close to the dominant one. At least through November, some of the beaches with *Tessaria* were occupied and others not at all, and no birds were known to return to the same beach the following October; it may be that all the beach vegetation is abandoned later in the rainy season (Foster 2007). As the waters rise on the Río Manu, flooding the *Tessaria*, Alder Flycatchers begin arriving in the Chaco Province of northern Argentina, where they establish individual territories of 0.04–0.06 ha along riverbanks in *Tessaria* and similar edge vegetation that they will occupy through March, even after the ground underneath floods (Areta et al. 2016).

Plovers and sandpipers have a wide range of territorial behaviors that can shift over the course of a winter or a day, depending on the relative benefits of defending a feeding area or flocking. At the Bahía Santa María in northwestern Mexico, where more than 350,000 Western Sandpipers (*Calidris mauri*) winter, some 10% of the global population, they feed in cattail marshes, mangrove flats, and brackish flats. While all three habitats have the same prey density, the brackish flats are most sought after, occupied by 365 birds per hectare, because these have the least danger from concealed raptors. Mangrove flats support 288 birds per hectare and cattail marshes 110 birds per hectare. In the brackish flats, 7% of the sandpipers are territorial; in mangrove flats 5%; and none in the cattail marshes. The individual territories are rectangular strips 5–9 m long. Because these feeding areas are tidal, the territories are defended only briefly, from 10 to 60 minutes (Fernández and Lank 2012).

In California, 11 wintering shorebirds are territorial in some form, and in Argentina 13, while others in the same habitat in both places are always in flocks. Just like the passerines already considered, shorebirds in stable environments such as Argentinean grasslands are likely to be territorial. There, American Golden-Plovers (*Pluvialis dominica*) and Buff-breasted Sandpipers (*Tryngites subruficollis*) each have a territory of 0.3 ha or less, which they leave in midafternoon to drink at local water sources and then leave again each evening to roost in flocks. Even when on their territories, the two species will join in a single dense flock at the sight of a predator and fly in circles until it leaves (Myers 1980b). White-rumped Sandpipers (*Calidris fuscicollis*) in the same area stay on their territory all day, departing after sunset to roost in flocks, sometimes several kilometers away, and returning before dawn (Myers et al. 1979a).

Some Sanderlings (*Calidris alba*) on coastal beaches will defend a territory, sometimes for several months, during the hours the tide level makes it most productive. The size of a territory may grow and shrink, depending on how many other territorial Sanderlings are nearby and on the energetic cost of defending the space. Since Sanderlings move elsewhere to feed when the tide covers their territory, they may have less incentive to defend it as vigorously as do species that never leave their territory (Myers et al. 1979b). In Bodega Bay, California, Sanderling territories on the beach are linear, at the edge of the water, with each bird typically defending about 50 m (Pitelka et al. 1980). When the tide drops on the beach, Sanderlings fly to nearby harbor sandflats; here a few may also defend a territory while most of the Sanderlings move about singly or in small groups. Occasionally, a Sanderling will follow a foraging Black Turnstone (*Arenaria melanocephala*), searching for food in the sand that the turnstone moves; the Sanderling treats the turnstone as its territory, defending the space around it against other Sanderlings. When scattered

On the Pacific coast of North America, some Sanderlings use Black Turnstones as their territory, moving with them on the shore and searching for food in the sand the turnstones move.

on a beach or flat, Sanderlings, too, will fly up in a tight group if a raptor flies over; should they all then land on one bird's territory, it does not bother to defend it before the birds disperse (Myers et al. 1979b).

Black-bellied Plovers (*Pluvialis squatarola*) winter on some of the same estuarine mudflats and sandflats used by Sanderlings, but they are entirely visual feeders, pursuing prey they see on the surface rather than probing underneath. This influences their territorial system. Each winter, the Tees Estuary in northeast England holds about 350 Black-bellied Plovers that come from Siberia. Some defend territories all through the winter, while others do so for a few hours, days, or weeks, and some not at all. A study that marked birds individually found that birds returned in successive years and repeated the same behavior, indicating that there must be benefits from each. The consistently territorial plovers claimed areas in the gullies or cracks in the flats that were sheltered from winds; they were able to feed during the most severe gales, while nonterritorial birds on the flats stopped feeding whenever the winds exceeded 25–30 knots. In cold winters, the territorial birds had the advantage; in milder winters, the birds that expended no energy defending their feeding area did as well. Short-term territories on the higher sandflats were most frequent during midwinter, when the greatest number of birds was present. At that time, the effort to exclude other plovers benefits territorial birds by reducing both competition and disturbance of prey that comes to the surface (Townshend et al. 1984).

FAMILY TERRITORIES

Families are, for a few birds, the unit that maintains and defends a winter territory. Dimensions reflect the distribution of the food and any other resource being defended. Family territories occur in nonmigratory species with any form of cooperative breeding, such as the Acorn Woodpecker (*Melanerpes formicivorus*), found from western North America to Colombia, a larder-hoarder that hammers hundreds of individual holes into selected trees, each to hold a single acorn that is a major part of the birds' winter diet. Groups of as many as 13 birds may live together, defending their storage trees against not only conspecifics but also many other species that might remove the acorns. When defending a tree against other Acorn Woodpeckers, each bird usually confronts one of its own sex. In California, territories may be 6 ha, but territory size varies widely across this species' range (Koenig et al. 1995).

Tufted Titmice are unusual among the members of the Paridae that have been fully studied in that young birds not only sometimes stay with their parents through the winter but also occasionally assist them in raising a brood the following spring. In winter-territorial groups composed of a core pair and young birds, the adults are more tolerant of their own offspring than of other young birds that join the group. But when these titmice are in a mixed-species flock with chickadees, all titmice are dominant over any chickadee (Pravosudova et al. 2001).

The Southern Black Tit (*Parus niger*) of southern Africa is a more regular cooperative breeder, perhaps because resources in its woodland habitat are relatively scarce.

Compared with the Black Tit's Eurasian ecological equivalent, the Great Tit, its year-round territories are much larger, averaging 3 pairs or groups per 100 ha, while Great Tit breeding territories are 28 pairs per 100 ha, and in winter Great Tits form mobile flocks of up to 200 birds. A study of Black Tits in the Transvaal of South Africa found one to three male helpers at 11 of 19 breeding units. All birds helped with territorial defense, especially in winter, when daylight declined by 25% and food resources diminished. The insects available then were much smaller; tits needed to collect eight times the number of prey items that they did in summer (Tarboton 1981).

Cranes and some waterfowl that migrate in family units also defend family territories. Common Cranes (*Grus grus*) do not mate until they are three or four years old and then remain paired for life, migrating and wintering together; young stay with their parents through their first winter. Of the 2,000–10,000 that winter in an area of lagoons and farmland in northeast Spain, 2% defend territories. These are all pairs that have young of the year; parents return to the site they used the previous winter. In years that a pair does not succeed in raising young, it does not defend a territory but may occasionally visit its territory of the prior year. The adult males are the active defenders, mainly against territorial neighbors, but not passing flocks. These territories have some advantages over the fields in which flocks feed. They are closer to the communal roost sites and include a source of water, while flocking cranes must travel beyond their feeding area to drink. Territorial adults need to spend more time in antipredator vigilance than do flockers, but they consume no less food. Their young are not involved in aggressive encounters, unlike birds of all ages in flocks, and therefore have more time to feed. Finally, because the cranes at this location were monitored for 11 winters, it was evident that marked territorial pairs had a higher average annual breeding success than marked flocking pairs, based on the number of young they returned with in subsequent winters (Alonso et al. 2004).

Some Common Cranes that have young of the year with them through the winter defend a territory, while adults without young move in flocks.

Barnacle Geese (*Branta leucopsis*) wintering in the Solway Firth of northern Britain feed in flocks walking across fields as they graze. They do not maintain a fixed territory, but families defend the area immediately around them against other geese. Parents will let one of their goslings feed on the same tussock of grass but will attack an unrelated bird coming within two goose lengths. A study that marked all members of many families found that pairs with young birds were no less fit at the end of the winter than geese that did not have any responsibilities for offspring; in fact, parents whose young stayed with them longest were most successful in the following breeding season. This may be because the young were also vigilant and repelled neighbors, so that the parents had more time to feed (Black and Owen 1989).

Scavengers usually have a home range but do not defend anything except the area around their nest and their feeding space at a carcass. Black Vultures (*Coragyps atratus*) remain in family groups throughout the year, and adults continue to feed their young until the April following their birth, even while they are incubating a new clutch. During their first winter, young vultures will not compete for food at a carcass when many adults are present. They stand away from it, waiting for a parent to come to them and regurgitate food. When a few months older, young birds join large feeding groups, where their parents may still aid them by threatening other adults that might try to drive them away (Rabenold 1986).

FLOATERS

Territories held by permanent residents or wintering migrants that become vacant are quickly occupied. The new claimant may be a bird that had a territory nearby or one that was a "floater." Among the Willow Flycatchers studied in Costa Rica were also a few floaters, quiet and submissive among the territorial birds until a territory emptied. The floater then claimed it, singing and displaying like its territorial neighbors, and some returned to the territory the following winter (Koronkiewicz et al. 2006).

"Floater" has been usefully defined as "an individual member of a largely territorial population who is not defending a territory, and whose movements encompass an area substantially larger than those of the average territorial conspecific" (Winker 1998). Floaters are present in many species in both the breeding and wintering seasons. They may be single or in groups, often in habitat of marginal quality or in the interstices between territories of high-quality habitat. They are not birds drifting over great distances or individuals temporarily leaving their territory to exploit a superabundant food resource like a tree in fruit.

In winter, individuals that have not established a territory at the beginning of the season have three alternatives (Brown and Long 2007). They can

- relocate to low-quality habitat and defend a territory there;
- relocate to low-quality habitat and become a floater; or
- remain in high-quality habitat as a floater.

The third option may sooner lead to occupancy of a vacant high-quality territory—but the odds are low if the initial holders in the better territories may be less likely to succumb to starvation or predation than those in the low-quality ones. Determining the actual scenario depends on marking every bird within the study area, so that a new claimant can be identified as either a true floater or a neighbor moving to a better territory.

In a population of Wood Thrushes (*Hylocichla mustelina*) wintering in lowland rainforest in southern Veracruz, Mexico, studied over six years, the sedentary birds did not move more than 150 m from where they were captured. They remained on-site through the winter and many returned in subsequent winters. Floaters moved more than 150 m, often within a few hours; they were never aggressive toward other Wood Thrushes and sometimes foraged within a few meters of other floaters at abundant sources of food like ant swarms. The sedentary birds had territories in wet lowland forest, while the floaters spent much time foraging in streamside vegetation and second growth around cornfields and pastures. Only 2 of 34 tracked territorial birds died from avian or mammalian predation, while 7 of 27 floaters were killed this way, perhaps because they spent more time traveling through exposed areas and never stayed long enough in places where they could find safe roosting sites (Rappole et al. 1989).

For the Wood Thrushes in Veracruz, the dynamics are even more complicated during the several periods each month between November and March when cold, wet weather systems, usually lasting two to six days, move in from the north. Then, Wood Thrushes from higher elevations descend to the rainforest, presumably because they can no longer find food at the elevations exposed to the worst of the weather. These birds effectively become floaters for a few days until the storms end and they return to their initial wintering site, where at least some may hold territories. Thus, some wintering Wood Thrushes regularly alternate every several days between territoriality and floating (Winker et al. 1990).

For many birds, floating behavior has hormonal aspects that match the experience of the floater Wood Thrushes. While testosterone, associated with dominance, increases in female European Robins in autumn when they establish territories (Schwabl 1992), corticosterone has been linked to floating behavior. It is released rapidly in the bloodstream in situations of stress that increase energy demands, such as attacks by predators and food shortages. It also increases locomotor activity, searching for food, and fattening and reduces aggression and territorial defense. Experiments with territorial and floater Hermit Thrushes found that floaters had increased levels of corticosterone when placed in contact with territorial birds; this subsided after the birds were separated (Brown and Long 2007).

For species with specialized winter habitats, the proportion of floaters may be an indication of population vulnerability or decline. If territorial Wood Thrushes return in subsequent winters to the territory they occupied before, birds born that year are more likely to become floaters, at least until they can find a vacant place. But, as increased conversion of land for farms and pastures degrades or destroys high-quality habitat, a greater portion of the local wintering Wood Thrush population, adult and first year, must resort to floating, with its lower survival expectations.

TERRITORIAL HABITAT SEGREGATION BY SEX AND AGE

More long-distance passerine migrants partition their distribution by habitat separation within the same region rather than by latitude. In the Greater Antilles, all within six degrees of latitude and crowded in winter with North American warblers, variations in elevation, soil, and rainfall result in different vegetation types dominating various parts of each island. These, combined with areas disturbed by clearance or hurricane damage, create a patchwork of microhabitats that are populated by wintering warblers in complex patterns determined by sex and age.

In the Dominican Republic on Hispaniola, male Prairie Warblers (*Setophaga discolor*) are more numerous in pine forests during the first part of the winter, while females are more common in the desert thorn scrub and washes. During the drier months of late winter, the sex ratios become more equal, with females still somewhat more numerous in the desert. Adult birds of both sexes are more sedentary, with 50%, slightly higher in males, returning the following year to their territories in both habitats. Adults dominate in all habitats, with more females and young birds living as floaters. Birds that spent the entire winter in pine forests were found to be in better condition by the end of the season than birds in the desert (Latta and Faaborg 2001).

These pine forests and desert habitats, as well as dry forests at intermediate elevations, also host Cape May Warblers (*S. tigrina*) through the winter. While the Prairie Warbler feeds on small insects, the Cape May is an unspecialized resource generalist that can exploit more foods in more of the island's habitats. In the desert, it feeds on arthropods, in the dry forest it concentrates on the honeydew produced by a scale insect (family Margarodidae) found on the tree *Bursera simaruba*, and in the pine forest it takes nectar and fruits from other plants. The Cape May's semitubular tongue, unique among warblers, enables it to feed on the honeydew as well as on nectar and the juice that exudes from fruits—an adaptation deployed only on southward migration and in winter, since none of these liquids are found in the conifer forests where it breeds. On Hispaniola, Cape Mays are most abundant, with males predominating, in pine forests, which have the most consistently reliable food sources; elsewhere, females predominate, with young birds of both sexes commonest in the dry forest. In the pine forests, 76% of individuals remain the entire winter, while only 24% stay in the dry forest and 33% in the desert, where birds have the lowest body mass of Cape Mays in any of the three habitats (Latta and Faaborg 2002).

At higher, wetter elevations on Hispaniola, rainforest and cloud forest are the winter home to 90% of the world's Bicknell's Thrushes (*Catharus bicknelli*). In the cloud forest, where arthropod prey is abundant, 74% of the birds are male; in midelevation rainforest, thrushes feed mainly on soft-bodied fruit and 53% are male. Both sexes are territorial at all elevations. Males in both habitats are in comparable body condition, but the females in cloud forest are in poorer condition than males and than females in the rainforest. The larger-bodied males may be better able to withstand nighttime cloud forest temperatures, which descend to 12° to 0° C (Townsend et al. 2012).

On Puerto Rico, there is a similar diet distinction based on habitats occupied by Black-throated Blue Warblers. Males are the majority in mature forests, where they feed on arthropods, while females are primarily in shrubby second growth with abundant fruit and nectar. The difference in diet does not affect body condition, but the birds in forest require larger home ranges than those whose primary food is at high densities. The forest is occupied by adults that return fairly early in autumn to their prior territories; the second growth is settled first by hatching-year birds, followed by adults that have occupied the sites before. Some birds of both sexes are floaters, especially in the second growth; they may be individuals of subordinate status unable to establish a territory or could be specializing in sources of fruit and nectar too widely dispersed to be defended in a territory. Because the forest has the least seasonal variation of the various habitats occupied by Black-throated Blue Warblers, its arthropod food is available continuously. This means birds with territories there are more likely to remain all winter than those in the second growth; males in forest can begin their premigratory fattening sooner and depart earlier, in order to arrive first at their breeding site (Wunderle 1995).

In Jamaica, American Redstarts occupy several habitats from wet limestone forest to coastal scrub. Habitats with more reliable rainfall, or those that, like mangrove forests, retain the most moisture as the dry season advances in late winter, harbor the most arthropods and are dominated by adult males. Hatching-year redstarts migrate sooner and arrive in Jamaica before adults; the young birds settle equally in mangrove and scrub habitats, but when adults return, often to territories they have held in the previous winter, many of the young birds are displaced from the mangroves. The females that succeed in maintaining a mangrove territory are larger than those in scrub, and birds of both sexes are more aggressive in defending mangrove territories than territories in scrub (Marra 2000).

When all the redstarts have arrived and settled, 61% of those in mangroves are male; in scrub habitats 76% are female. Among the males in mangroves, more have the black-and-orange plumage of birds in at least their second winter. Birds of both sexes in the mangroves retain their body mass through the winter, while some in scrub lose mass. The scrub residents that lose mass also have elevated levels of corticosterone by late winter, a sign of stress and of muscle catabolism resulting from lack of body fat. Females persist equally in both habitats through the winter, and males more in the mangroves, indicating that males in scrub habitat are more inclined to move. Similarly, 60–70% of birds of both sexes wintering in mangroves return there the following winter, while only 40–44% return to scrub habitats they have used before (Marra and Holmes 2001). As further evidence that mangroves are the superior habitat, females and young males that were moved experimentally from scrub to emptied mangrove territories maintained their body condition through the winter, departed sooner on migration, and returned the following winter in greater numbers than did scrub residents (Studds and Marra 2005).

Among many warblers wintering in the mainland Neotropics, males also predominate in larger-scale vegetation. In Mexico on the northern Yucatán Peninsula at the Sian Ka'an Reserve, Quintana Roo, and in adjacent areas ranging from pasture through scrub to medium-height forests, the Northern Parula (*Setophaga americana*), Magnolia Warbler, Hooded Warbler (*S. citrina*), American Redstart, and Common Yellowthroat (*Geothlypis trichas*) all fit this pattern. Parula and Magnolia males tend to inhabit taller forest, and females scrub. Male Hooded Warblers are eight times more numerous in forest and are only 13% of the birds in open and scrub habitat. Redstarts are numerous in all habitats with at least some woody vegetation, but only females are in open habitat, while 29–67% of the birds in scrub and semideciduous forest are male. Yellowthroat females are in pastures and fields and males more often in second-growth scrub. Other warblers in the area, including Black-and-white and Black-throated Green (*S. virens*), both essentially forest birds, show less disparity, and Yellow Warblers of both sexes are equally dispersed in coastal and inland scrub vegetation (Lopez Ornat and Greenberg 1990).

Experiments with Hooded Warblers in Quintana Roo that removed territorial birds—all adults—in both forest and shrubby habitat found that the replacements were all hatching-year birds; often, these were quite secretive during the first few days of their new occupancy, suggesting their previous experience was as floaters rather than as territorial birds moving to a better site. Female Hooded Warblers have varying amounts of male-like black feathering on the head and throat, but this—unlike the extent of black in hatching-year male redstarts—does not give them any advantage in encounters with males or with females that have less black (Stutchbury 1994).

Also different from redstarts studied in Jamaica, where male-dominated habitat provided more food and was sought by birds of both sexes, experimentally emptied male Hooded Warbler forest territories were not taken over by females from adjacent shrubby areas. In this species, there seems to be an innate sex-based preference for each habitat type, and territorial occupants in both, male and female, are equally aggressive to intruders of both sexes (Stutchbury 1994). Experiments with hand-reared Hooded Warblers that had had no experience in choosing natural habitats demonstrated that males preferred habitats composed of vertical elements—resembling forest—while females were more attracted to habitats with more oblique branches, closer to the growth pattern of shrubs (Morton 1990).

In Malaysia, Eastern Great Reed Warblers (*Acrocephalus orientalis*) also separate by sex based on vegetation structure, with females primarily in reed beds (*Phragmites*) similar to their nesting habitat and males at the edge of the beds or in scrub, sometimes at a distance from water. Males are 10% heavier than females and spend the winter in much thicker woody vegetation, where their feathers become more abraded than those of the females in the reed beds. The females, because they are lighter, can move more easily through the reeds, and the abundant small insects there meet their needs; most spend the bulk of the year in their winter territory—as many as 240 days—leaving only for a very brief breeding season in northeast Asia. This seems to be another case of sexual segregation by preference rather than by dominance (Nisbet and Medway 1972).

INTERSPECIFIC TERRITORIALITY

Where winter food resources are reliable but highly concentrated and permanent residents must contend with a seasonal influx of closely related species with similar feeding habits, interspecific territoriality reduces the competition. In the Mojave Desert of southern California, Costa's Hummingbirds (*Calypte costae*) are joined in winter by Allen's (*Selasphorus sasin*) and Rufous (*S. rufus*) Hummingbirds at a time when relatively few plants are in flower. In March and April, before the migrants have departed, bladderpod (*Isomeris arborea*) is the only source of nectar, and the territories of each individual hummingbird of the three species include three or four bushes. Where these are concentrated, 15 birds of the three species can each have a territory within an area of approximately 50 sq m; where the bushes are more widely scattered, the territories may be 10 times larger but still hold the same number of bladderpods (Cody 1968).

In an even harsher environment, the rocky desert west of the Gulf of Suez, four closely related wheatears defend individual territories against each other. Resident White-crowned (*Oenanthe leucopyga*), Mourning (*O. lugens*), and Hooded (*O. monacha*) Wheatears are joined in winter by Pied Wheatears (*O. pleschanka*). Each bird perches conspicuously and sings; by October there is little direct contact. But a migrant Northern Wheatear (*O. oenanthe*) that took up a territory that month was driven off five days later by two neighboring Mourning Wheatears. One of the White-crowned Wheatears also pursued and grappled with a female Blue Rock-thrush (*Monticola solitarius*) that established a territory in October, but the rock-thrush was not driven away. For the next few months, the two birds sang from the mutual edges of their territories (Simmons 1951).

Larder-hoarders that concentrate valuable food resources in a few locations are also likely to face competition from other species. Red-headed Woodpeckers (*Melanerpes erythrocephalus*) have individual winter territories in which they store acorns in dead trees. In Highlands County, Florida, they begin storing acorns in late summer and defend their storage sites, but not the oaks from which they collect the acorns, which may be as far as 100 m from their territory. The Red-headed Woodpeckers defend their hoards against Red-bellied Woodpeckers (*M. carolinus*), Northern Flickers (*Colaptes auratus*), Florida Scrub-Jays (*Aphelocoma coerulescens*), and Common Grackles (*Quiscalus quiscula*). The Red-bellied Woodpeckers have larger territories, overlapping those of several Red-headeds; Red-bellies sometimes succeed in stealing acorns from a cache but are more often driven off. At the start of the season, there are also fights between single Red-headed Woodpeckers and several scrub-jays; in due course, overlapping territories are established without further aggression. The woodpeckers also chase non–acorn eaters, from mockingbirds and American Robins (*Turdus migratorius*) to Red-shouldered Hawks (*Buteo lineatus*) and Great Horned Owls (*Bubo virginianus*) (Moskovits 1978).

Warblers are also often aggressive at rich but localized resources. In addition to the Yellow Warblers keeping Magnolias from the canopy, where insects are more abundant, Cape May Warblers similarly defend nectar sources in tree canopies while allowing other, smaller migrant warblers (Northern Parula and Prairie Warbler) and residents to remain in the understory (Staicer 1992). In the oak woodlands of Chiapas, western Mexico, the same sort of vertical stratification is dominated by "Audubon's Warblers" (*Setophaga coronata auduboni*), the western race of the Yellow-rumped Warbler. It seeks out infestations of a scale insect (family Margarodidae) that produces secretions of honeydew all through the day. In one woodland, each Audubon's Warbler defended one to five trees with the infestation in an area of 300 sq m through an entire winter. It chased smaller birds from an average distance of 8.5 m once every five minutes; 90% of the chases were directed at male Townsend's Warblers (*S. townsendi*), which in turn chased female and immature Townsend's as well as Wilson's (*Cardellina pusilla*) and Nashville (*Leiothlypis ruficapilla*) Warblers and some local resident species. These vertically stacked Audubon's Warbler territories each had at least one adult male Townsend's Warbler, which was dominant in the absence of an Audubon's Warbler; 75% of the territories also had a Wilson's Warbler and a female or immature Townsend's in the understory (Greenberg et al. 1994).

WINTER FLOCKS

Birds that feed on plant or animal matter that is not evenly distributed and occurs in larger quantities or over greater spaces than can easily be defended are often in flocks. These aggregations differ from the small groups of tits and other birds that have a collective territory. They vary from purely opportunistic gatherings at briefly available sources like a flowering or fruiting tree—where, as we have already seen, many warblers will congregate as long as the food lasts and then return to their territories or wander off individually—to birds that deliberately join cohesive groups. In addition to the birds that flock throughout the year, such as many herons, gulls, terns, swifts, swallows, and some blackbirds, other species that are territorial in the breeding season form flocks afterward, when their diet changes.

Among passerines, those that continue eating insects found on foliage or bark throughout the year are less likely to form large flocks in winter than are more omnivorous species that add fruit, nectar, or any other patchy and ephemeral resource to their diet. And, because omnivory is more frequent among larger birds, flocking occurs more often in larger species, even within families where the size differences are small, as among warblers. Finally, among Nearctic and Palearctic migrants to the tropics, far more Nearctic species switch to flocking behavior in the winter, because fruit and nectar are more widely available in Neotropical forests than in African savannas (Greenberg and Salewski 2005).

Studies of flocking birds in all parts of the world generally conclude that flocks provide two significant benefits: collectively, the birds in a flock are more likely to spot any predator, and flocks are more likely to find patchy sources of food than are individuals searching alone. Long before ornithologists tested and demonstrated some of the benefits experimentally and statistically, hunters using decoys to attract waterfowl and shorebirds knew that passing flocks recognized another flock on the ground or in the water as likely being in a safe place to land and feed.

We have seen that even when birds stay in small family units, each bird spends less time scanning for danger and can feed more. This is amplified when the number of scanners is multiplied severalfold in flocks, both in likelihood that danger will be detected and that it will be detected sooner. The tendency to flock is therefore stronger in regions with more predators. In Jamaica, where there are no resident bird-eating hawks, only 18% of finches, tanagers, and icterids move in families or flocks; in Costa Rica, which has many bird predators, 69% of birds in these families form flocks (Pulliam and Millikan

1982). Similarly, the regular absence or presence of a predator at a particular site can influence flocking behavior. In coastal California, Sanderlings defend feeding territories during winters when no Merlins (*Falco columbarius*) are present; in years when one or more Merlin has its own territory along the beach, the Sanderlings spend the winter feeding in flocks (Myers 1984).

Birds with a feeding method that reduces their ability to spot predators are especially likely to form or join flocks. While many shorebirds spread out to feed, Stilt Sandpipers (*Calidris himantopus*) and dowitchers (*Limnodromus* spp.), which all probe deeply into mud, often submerging their head, always feed and fly in tight groups on migration and in winter. Worm-eating Warblers (*Helmitheros vermivorum*) search dead, curled-up leaves for insects; they join mixed-species flocks of local residents that include birds that can spend more time scanning for danger. The warbler treats the flock as its territory, defending it against other Worm-eating Warblers.

For predators, flocks of any species are easier to locate than are solitary birds, and birds at the edge of a flock are most often pursued. In mixed-species flocks, the species that are most regular flock members are least likely to be at the periphery where they would be most exposed. Single-species flocks of shorebirds, pigeons, starlings, and other birds that take off and form compact masses moving in circles or erratic courses reduce the vulnerability of every bird.

In addition to giving each bird more time to feed if at least a few members of the flock are scanning for predators, flocks are more likely to locate food sources. Immature birds in their first winter, perhaps consuming a new type of food or in a new location, may learn to identify food from more experienced adults. In the same way, migrants joining local mixed-species flocks benefit from the knowledge of the residents. Birds that in winter roost communally and disperse by day in larger or smaller groups may follow the ones that depart early from the roost to return to a place where they fed the previous day.

For birds that feed on patchy and ephemeral resources, another benefit of flocking is that they can, by sheer numbers, overwhelm any defense by a territorial bird claiming that spot. Northern Mockingbirds can drive a single Cedar Waxwing (*Bombycilla cedrorum*), European Starling (*Sturnus vulgaris*), or American Robin from a shrub or tree bearing fruit, but when a flock of any of these species lands, the mockingbird can chase only one at a time while the rest of the group feeds, and the chased bird will return as soon as the mockingbird shifts to pursue another (Moore 1977). Resident territorial pairs of Common Ravens (*Corvus corax*) that discover a carcass seek to avoid attracting the attention of the many wandering immature birds that might join them at the scavenging site, but immature ravens that know the location of a carcass recruit others so that the normally dominant local adults cannot defend it against all the intruders (Heinrich 1989:310–311).

For some species, flocking, territoriality, and solitary behavior may all occur in the same region. In England, Great Tits form roving flocks in forests, but in suburban areas where food is more reliably available they tend to be territorial (Zahavi 1971). In a second-growth forest in Puerto Rico where several species of North American warblers winter, the three commonest are Northern Parulas, Cape May Warblers, and Prairie Warblers. Marked individuals return to the same territory in subsequent winters, but while Cape

Mays actively defend their territories and the smaller Prairies avoid the other warblers, Parulas range from territorial to gregarious, moving in flocks around a collective home range. At day's end, all three species leave the area where they forage to roost communally in dense trees surrounded by scrub or cattle pasture; these sites are presumably safer from nocturnal predators (Staicer 1992).

SINGLE-SPECIES FLOCKS

Eurasian Hazel Grouse (*Bonasa bonasia*) demonstrate the ecological factors that make flocking advantageous. In winter, Hazel Grouse eat catkins and buds of various deciduous trees. These are relatively scarce in dense Swedish spruce forests; there, the grouse are territorial, singly or in pairs. At the other end of their range, in Siberia, Hazel Grouse occur where larch and birch forests meet and along stream banks that have many deciduous trees. Here, food is more abundant and the grouse form flocks; in the larch-birch ecotone, there are 3 to 6 birds in a flock, and in riparian areas, up to 30. Throughout their range, Hazel Grouse forage in trees, but where they are alone, as in the Swedish spruce forests, they stay in the densest vegetation. In the more open Siberian woods, where food is more abundant but the birds are more exposed to predators, the number of vigilant birds in the flock reduces their risk. During the long hours of the northern winter night, Hazel Grouse normally roost in burrows each bird makes for itself in the snow, but when snow conditions are not suitable, they roost in flocks in trees. This must reduce each individual's risk of predation, even if the flock is easier for a predator to locate than more widely spaced birds (Swenson et al. 1995).

In Black-capped Chickadees, where both sexes and all ages have the same plumage and flocks moving through thick vegetation may lose visual contact, vocal communication is important for maintaining flock cohesion. A study focused on the variety of vocalizations in winter flocks of chickadees near Ithaca, New York, found that each member of a flock has distinctive elements in its *chick-a-dee* call; these include the frequency ranges, the total call duration, and the duration of the *dee* note. These attributes enable flock members to identify each other individually. At the same time, the calls of all members of a flock share certain characteristics, such as the duration of the *chick* note, the number of *dees*, and the intervals before the first *dee* and between *dees*. These help separate members of one flock from another. The aspects distinctive to each flock are at lower frequencies than the individual differences; lower frequencies travel farther and therefore can be heard by other flocks at greater distance. When the birds were brought into aviaries and the flocks reconfigured, the flock-distinctive calls of the members of each new flock converged within a month. This convergence facilitates cohesion in the new flock; in the wild it may help integrate new members into a flock (Mammen and Nowicki 1981).

Eastern Kingbirds (*Tyrannus tyrannus*), named for their pugnacious defense of their nesting territory against both conspecifics and much larger birds, change their behavior with their diet in winter. They migrate through Mexico and Central America to western Amazonia, where by mid-October they arrive in southeastern Peru in Manu National

Park in flocks of 25 to 250. Here, they feed in fruiting trees at river and lake edges, rather than in the forest interior. Their numbers overwhelm the several highly aggressive local flycatchers feeding in the same trees (Fitzpatrick 1980). A favorite food of Eastern Kingbirds is the fruit of *Didymopanax morototoni*, found in second-growth woodlands and other disturbed areas. Kingbirds go directly to the southern end of their winter range, at the southwestern end of the Amazon basin, and work their way north in synchrony with the fruiting of *Didymopanax*; by early March, they are in Panama, feeding on this tree in flocks of more than 100 (Morton 1971). The Eastern Kingbird, because it may travel farther in a day than resident flycatchers, may be a significant dispersal agent for *Didymopanax* (Morton 1980).

MIXED-SPECIES FLOCKS

The factors determining whether birds form a single-species flock or join a mixed one may be equal. In Costa Rican mangrove forests, Prothonotary Warblers were found in flocks 87% of the time; 48% of these were single-species flocks with an average of 4.5 birds. They also joined mixed-species flocks composed of residents and other wintering warblers; in these, the average number of Prothonotaries was 3.9, and Tennessee Warblers were the commonest flockmate. But since Prothonotaries were equally likely to form small single-species flocks as to join larger mixed flocks, increased predator detection seems unlikely to be a factor in their choice (Warkentin and Morton 2000).

Flocks that routinely contain several species combine features of territoriality with sociality. In both nonmigrant mixed-species flocks and those formed at winter locations, the flocks usually have a collective territory that they defend against other, similar flocks and, unlike the more casual assemblages of wintering warblers, flocks may have a fixed number of individuals of each species. Each species may have a different foraging method or preferred food, so that there is little competition among the flock members, while all benefit from the greater number of eyes scanning for predators and the reduced chance of becoming a victim. Some species are more regular members—"nuclear" species without which a flock may not form—and others are more casual joiners.

The nuclear species have an inherent tendency to form flocks, large or small, whether or not any other species join. Thus, they usually outnumber any regular followers or occasional joiners. Their number makes them easier for other species to find and follow. Often the members of the nuclear species are closely related, increasing the likelihood they will give alarm calls. Joiners benefit from this as well as from the knowledge the nuclear species, most often a permanent resident, is likely to have about food locales (Goodale and Beauchamp 2010).

Wintering migrant birds have different degrees of association with flocks composed primarily of resident birds, be they in temperate or tropical regions. A seasonal arrival may become an integral member of the flock, moving with it throughout the flock's home range and defending it against conspecifics. Alternatively, the winterer may establish its own territory and join the resident flock only when those birds move through it; the territory may overlap that of more than one mixed flock and the bird may associate with each as the flock appears; or, finally, two or more unrelated individuals may associate with the flock (Rappole 2013:144).

Resident Mixed Flocks

A "mixed" flock may contain as few as two species. In Latvia, Crested (*Lophophanes cristatus*) and Willow (*Parus montanus*) Tits move together in conifer forests, with four individuals of each species in the group. Their arthropod diet and foraging method are essentially the same, but the two use different parts of the trees. In groves of young pines, the dominant Crested Tits feed lower, in the less visible portion of the trees, while the Willow Tits are higher and more exposed. In experiments where Crested Tits were removed, the Willow Tits did not move to the seemingly safer part of the trees, but when Willow Tits were removed, the Crested Tits went higher in the trees. While more exposed, the upper and outer branches were better vantage points from which to spot predators; when Willow Tits are not there to do the job, Crested Tits must take it on (Krams 2001).

In spruce and pine forests of central Sweden, these two tits are the nuclear species, first and second ranking, respectively, when joined by Coal Tits (*P. ater*) and Goldcrests (*Regulus regulus*). Flocks usually average 3 Crested, 3 Willow, and 1.7 Coal Tits and 2.7 Goldcrests. Here, too, all four consume the same arthropods but differ in their foraging location. While arthropods are 20 times more abundant on the uppermost outer branches than the lower inner branches, the two dominant tits feed on the more concealed inner branches, the Willow Tits lower than the Crested Tits. The Coal Tits and Goldcrests feed on the exposed outer branches, but the risk there of predation, especially by Northern Pygmy Owls (*Glaucidium passerinum*), is greater. Each flock has fewer individuals of the species feeding in the most dangerous places than in the safer ones. Long before winter, the tits establish the hierarchy of safe places to feed: the part of the tree where each stores the most food matches the part where it continues to forage in winter (Suhonen et al. 1992).

In North America, where most forests have only one chickadee species and, if another parid is present, it is a much larger titmouse, the mixed flocks are characterized more by different foraging methods than by spatial partitioning. Typical flocks include Black-capped or Carolina Chickadees, Downy (*Dryobates pubescens*) and/or Hairy (*D. villosus*) Woodpeckers, White-breasted Nuthatch (*Sitta carolinensis*), Brown Creeper (*Certhia americana*), and Golden-crowned Kinglet (*Regulus satrapa*). Collective vigilance to increase chances of detecting predators may be a benefit to these flocks, but collective ability to find food may be more important. Where these birds have access to feeders, they spend less time in mixed-species flocks (Berner and Grubb 1985).

In European tit flocks, each species stores food long before winter and can resort to it on days when foraging is more difficult because of cold, rain, or snow. In North American mixed flocks, however, most birds except for the tits must find all their food each day. For them, more birds searching for food may be the incentive to join a flock. A study in central New York found that on sunny days in January the number of Black-capped Chickadees in a flock was the same when the temperature was near 0° C and when it dropped to −9°

C, but on the cold days, when finding food was more urgent, they were far more likely to be joined by the woodpeckers, nuthatch, creeper, and kinglet (Klein 1988).

Temperature and wind also affect where flocks forage. In Ohio, where the flocks in isolated woodlots included Downy Woodpeckers, Tufted Titmice, Carolina Chickadees, and White-breasted Nuthatches, the woodpeckers and nuthatches did not shift their place in the tree when temperatures dropped and wind increased, but the two tits, which feed on the more exposed twigs and outer branches, moved lower. When winds were weak, birds stayed on the windward side of the woodlot even when the temperature was −15° C. If winds were strong, however, all birds except the female Downy Woodpeckers, which spend more time on the trunk and thicker branches than do males, moved to the leeward side of the woodlot when the temperature was below −6° C. Since the Carolina Chickadee, the smallest and most susceptible of the flock members to cold and wind, was also the nuclear species, the others may have followed it to the leeward side rather than lose contact, even if they could have continued foraging in the more severe conditions (Dolby and Grubb 1999). A similar study in New Jersey found that when temperatures were −20° C to −29.9° C, the two tits moved from woodlands to adjacent old-fields where they could forage on or near the ground; they were not followed by the trunk-foraging woodpecker and nuthatch (Grubb 1977).

Mixed Flocks of Long-Distance Migrants and Residents

Among long-distance migrants, the likelihood of joining a local mixed-species flock varies by locale and by habitat. On Cuba, where far more hawks specialize in birds than on the smaller Caribbean islands, mixed flocks are larger and more widespread and involve more species. A survey there of several habitats found 56 species that participate in flocks, including 26 North American migrants. Two resident warblers, the Yellow-headed (*Teretistris fernandinae*) and Oriente (*T. fornsi*) Warbler, were, in different parts of the island, always the nuclear species. The average Cuban flock included 6.8 birds of 4.6 species other than the nuclear warblers. Migrants were part of 91% of the flocks, and most of these included 4–7 birds of at least 3 migrant species (Hamel and Kirkconnell 2005).

In parts of Mexico, migrants play a more central role. In Tamaulipas, northeastern Mexico, 37% of all forest species participate in mixed-species flocks, including the same proportion of migrants, 39%, or 16 species. Wintering Ruby-crowned Kinglets (*Regulus calendula*), never found singly, are the nuclear species in flocks in dry pine-oak forests, and Blue-gray Gnatcatchers (*Polioptila caerulea*) lead the flocks in tropical semideciduous forests, the habitat in which each is most abundant. In other habitats, resident birds are the nuclear species. The average flock contains 4.26 species, with more residents in the dry forests and more migrants in the tropical ones (Gram 1998). In Jalisco, on the western slope, where 27 species join flocks in lowland tropical deciduous forest, Blue-gray Gnatcatchers and Nashville Warblers are usually the nuclear species, found in 80% of the mixed flocks (Hutto 1994).

Deeper into the American tropics, in Costa Rica and Panama, where more of the natural habitat is mature forest with a closed canopy, many migrants, as we have seen, favor the more disturbed areas of dry scrub, dense edges and openings, and broken canopy. Only about 20 of the migrants that regularly winter in forest habitats here associate with resident flocks. Of these, 7 join flocks in closed-canopy forest and 13 in the more open areas, where the flocks are composed of different resident species. The flocks in open areas move more slowly, return more often to the same places, and stay in the migrants' preferred habitats more than do forest interior flocks (Powell 1980). On Barro Colorado Island, in central Panama, small resident members of flocks supplanted and chased the larger Magnolia and Chestnut-sided Warblers, and most interactions of Tennessee and Bay-breasted Warblers (*Setophaga castanea*) in the flock were with other warblers. The two commonest warblers in these flocks, Bay-breasted and Chestnut-sided, each associated most closely with certain of the residents—Chestnut-sideds with the Dot-winged Antwren (*Microrhopias quixensis*), Bay-breasteds with a small flycatcher and tanager and a resident gnatcatcher (Greenberg 1984:43–47).

In northern South America, most of the wintering passerines are at higher elevations. A total of 63 species participate in mixed-species flocks in Venezuelan shade coffee plantations between 675 and 1230 m, with an average of 13 species in any flock. American Redstarts and Bay-breasted, Blackburnian (*Setophaga fusca*), and Cerulean (*S. cerulea*) Warblers are the most frequent of the migrants, in that order. Flocks contain 8–40 birds, with local species as the nucleus (Jones et al. 2000). In the Central Andes of Colombia, 68 species, including 19 migrants from North and Central America, are part of flocks in oak woodlands surrounded by pasture at 1800 m. Nearly half the birds in any flock are migrants, and Blackburnian Warblers constitute 57.2% of all the migrants. These flocks seem to have no nuclear species or patterns of preference for joiners. The resident members in this fragmented habitat are widespread generalists, species with extensive ranges from Central America to Bolivia or Argentina, rather than forest specialists. During the months the migrants are present, the flocks forage lower in the trees than when residents have the woodland to themselves. This may be because the additional numbers force some birds out of the preferred canopy zone, or because their presence makes the lower, more exposed levels safer (Chipley 1976).

Most of the wintering passerines in the Neotropics seem to be joiners rather than nuclear species and may drop in and out of mixed flocks, leaving them, for example, to feed at a source of nectar or fruit while the essentially insectivorous residents continue moving through their collective territory. Some migrants, however, treat the flock as their own territory. A few with specialized feeding techniques, like the Worm-eating Warbler already mentioned, benefit from the presence of more vigilant birds. Others that make a flock their territory are Golden-winged Warblers, which also probe into clumps of dead leaves in vine tangles, and Black-and-white Warblers, which search the bark of tree trunks and major branches. All three warblers defend the flock against conspecifics (Tramer and Kemp 1980). For them, the predator-scanning benefits must be more significant than whatever guidance to food sources the resident birds can provide, because the other members of the flock do not forage in the same substrate. Several other more generalized foliage gleaners wintering in various parts of the Neotropics also defend flocks, including vireos, other warblers, and the Summer Tanager (*Piranga rubra*); they must gain from the residents' local knowledge of food sources, especially if food is more thinly distributed than in an area the wintering bird could efficiently find and defend on its own (Rappole 1995:44–45).

DOMINANCE

Except for the few birds that are still looking after their young in winter, every individual is potentially competing with all others of its species until the following breeding season. Birds that defend winter territories substantially reduce their competitive interactions; all territories may not be equal in quality, and there may be a hierarchy based on age, sex, and size among the claimants, but once established, the territories are accepted by their neighbors, at least until a vacancy occurs. Among birds that flock or depend on resources that are not evenly distributed or of equal quality, however, the energetic drain of constant interaction within a flock is usually reduced and simplified by establishment of a hierarchy of dominance. And since rank correlates with access to food and other vital resources, dominance has a significant impact on winter survival.

Age and sex are the usual determinants of dominance. Research demonstrating the patterns and their ecological correlates is easiest with species in which males and females, as well as adults and young, have different plumages. Finer details come from studies where the population has been individually marked. The Eurasian Oystercatcher (*Haematopus ostralegus*) is an ideal subject for research of this kind because large numbers can be observed simultaneously on coastal estuaries and mudflats, their long legs can carry highly visible individually color-coded bands, and the birds take five years to reach adulthood, so there are more different age classes that can be recognized by plumage variations. As in many shorebirds, female oystercatchers are larger than males.

On the Exe Estuary on the south coast of England, subadult oystercatchers are present year round, but 80% of the young birds leave the preferred mussel beds in autumn when adults arrive; the young return from inferior parts of the estuary when the adults depart

Adult Eurasian Oystercatchers vary in their aggressiveness, with some dominant birds attacking near neighbors as often as every five minutes when they are feeding at low tide.

in spring. There is generally low mortality in winter, but it is sharply skewed; for adults, it is less than 2%, but 12% for juveniles in their first winter. In contrast, during spring and summer, adult mortality in their breeding area is higher than for the subadults remaining at the Exe Estuary. In autumn, adults not only prompt most young birds to move away from the flats with the most mussels, but among them some adults are more aggressive than others. The most aggressive oystercatchers attack another every five minutes when feeding at low tide in the best zone for mussels, often stealing a mussel the other bird has just taken from the mud. The aggressive birds are themselves seldom attacked, while unaggressive birds are attacked frequently and usually lose the encounter. The dominant birds gain most from their aggression where the density of oystercatchers is highest, because it is easier to steal from or drive away a bird that is close rather than farther away. The subdominants consume less, not because mussels are stolen from them, but because they find fewer. It may be they are too distracted, or they are forced into less profitable areas (Goss-Custard and Durell 1984).

Oystercatcher roosts further reveal the dominance hierarchy. On Texel and Vlieland, two islands along the Dutch Wadden Sea that in winter harbor as many as 40,000 and 25,000 oystercatchers, respectively, there are several roosts, ranging from large to small along each coast. The largest roosts are nearest the largest feeding areas, with gradually sloping shores that have the greatest tidal fluctuation and therefore the most hours available for feeding. Most birds are faithful all winter to a single roost, feeding on the flats nearby. The largest roosts on each island were found to have the greatest number of adult females, the midsized roosts more subadults, and the smallest roosts the most males and birds in their first year. Physical condition also correlated with roost size: birds with external tumors, missing toes, or malformed bills were commonest at the smaller roosts, no matter their age. Mortality was higher at the small roosts, but most notably during a period of two weeks with strong winds and temperatures below 0° C, when 36–39% of the birds found dead at the small roosts were below normal weight and 26.6% had abnormalities (Swennen 1984).

Flocking birds that remain in families through the winter also have hierarchical interactions with nonfamily members in the flock. Snow Geese (*Chen caerulescens*) wintering in southern Louisiana salt marshes and rice fields may be single, in pairs, or in families with one or more young birds. Families with more young win encounters over those having fewer young; families of any size, even with only a single parent, prevail over pairs without young; and pairs dominate solitary birds. As with the oystercatchers, individuals with any injury are most subordinate. Disputes are food related; the geese protect the area in which they are foraging against birds not in their group. In salt marshes, interactions average 4.7 per hour and in rice fields 8.7 per hour. The birds are more dispersed in salt marshes and may continue pulling up rhizomes from a single spot for more than 20 minutes, but the interactions here are more aggressive, reaching the level of fights rather than occasional pecks. Adults with young initiate the most conflicts, having the greatest need to defend space for themselves and their family. In most cases, the initiator of any aggression usually wins; that bird is likely to have assessed its chances, based on the status of its neighbor, before it instigates anything (Gregoire and Ankney 1990).

The little groups of chickadees and titmice so visible in winter have a distinctive social structure that is maintained when they are joined by other species. Some of this structure can be seen even without individually marking each bird. The core pair of adults is dominant. There is a linear dominance determined by sex and mate. Females have the rank of their mate, but males dominate over most or all females. In Carolina Chickadees, the alpha female dominates all birds except her mate—but only when in her mate's presence. In Tufted Titmice and Black-capped Chickadees dominance among those that join the core pair is determined by when the birds arrived; earlier joiners outrank later ones. Later still, the occasional unattached young Black-capped Chickadee visiting different groups in winter is looking for vacancies in the higher ranks that it might fill. All members of the group will help the core pair defend its territory and do not usually associate with neighboring groups. This, however, may break down at artificial sources of superabundant food, such as feeders, where birds of several nearby groups may visit sequentially or simultaneously without aggressive interaction (Matthysen 1990).

Similarly, in Willow Tits in central Norway, winter flocks usually contain a resident pair of adults joined by two unrelated juvenile males and two juvenile females that each form a pair. The resident adults are dominant over all juveniles of their sex; the ranking among the young birds is determined by size, weight, and seniority in joining the group. If the dominant among juvenile males dies, its mate sinks to the lowest position in the flock within one day. Occasionally, an adult female is the dominant bird (Hogstad 1987a). Average overwinter survival is 74% for adults, 32% for juveniles. Adult survival is enhanced because they feed in the parts of trees most sheltered from predators; they force the juveniles to feed in the more exposed branches. In warm weather, the flock may split temporarily, with the juveniles moving within the collective territory to better feeding spots; they return when the weather turns cold. When together, all members of the flock have more time to feed because each bird needs to spend less time in vigilance. The occasional all-juvenile flock is restricted to poorer habitat, and in experiments where the adult pair is removed from a flock, the juveniles are less likely to survive the winter (Hogstad 1989).

In North America at far lower latitudes, dominant Tufted Titmice and Carolina Chickadees have fewer fat reserves than subordinates. While in the longer nights and colder temperatures of coniferous Norwegian forests more fat may be an advantage to the dominant Willow Tits, in leafless deciduous woods, where shelter from hawks is limited, maneuverability by day may be more valuable. Since the dominant birds have priority access to food, they have less need than do subordinates to put on extra reserves to buffer themselves against shortfalls (Pravosudov et al. 1999). When these two species are in a flock together, the titmice are always dominant; all members of the flock are less vigilant than when alone, but subordinates of each species are still the most vigilant (Pravosudov and Grubb 1999). When a predator is sighted by flocks of titmice and chickadees, birds may freeze for as long as 15 minutes after the alert, and the dominant titmice wait an average of 84 seconds longer than subordinates to begin moving again; this may be another factor in their higher survival rate (Waite and Grubb 1987).

Sparrows that winter in small flocks and come readily to feeders also lend themselves to research on dominance. A three-winter study of individually marked White-throated

Sparrows (*Zonotrichia albicollis*) in North Carolina found that in addition to the usual dominance in sparrows of males over females, of older birds over first-year birds, and of larger birds over smaller ones, a key factor in the flocks' hierarchy was the place where an aggressive encounter occurred. Each of the marked birds in these loose flocks living in a long hedgerow had more success in aggression when it was near the center of its range in the hedge. When birds were farther away, they had more confrontations with unfamiliar birds that, if at the center of their own range, were more likely to dominate. Birds of low status fared equally poorly at the core or the periphery of their range. The ultimate determinants, in addition to placement, were size and age; birds returning for their second winter made large gains in status, less the following one, and little after their third winter. Return to the same small range within the study area hedge was not a significant factor, nor was the arrival date of the bird returning for the winter or any of the normal plumage variations such as the brightness or extent of yellow on the crown and whether the crown stripes were black and white, black and tan, or brown and tan (Piper and Wiley 1989).

The Dark-eyed Junco (*Junco hyemalis*) is so closely related to the White-throated Sparrow the two species sometimes hybridize, but the junco has a different winter ecology. Most form small, cohesive flocks, rather than the sparrow's loose aggregations or small territories. Juncos are more mobile, feeding in open fields, not woodland thickets and edges, so the predator detection benefits of a flock may be more important to them. Seeds in fields are also likely to be more evenly distributed, so individual juncos may not be as aggressively competitive during most winter conditions once the flock hierarchy has been established based on sex, age, and arrival or prior occupancy. Birds that join a flock later have the lowest rank. Various studies have found that juncos' normal distance from one another during feeding is a meter or less—but when food is scarce and concentrated they dispense with the threats or pecks that keep subordinates at a distance and tolerate closer proximity, prioritizing feeding over dominance (Sabine 1959).

While most juncos join the winter flocks feeding in fields, some, generally smaller birds that may have found themselves lowest in the hierarchy, shift to woodlands, where they move about either singly or in small groups. Here, the birds are each at a greater distance from one another, so there are few aggressive encounters. Later in the winter, if they have exhausted the food supply in the woods, they may join flocks, or resort to bird feeders if these can be found. But, as with the subordinates that remain in the flocks, they tend to have less fat, indicating that they have had less access to food; they also have larger adrenal glands, a symptom of stress. In North Carolina flocks where each bird was marked and its rank known, the individuals that most often died following ice storms or other conditions that significantly reduced food availability were subordinates (Fretwell 1969).

Experiments with juncos in Utah in which flock members were removed for a week found that when reintroduced, the birds quickly regained their original rank after some aggressive interactions. Birds from other flocks that attempted to join were subject to more aggression and acquired only subordinate rank. The plumage characters usually associated with high rank, such as a darker head and more white—most prominent on adult males—did not advance late joiners. While in juncos more males winter north of most females, there is always some overlap, and dominant males are more tolerant of females in the flock than of other males; the males gain as much predator vigilance from females without spending energy on aggression (Balph 1979).

There are some costs to maintaining the dominant position in a winter flock. Among the Norwegian Willow Tits, the dominant male is always the most aggressive in territorial encounters with other flocks. That drains energy and increases the risk of exposure to predators. Perhaps as a consequence, this male has the highest metabolic rate within the flock. The metabolic rates of the other members correlate with their rank in the linear sequence among the sexes and ages. If the dominant male is removed, the higher-ranking juvenile male becomes the dominant, and its metabolic rate increases; no other birds in the flock increase their rate. Nor do they if any female is removed from the flock (Hogstad 1987b).

Dominance within a group of birds as it is forming in autumn or winter may be established by a mix of visual and social signals reinforced by aggressive interactions. Every member of the flock benefits from minimizing the chases, attacks, fights, and other encounters that are energetically draining and take time away from foraging and vigilance. Plumage variation can help flock members identify each other and recognize their actual or potential place in the hierarchy. This may be especially useful where flocks are unstable and newcomers need to be integrated frequently. While individuals of some flocking species composed of permanent residents like tits have close to identical plumage—and these may be the most stable of winter groups—among looser flocks of migrant juncos and sparrows, birds may have a wide range of shades, especially on the foreparts, which birds see when they are confronting one another.

In Harris's Sparrows (*Zonotrichia querula*) the extent of black on the crown and throat is especially variable. Research with marked birds in Kansas found the familiar hierarchy, reflected in the amount of black, of adult males, juvenile males, adult females, and juvenile females. But the most frequent aggressive encounters were not between birds ranked near

The extent of black on the crown and throat determines the rank of Harris's Sparrows in a flock.
Adult males usually have the most black, juvenile females the least.

each other, but between those with the blackest heads and those with the least. In these cases, where the rank disparity was already established, the purpose may have been to drive the most subordinate birds out of the flock so there would be less competition for limited food resources (Rohwer 1975).

Plumage variation is also a factor in whether a winter-territorial bird can establish itself in better habitat among higher-ranking neighbors, and in whether it becomes territorial at all. Male American Redstarts take two years to acquire the adult black-and-orange plumage. Some juvenile males have more black on the head and throat than others; in Jamaica, these were able to defend territories in superior mangrove forests populated by adult males, while juvenile males lacking any black settled in the dry scrub habitats used also by females of all ages (Germain et al. 2010). Among Yellow Warblers juveniles of the Alaskan and western Canadian subspecies have more olive plumages than do adults and the bright yellow birds of eastern North America. Each autumn, these juveniles are the last to arrive in Panama, where the more yellow individuals have already established territories; the duller juveniles avoid competition and instead join mixed-species flocks of resident and wintering species (Morton 1976).

In the finch genus *Haemorhous*, females are dominant over males. Juvenile male House Finches (*H. mexicanus*) are distinctive in the genus in already having some red feathers characteristic of adult males; the amount of red increases as the birds age. At a feeder in Connecticut, the all-gray females were dominant, and young males with the least red were dominant over males of all ages with more red. In encounters between birds of various plumages, gray House Finches displaced red ones 45.6% of the time, while red birds displaced grays only 7.6% of the time. In encounters between reds, those with less displaced those with more red plumage 61% more frequently. For young males, there may be an advantage to "fooling" both females and adult males, reducing aggression from females and winning contests with other males. Evidence that dominance at a feeder has real survival value comes from the fact that when there was cause for alarm and birds flew away, in 66.9% of cases the only bird to remain was red, indicating it was willing to take more risks to secure food (Brown and Brown 1988).

INTERACTIONS AND COMPETITION IN THE WINTER RANGE

COMPETITION WITH RESIDENT CONSPECIFICS

Partitioning by habitat and diet, among both resident and migrant species, substantially reduces the likelihood of direct competition between species when an influx of wintering birds enters a region it occupies only seasonally. Where migrant populations settle for a few months in the range of resident conspecifics, however, competition may be more direct. In some species, this is avoided when a local breeding population itself migrates just as others are arriving. In West Africa, the first Hoopoes (*Upupa epops*) from Eurasia arrive south of the Sahara in late August, when the southern race is departing at the end of its nesting season; these return to breed at the beginning of the following rainy season, when the northern Hoopoes are leaving. In East Africa, the local Hoopoes do not migrate and may compete with wintering ones from farther north (Moreau 1972:195). Among most species with overlapping winter populations, residency and age usually give local individuals the advantage over migrants, but in some cases size may lead to dominance.

In southwest Spain, Red Kites (*Milvus milvus*) live year round in the marshes of the Guadalquivir River. From November to March, migrants from central Europe also occupy the area, outnumbering the residents during the peak months, November–January, by four to one. The kites are generalized predators and scavengers, with residents taking more insects in winter and migrants more often scavenging on larger items such as dead geese. During winter, the residents spend more time in forests, roosting at their nest to defend it against other residents, and migrants forage gregariously over the marshes; they also roost in trees but shift sites every few days. Residents, familiar with their territory, do not usually begin foraging until midmorning, while wintering birds depart their roost at dawn to search over a wider home range in the marshes—which in winter are vacant of competitors because the local Black Kites (*M. migrans*) have left for Africa (Heredia et al. 1991).

Elsewhere in southern Spain, in areas with a mix of forest and shrubland, resident European Robins are joined by migrants. The resident birds have each established their individual winter territories in forest before migrants arrive. Only a few of the migrants are able to find a space in the forests, and most settle in shrubland, which has more fruit, but fewer invertebrates or places to shelter from predators. The greater desirability of the forest is apparent in that no resident birds occupy shrubland before migrants arrive, and the migrants able to settle in forest are the first to arrive and the last to depart. Over the course of the winter, some 35% of the original forest birds shift to shrubland; these are mostly juveniles, replaced in the forest by adult migrants. While food is more abundant, if less varied, in the shrubland, forest birds are heavier; this may, however, reflect dominant individuals, resident and migrant, successfully defending territories that have additional benefits such as shelter from predators (Telleria and Perez-Tris 2004).

In the same landscape, resident Blackcaps (*Sylvia atricapilla*) are joined in the forest by equal numbers of adult migrants, while the shrubland is occupied 95% by migrants, with 80% of all individuals there being juveniles. Over the course of the winter, the forest population is steady, but in the shrubland the number of Blackcaps fluctuates in different winters with the amount of fruit available, suggesting there may be large-scale movements in response to fruit abundance and decline (Telleria and Periz-Tris 2007).

In the Venezuelan Llanos, a mix of seasonally flooded grasslands and gallery forest along waterways, four resident vulture species each search for carrion by using different flight methods, concentrating in certain habitats, and favoring carcasses of different sizes. From October until April, during what is mostly the dry season, they are joined by wintering Turkey Vultures (*Cathartes aura meridionalis*) from North America that compete most directly with the smaller resident subspecies, *C. a. ruficollis*. During those months, the resident Turkey Vultures, which had been foraging over both savannas and forest, restrict themselves almost entirely to gallery and semideciduous forest, while the wintering birds favor the open and semiopen areas and readily come to road kills and carcasses near buildings, which the residents avoid at all seasons. Where they gather at the same carcass, wintering Turkey Vultures displace residents 86% of the time. The winterers also displace resident birds at roost sites—towers and adjacent dead trees; in late October and November, residents occupy only the lower rungs of local towers, and by December no residents roost on or near towers. During the course of their stay, the body condition of wintering Turkey Vultures improves while that of residents declines, suggesting that residents forage in habitats with less carrion. After the wintering birds depart, resident Turkey Vultures return to searching for carrion in open habitats, and more of the local Lesser Yellow-headed Vultures (*C. burrovianus*) move into the area (Kirk and Currall 1994).

INTERACTIONS WITH RESIDENT SPECIES

From the evolutionary point of view, most of the birds wintering in the tropics are in fact returning "home" where they have a long-standing place. The idea that they may be competing for resources or driven to intrinsically poorer ones in marginal habitats has been substantially dispelled by research in the Neotropics and Africa. Some wintering species use habitats that residents do not favor, because these are not useful to residents all through the year, while other winterers sharing habitats used year round by residents specialize in food resources not exploited by the residents.

In the Neotropics north of the equator, birds arrive at the end of the rainy season or the beginning of the dry season, when less food of all kinds is available. Some resident birds that nested in forests earlier in the year now also face the same decline and shift to using the disturbed areas that receive more sun than the forest and have more flowering and fruiting plants with their attendant insects; for them, too, the breeding season, with its pressure to find protein-rich arthropods to feed their young, is over and they can diversify their diet. These birds may avoid disturbed areas when nesting if second growth harbors more predators than the forest (Gates and Gysel 1978). But at the same time that a few forest residents may temporarily exploit edge habitat, certain migrant species are establishing individual territories and their distinctive niche in the forest, where the density of resident birds remains far greater.

During the breeding season, resident Neotropical passerines comparable in body size to the migrants tend to feed themselves and their young large, soft-bodied insects greater than 5 mm long. These are primarily lepidopteran larvae and orthopterans—crickets, katydids, grasshoppers, and the like—and are commoner in forests than in disturbed habitats, especially during the local breeding season, which peaks between April and June, when migrants are gone. To exploit them, the resident birds have noticeably longer and narrower bills than do their relatives from higher latitudes that prey on smaller insects throughout the year. These long, narrow bills can close more rapidly on fast-moving orthopterans, while the shorter, wider bills of migrants from North America have no difficulty picking immobile caterpillars, the principal prey they feed their young, from leaves and branches; the broad bill in fact makes caterpillars easier to hold and to beat as they struggle. As adults, these migrants consume primarily smaller insects than they feed their young (Greenberg 1981). The birds are therefore well suited to the disturbed habitats favored by many migrants; these have few large arthropods. The wintering birds there seek out smaller prey, especially spiders and homopterans—sucking insects such as leafhoppers, aphids, and mealybugs (Greenberg 1995).

In West Indian forests, large arthropods are scarcer than in the mainland tropics—but even there, a study in Jamaica found that during the resident birds' breeding season the biomass of these arthropods increases by 560%, while the biomass of small arthropods increases only 10%. At the end of the breeding season, when large arthropods decline, two of the insectivores, the Gray Kingbird (*Tyrannus dominicensis*) and the Black-whiskered Vireo (*Vireo altiloquus*), both with significantly longer bills than their northern congeners, leave the island to winter in northern South America, where they feed substantially on large fruit. The resident passerines that remain have moderate bills, and in winter they also expand their diet to include local fruit (Johnson et al. 2005).

In central Panama, food is scarcest in October, at the height of the rainy season, when wintering birds are arriving and other migrants, continuing to South America, are also abundant. During light rain, all passerines continue feeding or foraging, but when it grows more intense most residents decrease their activity. Migrants, meanwhile, become more active, stopping only when rain is heavy. This suggests that wintering birds are at a disadvantage and need to continue searching for food during unfavorable conditions,

Wintering migrants like this Bay-breasted Warbler in Panama often continue feeding in the rain while the more experienced, resident birds have stopped.

even if they are losing energy while flicking their wings and shaking water off their body. In mixed-species flocks and in other encounters at food sources, the migrants are usually subordinate to residents, which are larger (Leck 1972).

A study in Soberania National Park, Panama, that compared the diets of migrants, winterers, and residents in forest understory from September to May found substantial differences in the arthropod prey of the North American birds and the locals. In second-growth humid forest, the migrants and wintering birds ate some fruit but took more hard-bodied arthropods of low nutritional value—beetles and ants—and nonflying termites, centipedes, and millipedes, which all have distasteful or toxic chemicals. Residents feeding on the same substrate ate some of these but fed most on more nutritious spiders, insect pupae, alate ants, and larger orthopterans and lizards. Migrants and winterers took berrylike fruit while residents ate larger fruit. The North American birds—flycatchers, thrushes, and warblers that in their breeding habitat feed at such different heights in forest—were all near the ground in the understory and had diets more similar to those of one another than did any of the residents of similar size. Only in April and May did the winterers and migrants, then building fat reserves, avoid beetles and ants (Poulin and Lefebvre 1996).

Relatively few North American passerines winter south of the equator. Most between Ecuador and Bolivia are found at the western edge of the Amazon basin and in the lower and middle elevations of the eastern slope of the Andes. The few wintering species in the lowland forests are thinly distributed, but because the forest is the dominant habitat, they may be quite numerous. In these latitudes, northern migrants arrive at the beginning of the rainy season, when insects and fruits are at their highest levels, and there is little likelihood of competition with residents, with altitudinal migrants that descend from higher in the Andes, or with austral migrants that in October or November have not yet departed (Pearson 1980).

In the Neotropics, the most likely evidence of competition or displacement of resident birds by winterers would come from warblers, with 47 species wintering from the Caribbean islands and Mexico south to Bolivia, in habitats where their closest ecological counterparts are small flycatchers, gnatcatchers, greenlets, and other warblers. The migrant warblers, as we have seen, have already reduced competition among themselves, since closely related species winter in different regions, elevations, and habitats and partition the places they share by differing in social habits and feeding specialties.

In the Caribbean, the warblers of endemic genera predominantly inhabit undergrowth, while the migrants and their closely related endemic *Setophaga* congeners are arboreal foliage gleaners (Keast 1980). The retreat of resident Yellow Warblers in Jamaica to mangrove forests when North American Yellow Warblers arrive, and their return to scrub habitat after the winterers leave, is the only known case of significant habitat displacement (Lack and Lack 1972). This may, however, represent residents shifting to their own advantage, not being pushed out. The mangroves are likely to be the superior habitat—given the preference some wintering warblers also show for it—and the scrub may again become productive enough for nesting birds when the rains resume in spring. On Hispaniola, resident warbler species occupy only 30% of the island's habitat, while wintering warblers are common at all elevations; presumably the areas unused by residents do not meet their needs during the breeding season, when these places have been vacated by the migrants (Terborgh and Faaborg 1980).

In western Mexico, warblers wintering in the highlands join local mixed-species flocks while those in lowlands fit the familiar pattern of concentrating in disturbed areas; there, areas that support 64 individuals per hectare in winter hold only 1.7 per hectare in summer. At none of these elevations are resident birds detrimentally affected by the seasonal addition of so many others. In fact, the presence of wintering birds in the mixed-species flocks in the highlands, where the winterers compose 30% of some flocks, may provide additional vigilance, a benefit to the resident members during the season when there may also be more raptors present (Hutto 1980).

Farther south, in Campeche, where seven warblers winter during the dry season (November–April) in second growth with several resident insectivores of comparable size, only one endemic warbler, the Gray-throated Chat (*Granatellus sallaei*), shifts its usual feeding position to lower in the crown of trees (Waide 1981). In Chiapas, resident Rufous-capped Warblers (*Basileuterus rufifrons*) in shade coffee farms similarly spend more time in the understory when mixed flocks of migrants are in the canopy. At this time of year, large arthropods decline by 58% in the canopy even in the absence of birds feeding on them, so the presence of additional birds searching for what arthropods remain is an incentive to shift to the understory, where the abundance of large arthropods, though less, does not change between wet and dry seasons (Jedlicka et al. 2006).

In the Santa Marta highlands of Colombia, the Blackburnian Warbler is the commonest wintering species. It and other migrant warblers join local mixed-species flocks, where they are usually subordinate to resident birds. In this region, the breeding season for many passerines is December–May, and the mixed flocks then have the fewest resident members. The remaining residents in the flocks may therefore benefit from the addition of wintering birds, which, even if these are competing for food, may aid in predator detection and, by their numerical presence, reduce the odds of any one bird becoming a victim. At the end of the breeding season more local birds rejoin the flocks; by then, the migrants have departed (Johnson 1980).

Elsewhere in the Colombian highlands, in a region where no original forest survived in the early 1970s, a 30-ha patch of oak woods 5 km distant from any other woodland was an island refuge that harbored 49 resident species and 19 migrants, including 17 passerines, a fraction of the original avifauna. The remaining resident birds were all widespread generalists found from Central America to Bolivia or Argentina; in winter, 50% of the passerines in the oak woodland were migrants and 57% of these were Blackburnian Warblers. At this time of year, more resident birds form mixed-species flocks than at other seasons; these are sometimes joined by migrants. The added presence of so many migrants puts more pressure especially on the two resident warblers and one vireo that are ecologically most similar to the Blackburnian, which, like them, prefers the canopy and outnumbers them. Singly or as part of mixed-species flocks, these three closest ecological counterparts shifted to feed lower in the woods, which are more open and may provide less food. There were more aggressive encounters at lower levels, further indicating more potential competition (Chipley 1976).

The mixed-species flocks following ant swarms are another place to look for potential competition during the months migrants from North America are in the region. A study at several locations in the Cordillera de Tilarán, Costa Rica, however, found that the size and species composition of resident flocks were unchanged during the months they were joined by individuals of as many as 11 migrant species. The migrants were smaller on average and were less likely to capture prey. They more often perched on vegetation over the ground and sallied to catch items, more rarely coming to the ground to pick something up or walking or hopping on the ground, as did many residents (O'Donnell et al. 2014). Only resident followers checked ant bivouacs morning and evening to determine where the swarm would next move, and migrants fed only at swarms that already had resident followers (O'Donnell et al. 2010). On Barro Colorado Island in lowland Panama, migrant followers are regularly supplanted by residents, even smaller ones, and are usually at the periphery of the area in advance of the ant swarms, where they have access to fewer fleeing arthropods. Most aggression is between residents, especially among conspecifics, and migrants usually chase only other migrants (Willis 1966).

In Africa, most migrants winter between the Sahara and the equator during what for many resident birds is the breeding season. Even in the very dry habitats on the southern end of the desert belt, several resident larks nest while two Eurasian lark species are present (Moreau 1972:116). Farther south, migrants favor seasonal savannas and open woodland, using temporary and locally abundant food sources generally unexploited by the residents. Local songbirds prefer small ground-dwelling insects, while the larger insects in trees and shrubs that are favored by migrants are more seasonal. In Kenya, a comparison of eight resident and wintering small thrushes found that the African species have shorter, more rounded wings than the migrants, they forage more slowly, and they pursue fewer insects by aerial maneuvers (Leisler 1992). Of course, highly migratory birds generally have longer, more pointed wings than closely related sedentary species; the longer wings required for long migratory flights could be considered a preadaptation for pursuit of aerial prey—just as the anatomy required for the pursuit of aerial prey could have been a preadaptation enabling some African birds to migrate. The two alternatives may be impossible to separate, given how many Palearctic species that winter in Africa are likely to have originated there.

A comprehensive literature review concluded that direct competition between African resident and wintering species has been less demonstrated than in the Neotropics (Salewski and Jones 2006). One study found that among 12 small ground-foraging insectivores that occur together in East Africa, resident species always dominate over Palearctic winterers, irrespective of size, but dominance in the winterers themselves is always determined by size; East African residents, however, are not always dominant over austral migrants from farther south in Africa (Leisler 1990). Among the few Eurasian birds that winter in African forests, Pied Flycatchers (*Ficedula hypoleuca*) and Willow Warblers (*Phylloscopus trochilus*) in Comoé National Park, Ivory Coast, were found to be more flexible in foraging behavior than residents in the same guilds, but they had similar foraging speed and intake rates and shared the same microhabitats (Salewski et al. 2003). Other birds wintering in Africa avoid competition with close relatives through niche and latitude partitioning. Common

Common Swifts reduce competition with local African species by foraging at higher altitudes. They may also spend the night on the wing, thereby avoiding competition even for roosting sites.

Swifts (*Apus apus*) and House Martins (*Delichon urbicum*) from Eurasia both hawk for flying insects at much higher altitudes than their local congeners, often beyond sight; both are believed to sleep on the wing, thus never encountering potentially competitive residents at either feeding or roosting sites. In the dry belt across northern tropical Africa, White Storks (*Ciconia ciconia*) arrive when local Abdim's Storks (*C. abdimii*) are departing for the south at the end of their breeding season. At the southern end of the White Stork's winter range, the two species are found together; here, White Storks concentrate on locusts when these are available, while Abdim's Storks take other orthopterans, but in years without locust swarms their diets overlap (Moreau 1972:170, 120, 241).

INTERACTIONS BETWEEN NORTHERN HEMISPHERE AND AUSTRAL MIGRANTS

Few austral migrants migrate as far as do many Northern Hemisphere birds. Most of the places they may overlap are south of the equator. The one significant exception is the oceanic shearwaters and petrels that breed at high latitudes in the Southern Hemisphere and winter where it is summer in the northern Pacific and Atlantic. Here, they outnumber by tens of millions those of the few related species that breed at similar latitudes of the Northern Hemisphere during this season. Eight austral petrels and shearwaters winter in high latitudes of the North Pacific, where there are only two breeding species; in the Atlantic, three austral species reach the same latitudes, where there are five local breeders. There is no evidence that food is a limiting factor for any of these species; they feed in different proportions and on different sizes of fish, squid, and crustaceans, either caught or scavenged, with some species mainly plucking food off the water surface, others reaching into it, and a few diving as far as 67 m in pursuit of prey. The oceans are further partitioned as different petrels, shearwaters, and storm petrels concentrate in waters of various temperatures, depths, and distances from land (Brooke 2004:122–139).

Among landbirds, a few species of tropical origin have breeding populations that have expanded both north and south from the equator and that return toward the equator in winter. Southern populations of 10 African species of cuckoo, three kingfishers, and a nightjar move north to the equator at the end of their breeding season just when the northern African populations of these species that wintered there migrate north to breed. In South America, two flycatchers, the Tropical Kingbird (*Tyrannus melancholicus*) and Fork-tailed Flycatcher (*T. savana*), follow the same pattern (Dingle 2008).

Elsewhere in South America, ecologically similar Nearctic and Neotropical flycatchers replace each other with minimal overlap. In southeastern Peru at 8° S, Eastern Kingbirds (*T. tyrannus*) arrive by early September with the first burst of new fruit in canopy trees. The Vermilion Flycatcher (*Pyrocephalus rubinus*) and Crowned Slaty Flycatcher (*Empidonomus aurantioatrocristatus*), also canopy feeders, are then departing to nest in Bolivia and Argentina (Pearson 1980). Farther south in Peru, at 12° S, the Vermilion Flycatchers have left by early November, when Eastern Wood-Pewees (*Contopus virens*) take their place sallying for insects at forest openings near lakeshores and at the edge of rivers; the Vermilions return in early May, three to four weeks after the pewees have departed (Fitzpatrick 1980).

The most complex interactions between Northern Hemisphere and austral migrants are among shorebirds that winter in coastal Buenos Aires Province, Argentina, at latitudes 36–38° S. Some 30 shorebird species live in the region over the course of the year, using sandy beaches, tidal mudflats, and interior wetlands and grasslands. North American winterers, mostly sandpipers, arrive between mid-August and November. Some stay into April, by which time the Patagonian migrants, all plovers, have arrived; some of these remain until late September, creating a second overlap period (Myers and Myers 1979). Neither the Patagonian Tawny-throated Dotterel (*Oreopholus ruficollis*) nor the Rufous-chested Dotterel (*Charadrius modestus*) is territorial, so during the beginning and end of the austral winter they may be at a disadvantage when territorial northern migrants are also using the same grassland habitat (Isacch and Martinez 2003).

The overlap is not only between species that nest in the Arctic and Patagonia, but also with resident shorebirds and a few local breeders that migrate north in austral winter to Brazil and Paraguay. The eight nesting species include several in families unrelated to the migratory contingent of plovers and sandpipers; their very different ecologies spare them competition with the wintering birds, which at all seasons vastly outnumber all the residents. The local nesting season begins at the same time migrants arrive from North America. Southern Lapwings (*Vanellus chilensis*) may then be affected in two ways: they live at densities of 1 bird per hectare, while American Golden-Plovers and Buff-breasted Sandpipers, which feed in a similar manner, occupy the same grasslands at densities of

Resident Southern
Lapwings and wintering
Buff-breasted Sandpipers
on Argentinean grasslands
may compete for food.

15–25 per hectare. The absence of these birds might leave much more food for lapwings and their young. The lapwings are not aggressive except when other birds are close to their nest or young; adult lapwings are so much larger than the wintering shorebirds that they may not take the same food, but the lapwings may seek to drive away any bird that could attract the attention of predators, and their small young may seek the same items as nearby plovers and sandpipers (Myers 1980a).

Many first-year North American shorebirds remain to spend the austral winter in Argentina, thereby increasing potential competition for food when it is scarcest. Then, Two-banded Plovers (*Charadrius falklandicus*), wintering from farther south, and first-year Nearctic Hudsonian Godwits (*Limosa haemastica*) feed primarily on small polychaetes in mudflats. While the godwit probing deep into the mud with its long bill and the plover pecking at prey it sees on the surface do not likely compete directly, they reduce the polychaete population significantly, perhaps explaining why White-rumped Sandpipers, which fed on them during the austral summer, do not overwinter (Martinez-Curci et al. 2015). In the early austral spring and late austral summer, when white-rumps are present, wintering Two-banded Plovers and resident Rufous-chested Dotterels repeatedly fight with them (Myers 1980a).

SUMMARY

In winter, when most birds focus only on individual survival, birds exhibit the widest range of spatial and social organization. These may be very different from their use of space and degree of sociality when they must pair and raise young. Some birds that were aggressively territorial in the breeding season join others in flocks, while many colonial nesters disperse widely and solitarily. Other birds continue to defend the same territory through the year or create and defend a new one at their winter destination, to which they may return in subsequent years. The spatial and social patterns are a response to the availability of the resources—principally food and shelter—that each bird needs over the course of the winter. Species that feed on evenly dispersed items within a defensible range are likely to establish a territory, which may last for the entire winter or for only briefer periods, depending on local conditions. Birds that feed on highly mobile foods, such as swarms of aerial insects and schools of fish, or on resources that are patchy in time and space, like fruit, cannot defend an area large enough to supply their needs through an entire season and cannot alone defend any food source when it becomes available. For them, flocking can both help locate these sources and reduce any individual bird's danger from predators when it is feeding.

The scarcity of resources at high and middle latitudes as a result of cold and shortened day length, and in some tropical regions where wintering birds are present during the least productive months of the year, means that birds are competing more intensely not only with conspecifics but also with other species they do not encounter during the breeding season—including permanent residents, fellow migrants, and even migrants from other hemispheres where winterers from the north and south overlap. Direct competition among species is often reduced by the preference of wintering birds for marginal habitats that are productive only seasonally or that produce food inadequate for permanent residents and are thus avoided or little used by them. Among winter-territorial birds, older individuals and often males, especially where these are larger, are able to acquire and defend the best sites; many females and most younger birds may be limited to inferior habitat and may be in poorer condition by the end of winter. For lack of an adequate territory, some of these birds may join opportunistic flocks or drift around territories on the alert for a vacancy. Where several species have very similar ecological niches, they may defend their territory against these as well.

Flocks may consist of a mobile cohesive group that ranges widely in search of food or may form spontaneously at any locale where food is abundant and ephemeral. Flocks may contain more than one species when these harvest the same resource, or where several species work through the same habitat with each searching in a different place or for different food. Often these mixed-species flocks form a territory that they collectively defend, each bird against other conspecifics. In sedentary single-species flocks there is usually a hierarchy of individuals based on age, sex, plumage variation, and the time each bird joined the group. In multispecies flocks the ranking is by species as well—with permanent residents enjoying dominance by being able to feed where least exposed to danger, and often on food of higher quality.

CHAPTER 5

SURVIVAL

I F REPRODUCTION IS THE primary goal of any animal's existence, survival must be its goal during the nonreproductive seasons. The two major challenges to survival through a winter are securing enough food, often in less favorable or less familiar environmental conditions, and avoiding predation. At higher latitudes, winter weather particularly affects the amount and availability of food, and extreme weather events can have still greater impacts over the short term. At all latitudes, however, both permanent residents and migrants may need to shift their diet or foraging methods to meet their daily requirements. Since birds, especially small ones, have little capacity to live very long without food, even periods of a few days when food is impossible to secure will affect a local population.

Larger birds can withstand longer periods of food deprivation; their bodies retain heat longer, their feeding method may not require as much energy as do those of small birds constantly on the move, and their food, when they secure it, often comes in larger portions, so fewer feedings are required. Tits can store only half their daily metabolic requirement as fat (Pravosudov and Grubb 1997). A European Blackbird (*Turdus merula*), the size and ecological equivalent of the American Robin (*T. migratorius*), can survive 2.2 days without food with normal movement, and 4.6 days if it is inactive (Biebach 1977). Following the severe British winter of 1978–1979, populations of woodland and farmland birds weighing less than 10 g dropped by 40%, while birds weighing more than 100 g were not noticeably affected (Cawthorne and Marchant 1980).

The northern limits of the usual winter range of any Northern Hemisphere bird may be a useful indicator of the factors that determine survival. A study using Christmas Bird Count data for 148 North American landbirds found the strongest correlations between northernmost latitude of winter range and average minimum January temperature (60.2% correlation), mean length of frost-free period (50.4%), and potential vegetation (63.7%). When species were grouped into different ecological guilds—raptors, bark gleaners, foliage gleaners, and seed eaters—the influence of each of the three factors was seen to vary by group. Foliage gleaners showed the strongest correlation with average minimum January temperature and mean length of frost-free period (86.4% and 68.2%, respectively), while for raptors the correlations were only 54.5% and 36.4%. This may be because raptors, being larger bodied, can survive colder weather, and their winter distribution may be more directly linked to the presence of their prey, which are usually smaller bodied and therefore likely to be less cold tolerant. Of the four guilds, the northern limit of bark gleaners was the most closely correlated (76.9%) with potential vegetation, given that these birds require trees. For many of the seed eaters, the northern limit was where ambient temperatures did not require the birds to raise their metabolic rate more than 2.4 times their basal metabolic rate (Root 1988).

Reliable data on winter survival rates are difficult to obtain from individual wild birds. These must be marked, and if one disappears, there must be evidence that it actually died rather than simply moved beyond the study area. And many research projects focus on annual survival, which does not separate winter mortality from that of other seasons. Finally, studies should continue as long as any of the marked birds are alive, because age affects winter survival; young birds usually have lower survival rates, especially during their first winter, but different age classes of adults may also vary.

For long-distance migrants, annual survival rates, if high, are of course a good indicator of winter rates. Some Pacific Golden-Plovers (*Pluvialis fulva*) wintering on Oahu, Hawaii, that were first marked as adults lived another 17 or 18 years; this 20-year study concluded that mortality was equally likely during the breeding and nonbreeding seasons, but since the breeding season, including migration to and from Alaska, is only four months, it must have proportionally more hazards than the birds face during their long winter stays on Oahu (Johnson et al. 2001). In Bristle-thighed Curlews (*Numenius tahitiensis*) wintering on Laysan Island in the northwestern Hawaiian chain, 92.4% of first-year birds, which do not leave Laysan until the end of their third winter, survived their first year there. Eighty-three percent of first-year birds marked in 1988 were still alive in the spring of 1991, and annual survivorship for adults, which did migrate, was more than 85% (Marks and Redmond 1996).

Among long-distance passerine migrants, actual winter survival has been tracked for only a few species. On a monthly basis, the average survival rate over a winter in Jamaica for Black-throated Blue Warblers (*Setophaga caerulescens*) was 99% for both males and females, including adults and first-year birds, which often winter in different habitats on the island (Sillett and Holmes 2002). In Mexico, monthly survival rates for Wood Thrush (*Hylocichla mustelina*), Hooded Warbler (*S. citrina*), Kentucky Warbler (*Geothlypis formosa*), and Ovenbird (*Seiurus aurocapillus*) averaged 89–93% (Conway et al. 1995). Willow Flycatchers (*Empidonax traillii*) wintering in northwestern Costa Rica have a monthly survivorship averaging 98% and an overwinter survivorship rate of 88% for the six months from mid-September to April that they are present (Paxton et al. 2017). Eastern Great Reed Warblers (*Acrocephalus orientalis*) wintering in peninsular Malaysia have a 71% seasonal survival rate (Nisbet and Medway 1972).

EFFECTS OF WEATHER

Variations in winter weather affect survivorship. A British study from 1962 to 1988 found that temperature alone had little effect on survival of resident passerines in forests and farmlands, but that snow cover had a significant impact, especially on ground feeders. The inaccessibility of food under snow was the key factor, not the frozen ground, glazed vegetation, prey inactivity at low temperatures, or increased energy requirements (Greenwood and Baillie 1991). A 20-year review that focused on Winter Wrens (*Troglodytes troglodytes*) and Eurasian Treecreepers (*Certhia familiaris*) in one forest in Nottinghamshire, England, concluded that wren mortality was highest during winters with the most days of snow, and treecreeper mortality was greatest in winters with greatest total rainfall. Wren annual survival at four sites was between 32% and 42%, and treecreeper survival was 44%. Both species had especially high mortality during two notably severe winters (Peach et al. 1995). Finally, a 19-year study in Great Britain found that mortality in Eurasian Sparrowhawks (*Accipiter nisus*) was greatest in years with more rainy days between October and April. This pattern was constant from a location in Scotland where the number of rainy days varied from 106 to 138 in different winters to another in England with 87–117 rainy winter days (Newton et al. 1993).

Migrant birds, with less need to remain on their territory than these British landbirds, can respond to unfavorable local weather conditions by temporarily moving elsewhere. As many as 66% of individually marked Dunlins (*Calidris alpina*) at Bolinas Lagoon, on the central coast of California, left there during periods of heavy rain; some were found 14 km inland in San Francisco Bay, others up to 140 km away in the Sacramento Valley. Of the birds of known sex, all those that went inland were males, which have shorter bills than females and may not be able to feed as well on the coastal beaches when water may be higher than normal. Similarly, short-billed male Eurasian Curlews (*Numenius arquata*) on the Tees Estuary in England move inland during cold spells to feed in fields; they may be less able than the longer-billed females to reach invertebrates that burrow more deeply to escape the cold (Warnock et al. 1995).

In Eurasian Oystercatchers (*Haematopus ostralegus*), males have the advantage in cold weather. Individuals of each sex feed consistently using one of three methods: they open mussel shells by hammering them or by using their bill to pry the shell apart when the mussel is gaping, or they feed on polychaete worms and clams. Most males are hammerers, most females and young birds feed on worms and clams, and an even number of prying birds are male and female. A study conducted during the mild winters between 2002 and 2006 found no difference in winter survival among the birds using each technique, but lower survival of females in all three because of dominance by males. The general rates of survival among the worm and clam feeders during these years were higher than in earlier studies because in milder winters, the worms do not reduce their activity and the mud in which they live does not freeze, so oystercatchers can get to them (Durell 2007).

The occasional year with unusual cold can have more substantial impacts. In central and southern England, the winter of 1962–1963 was the coldest since 1740 and possibly the snowiest for 150 years. Almost everywhere, rivers, lakes, and beaches were frozen, and freezing fog coated trees with ice. More than 15,000 bird carcasses were found, ranging from loons, grebes, and waterfowl through shorebirds and gulls to passerines of all sizes. The three species most found were Woodpigeon (*Columba palumbus*), European Starling (*Sturnus vulgaris*), and Redwing (*Turdus iliacus*), in that order. Many birds not previously known to visit feeders came to them, including various gulls, doves, corvids, and thrushes. The following spring, the species that had most declined were Eurasian Kingfisher (*Alcedo atthis*), Grey Wagtail (*Motacilla cinerea*), Goldcrest (*Regulus regulus*), European Stonechat (*Saxicola rubicola*), Winter Wren, Barn Owl (*Tyto alba*), and Common Snipe (*Gallinago gallinago*), demonstrating that all types of foragers were affected. Conversely, tits, vulnerable to cold because of their small size, were not notably reduced; the prior autumn had been a good year for beech mast and acorns, so tits that cached food would have had enough (Dobinson and Richards 1964).

Short bouts of extreme weather also have an impact. In North America, winter temperatures can drop as much as 30° C in 24 hours when a warm air mass meets a polar air mass. If water bodies freeze rapidly, large numbers of waterfowl can be frozen in. In January 1998, freezing rain fell over Quebec and the northeastern United States for five days, followed by several days below −10° C. Birds that forage in trees were especially affected—Hairy (*Dryobates villosus*) and Downy (*D. pubescens*) Woodpeckers, Blue Jays (*Cyanocitta cristata*), Brown Creepers (*Certhia americana*), and Black-capped Chickadees (*Poecile atricapillus*) in the region all dropped 20–40% on the subsequent winter's Christmas counts (Elkins 2004:197–198). A storm on February 9, 1939, in Urbana, Illinois, killed about 4% of the 25,000 "blackbirds" roosting in an extensive grove of coniferous and deciduous trees; winds of up to 77 km per hour killed some birds, and others froze to death when the temperature dropped below 0° C during the rain. While Common Grackles (*Quiscalus quiscula*) and Brown-headed Cowbirds (*Molothrus ater*) were in the more sheltered conifers, they suffered higher rates of mortality than the European Starlings in the leafless trees; presumably, they are less resistant to cold and freezing rain (Odum and Pitelka 1939).

Birds that winter at sea at high latitudes where even normal conditions may seem daunting are also affected by extreme weather. Some Northern Fulmars (*Fulmarus glacialis*) winter in the ice-free waters around Arctic islands, where January air temperatures can descend to −28° C with strong northerly winds, but the exceptional winter of 1962 probably killed thousands in the North Sea, where they could not feed in seas with gales generating waves over 5 m high (Elkins 2004:213, 217). Spectacled Eiders (*Somateria fischeri*) winter in the Bering Sea south of St. Lawrence Island in an area that is usually covered at least 90% with sea ice at least half the time from January through April. Between 1990 and 2002, nine winters had more days than normal of ice coverage, compared with only three winters during 1979–1989. The winters with more ice were followed by nesting seasons in Alaska with fewer birds, but the specific impacts of the extensive ice on survival are not clear (Petersen and Douglas 2004). European Shags (*Phalacrocorax aristotelis*) on the east coast of Britain suffered severe mortality during an extended period of onshore winds in February 1994. In one well-studied colony in Scotland, overwinter survival dropped from the normal 88% to 14.7%. Birds in this colony that were less than 7 years and more than 16 years old died in greater numbers than middle-aged birds, while in most years survival declined significantly in birds older than 13 years (Harris et al. 1998).

While high latitudes may be more challenging to many birds during winter, low latitudes also have their hazards. Significant numbers of Ruddy Turnstones (*Arenaria interpres*) wintering in Australia settle in Broome, on the north coast of Western Australia at 18° S, while others travel 2800 and 3300 km farther, to Nene Valley on the south coast of South Australia (38° S) and King Island, Tasmania (40° S), respectively. Blood samples taken from turnstones at the three sites found that the birds wintering at Broome had higher levels of the stress hormone corticosterone as well as greater incidence of blood parasites such as avian malaria. Their daily energy expenditure rate was lower, as expected in a location with higher temperatures and less wind. The two more southerly wintering populations had higher fuel deposition rates but, in addition to the benefit of less disease, were in regions with fewer avian predators; this may have contributed to their lower stress-hormone levels. Turnstones in each of these wintering areas returned to the same one in successive years. Actual winter survival rates at the three locales were not determined, but it is clear that the turnstones encountered different potential hazards at latitudes with very different climatic conditions (Aharon-Rotman et al. 2016).

DIET SHIFTS

From the highest latitudes to the equator, many wintering birds must shift their diet and foraging methods to harvest whatever foods are available at that season. The shift may be gradual or abrupt, depending on the scale of difference between habitats used in different seasons and the speed at which birds move from one region to another. Ross's Gulls (*Rhodostethia rosea*) nesting on tundra marshes and deltas of rivers flowing into the Arctic Ocean feed primarily on insects. In August, at the end of the breeding season, they immediately move north to the pack ice, where they will remain until June, feeding at sea and at the edge of the ice. Here, their diet changes entirely to include zooplankton that thrive under the ice floes, juvenile Arctic cod (*Boreogadus saida*), and walrus feces (Divoky 1976). At the same latitudes, Ivory Gulls (*Pagophila eburnea*) consume fish and crustaceans in summer; wintering on pack ice, they probably get much of their food by scavenging carcasses left by polar bears (*Thalarctos maritimus*) (Bateson and Plowright 1959).

Birds may shift their diet less abruptly or radically over the course of the winter because their initial source has diminished in quantity or value. As many as 350,000 Eurasian

Ross's Gulls wintering on the Arctic Ocean ice pack shift from their summer diet of insects to feed on fish, zooplankton, and walrus feces.

Wigeon (*Mareca penelope*) live on the coast of the Netherlands during a mild winter. Between September and November, most graze on salt-marsh plants, especially *Salicornia*. As these become depleted, they move inland from December through March to feed on freshwater grasses (van Eerden 1984). The benefits of grass were demonstrated in southwest Scotland, where wigeon also graze on grasses at the edge of ponds. They return repeatedly to the same small swards, where their regular cropping increases leaf production by 52%. By the end of winter, the grazed grass has 4.75% more protein than ungrazed grass; the increase is due to the continued stimulus to new growth, not to fertilization from droppings (Mayhew and Houston 1999).

The amount of energy required to harvest any resource is another important factor. When preferred food is highly available, a hunting or foraging method that requires substantial energy expenditures may still provide more energy than less costly methods of obtaining other items. When this resource is scarce, birds may shift to a food that has less value but also requires less energy to secure (Evans 1976). This may, however, require anatomical adjustments. In an experiment where Red Knots from the Dutch Wadden Sea were shifted from high-quality trout food to low-quality hard-shell blue mussels (*Mytilus edulis*), over 30 days the birds increased the height and width of their gizzard, 66% and 71%, respectively, in order to crush the shells so they could digest the flesh. Associated with this was a 15% loss of body mass and an 11% reduction in the size of their pectoral muscles, which reduced the birds' ability to endure cold, since pectoral muscles are the main source of shivering heat production (Vézina et al. 2010).

For some birds, the shifts during the season of greatest scarcity or increased competition are likely responsible for the evolution of anatomical features such as bill length and shape that best enable birds to secure foods at their wintering site but are less necessary in the breeding range. Tundra-nesting sandpipers that are visual feeders take insects and berries from the surface of plants; they shift their diet substantially when they move south to coasts and wetlands, where many become tactile feeders. The fine species partitioning of wintering habitats—by latitude, preferred substrate, bills of various lengths and shapes that enable different species to probe to different depths, and the size differences within many species by sex—all suggest that the diet in the nonbreeding season has influenced their morphology more than has their generalized superabundant Arctic summer diet, much of which can be obtained equally well by birds with bills of any type. Of all the sandpipers, the Eastern Curlew (*Numenius madagascariensis*) has the longest bill, in males 128–170 mm, in females 154–201 mm. Its summer and autumn diet includes insect larvae, frogs, crabs, and berries (Johnsgard 1981:369). In winter, the curlew puts the bill to more specialized use; on the south coast of Australia it probes in sand and mud for shrimp and crabs and is able to reach farther than any other shorebirds that may be on the same flats. Males feed most often on mudflats, where, in addition to probing, they take crabs on the surface, while the longer-billed females probe in sandy areas of intertidal flats for shrimp and crabs that burrow more deeply (Dann 2014).

Similarly, the size of the bill of the Worm-eating Warbler (*Helmitheros vermivorum*), comparatively long for a warbler, is due to its winter feeding habit. While it seeks out large arthropods throughout the year, in North America, these are mainly caterpillars living on leaf surfaces and active by day, but in Central America and on Caribbean islands, these include more types of insects and spiders that are mostly nocturnal and spend the days concealed in curled-up dead leaves. In summer, 75% of Worm-eating Warbler foraging maneuvers are therefore directed to living foliage; in winter, 75% of maneuvers are directed to dead curled leaves. The warbler's long bill enables the bird to probe more deeply and open a curled leaf more widely. In September, when Worm-eating Warblers are arriving, they spend only 50% of their time investigating curled-up leaves; evidently, it takes some time—perhaps especially for juvenile birds—to make the shift to the most profitable potential food sites (Greenberg 1987).

In eastern North America, two otherwise insectivorous birds have evolved the capacity to digest the fatty acids in waxes coating bayberry fruit (*Myrica* spp.). They exploit a niche that both eliminates competition with close relatives and enables them to winter much closer to their breeding range. On the Atlantic coast, the Yellow-rumped Warbler (*Setophaga coronata*) winters from southern Nova Scotia and central Maine to Panama, and the Tree Swallow (*Tachycineta bicolor*) from Massachusetts to Central America, matching the range of *Myrica*. Both shift from an insect diet to one dominated by bayberry fruit, which ripens between August and October, as these birds are migrating south, and persists well into winter. No other warbler or swallow can assimilate the waxes coating the fruit; Yellow-rumps assimilate more than 80%, but experiments with Yellow Warblers (*S. petechia*) found these absorbed less than 5%. Yellow-rumped Warblers also consume fruit of poison ivy (*Toxicodendron radicans*), which often grows with bayberry, but the birds will lose weight on an exclusively fruit diet; they and the Tree Swallows also take insects when they can (Place and Stiles 1992).

Winter food scarcity driven by increased competition may prompt very rapid shifts to new feeding methods. Between the early 1990s and 2010, the population of the Icelandic

Tree Swallows are unique in their family for having evolved the ability to subsist substantially on bayberry fruit in the winter.

Black-tailed Godwit (*Limosa limosa islandica*) grew dramatically, and the number of birds wintering on the Atlantic coast of France rose from 4,000 to 27,000. Near La Rochelle, godwits displaced from the traditional local wintering areas began using a new site, the wetlands of the Ile de Ré, where, instead of mollusks, the birds' diet is now as much as 94% rhizomes of *Zostera noltii*, a seagrass. The Ile de Ré is the most densely populated of the several local wintering sites, suggesting that its carrying capacity with an alternative food source is greater than that of other places nearby where godwits continue to feed on bivalves (Robin et al. 2013). What impact this shift to a vegetarian diet may have had on anatomy, individual fitness, or overall survival has not yet been studied.

Some birds may alter their feeding method more than their actual food source at different seasons and latitudes. Thick-billed Murres (*Uria lomvia*) nesting on Bjørnøya, Svalbard, at 74° N migrate in winter to the Atlantic between Greenland and the British

On the Atlantic coast of France, where the wintering population of Black-tailed Godwits has grown substantially in recent decades, some birds have shifted their diet from mollusks to rhizomes of seagrass.

Isles, while the Common Murres (*U. aalge*) on Bjørnøya disperse more locally into the Barents, Norwegian, and White Seas. In the breeding season, the usual diving depth of both species is 20–50 m, but in winter, when their prey is farther from the water surface, Thick-billed Murres dive 67–182 m and Common Murres 55–136 m (Fort et al. 2013). Whatever adaptations murres have evolved to withstand the pressure of diving to great depths have therefore been in response to their wintering feeding needs.

All through the year, Bald Eagles (*Haliaeetus leucocephalus*) swoop down to water to catch fish in their talons. When ice covers lakes, they must use other methods. At Mormon Lake, 34 km south of Flagstaff, Arizona, eagles can be seen jumping up and down on ice to break through and create a hole. They may next remove more ice by grasping a piece in their toes and pulling it away. Then, they watch the hole and grab a fish in their talons. The eagles also extract dead fish that have been frozen in ice, as well as walk on partially submerged ice at the lake edge to reach deeper water from which they take fish and snails, an unusual item in eagle diets (Grubb and Lopez 1997).

For most passerine migrants, the transition over the seasons from one diet to another is gradual, matching the pace at which birds move from breeding to wintering sites. In spring and summer, especially when feeding young, these birds can concentrate on abundant preferred foods, such as soft-bodied insects, and may continue selecting the preferred items even when these diminish in abundance. By the time the young fledge, both they and their parents begin expanding their diet. Passerines then seek out fruit, which has lipids that may help fuel molt as well as future migration. Once on migration, birds stop in regions with different vegetation and food sources; here, too, in order to replenish fat supplies rapidly before the next flight, their diet may be broader and more opportunistic than when nesting. By the time they reach their winter destination, it is not, therefore, a significant behavioral shift for most migrant birds to continue searching for locally superabundant or ephemeral resources that may be different from what they consumed in the nesting season.

Many migrants, like the warblers that spread over several ecosystems on Caribbean islands, occupy a broader spectrum of habitats and consume a wider variety of foods in winter than in their breeding range. They shift the balance among several foraging techniques they use in both breeding and winter ranges. This is likely in response to differing availability of prey that can be captured most efficiently using each technique. The competition with residents and other winterers may also encourage the seasonal occupants to broaden their range of feeding methods. American Redstarts wintering on Caribbean islands hawk for flying insects more often and do less hovering and foliage gleaning than in their breeding range (Bennett 1980). They also fan their tail more often, which may help flush prey from a hiding place so it can be pursued in flight. The shift in method reflects a diet that includes more flying insects of smaller size than the caterpillars and larger flies they take in summer. In winter, redstarts forage at greater speed but capture fewer items, indicating that food is scarcer and the birds must expend more energy to secure items of less value (Lovette and Holmes 1995). In southwestern Jamaica, where by spring most leaves fall from trees in scrub habitat, female redstarts wintering there shift from near-perch maneuvers to more aerial ones in pursuit of insects, while the redstarts in richer mangrove forests, although already pursuing more insects in flight than during the breeding season, do not increase this maneuver over the course of the winter (Powell et al. 2015). In Maine and North Carolina, Black-throated Green Warblers (*Setophaga virens*) gather more prey in maneuvers involving flight—by sallying, hawking, and hovering—than by picking items while perched; in Panama, they not only spend more time foraging higher and closer to the edge of forest foliage but also capture more arthropods while hanging, clinging, or peering under foliage (Rabenold 1980).

A food source unique to the tropics is the horde of arthropods and other small animals fleeing advancing columns of army ants. The ants are exploited year round by many resident forest birds, which follow them as the ants flush prey; the birds do not consume the ants themselves. Some birds, nearly all living in forests below 1000 m, are obligate, or "professional," ant followers, and many additional species opportunistically join the professionals from time to time. Some migrant and wintering birds from North America also join these mixed flocks. Two species of ants, with different ecologies, are exploited by the residents and migrants. *Eciton burchellii* is a large ant that flushes mainly cockroaches, spiders, and other large arthropods; its columns move about 15 m per hour and 100–200 m per day in two-week periods with a different bivouac each night, followed by two-week stationary periods when any raids radiate from the temporary nest while eggs and pupae are produced inside it. *Labidus praedator* is a much smaller species that flushes more small arthropods while it moves in a regular course over a restricted area; it is not, however, as regular aboveground, especially in the dry season, so resident birds follow *Lapidus* swarms less frequently, leaving this niche open to wintering birds (Willis 1966).

Wintering birds following *Eciton* have been found as far north as Guerrero, Mexico, which is beyond the range of any ant-following obligates. Here, of seven species found in association with the ants, were three migrants—Hermit (*Catharus guttatus*) and Swainson's (*C. ustulatus*) Thrushes and Wilson's Warbler (*Cardellina pusilla*) (Greene et al. 1984). In Central America, the most frequent North American followers are the Wood (*Hylocichla mustelina*), Swainson's, and Gray-cheeked (*Catharus minimus*) Thrushes, and the Kentucky Warbler (Willis 1966), while in the Cordillera de Tilarán in the highlands of Costa Rica, seven warbler species in addition to the Kentucky follow ant swarms at least occasionally (O'Donnell et al. 2014). In South America, most migrants winter at higher elevations and in more disturbed habitats than either ant species uses.

Some migrants revert in winter to a food source they use *except* when breeding. During spring and summer, nesting Swainson's Hawks (*Buteo swainsoni*) feed both themselves and their young primarily small mammals, while nonbreeding birds in the region consume large insects like grasshoppers (England et al. 1997). In Argentina in winter, the major food for the entire population is insects, including locusts and irregularly superabundant resources such as migrating dragonflies, which the hawks rarely eat in North America. In Buenos Aires Province, dragonflies may suddenly appear in swarms of millions, with the first Swainson's Hawks arriving a few minutes later, stooping at them, catching a dragonfly in their talons, and transferring it to their bill. In one aggregation of more than 5,000 hawks seen pursuing dragonflies, the birds left as the dragonflies continued north. When strong winds prevent dragonflies from migrating and they cluster on the lee side of vegetation, Swainson's Hawks come to the ground and pluck them off (Jaramillo 1993).

Similarly, Eleonora's Falcons (*Falco eleonorae*) breed on islands in the Mediterranean and off North Africa in late summer and autumn in order to feed southbound migrant passerines to their young. The entire population winters in Madagascar, where the falcons are more densely concentrated and there are far fewer small birds. There, they prey exclusively on large insects captured in flight. They return to breeding sites as early as late March and continue to feed on flying insects until the next southward migration of songbirds (Walter 1979:144, 284).

Finally, in addition to the nonmigratory birds that store food to retrieve during the winter, a few species also directly manage their winter food supply, for the current year as well as for future years. In Siberia, Black-billed Capercaillie (*Tetrao urogalloides*), poor fliers and therefore sedentary, feed for seven winter months exclusively on larch twigs. Females, weighing 1.5–1.8 kg, move in groups of 2–14 higher in the mountains, where there are few predators; they further reduce their exposure to danger by feeding in the canopy, usually in a single tree each day, and spend as much as 22 hours in snow burrows, sometimes emerging only in the morning. The males, at 3.0–3.5 kg, maintain individual territories in the lower foothills, foraging mainly on the ground because they are too heavy to reach the outer twigs on trees. They eat twigs from the larches that protrude over the snow, visiting 75–125 trees per day, taking a few twigs from each. Their regular trimming prevents the trees from growing vertically, while increasing the number of lateral shoots. These "gardens" last for generations; some of the trimmed larches are 30–60 years old (Andreev 1991).

In northern Alaska, where Willow Ptarmigan (*Lagopus lagopus*) and Rock Ptarmigan (*L. mutus*) feed through the winter on the buds of feltleaf willow (*Salix alaxensis*) on the twigs that protrude over the snow, the birds' consistent browsing also "cultivates" the willows so that these generate more buds. When the terminal buds are consumed, the willows put forth more woody shoots in the following growing season, creating a broom-like appearance. The buds on these shoots are more numerous, albeit smaller, and have fewer tannins. These browsed twigs, however, produce no catkins and do not grow as tall as they might otherwise. But the fertilizing these plants receive from ptarmigan excreta may provide some benefit to the willows that in turn enhances productivity for ptarmigan in successive winters (Christie and Ruess 2015).

IMPACTS OF SUPPLEMENTAL FEEDING

Experiments comparing the survival rates of birds under natural conditions with those offered supplemental food have shown that the physical, ecological, and behavioral adaptations birds have evolved to cope with winter food availability enable most of them to survive the season's normal variability. During periods of extreme weather, birds, especially those at the coldest edge of their winter range, may be carried over by feeders. Several studies of Black-capped Chickadees, the smallest birds to remain common in northern North America all through the winter, have demonstrated that except in extended periods of extreme cold, the birds can secure all the food they require without supplemental sources like feeders. In central Pennsylvania, a comparison of a chickadee population in a forest with no access to feeders, another in a forest with feeders, and a third in a suburb with feeders found that 81% survived the winter without feeders and 93–94% in the two sites with feeders; the survival rates at each of the sites were the same during both cold and moderate periods (Egan and Brittingham 1994).

In Wisconsin experiments, chickadees resorted to feeders most during periods when the temperature was less than −18° C for five days or more; feeders were relatively unimportant during mild or average winter weather. During those periods of intense cold, the birds in the flocks with no access to feeders were lethargic and sat motionless facing the sun with their feathers puffed out; for them, the energy depleted in foraging may have been greater than what would have been gained by finding food. Overwinter survival in these flocks was 37%, while 67% of birds using feeders survived (Brittingham and Temple 1988). Availability of feeders does not, however, create long-term dependency. No survival difference was found in two Wisconsin chickadee populations where one group had never been exposed to feeders and the other had had access to them in the previous winter, but during the study winter had to find all its food at natural sources. For the latter group, only 21% of its daily energy requirements had come from the feeders in the prior winter when these were available (Brittingham and Temple 1992).

A study focused on winter fat reserves monitored seven species in south-central Kansas—Dark-eyed Junco (*Junco hyemalis*), Harris's Sparrow (*Zonotrichia querula*), Song Sparrow (*Z. melodia*), American Tree Sparrow (*Spizella arborea*), Northern Cardinal (*Cardinalis cardinalis*), Black-capped Chickadee, and Tufted Titmouse (*Baeolophus bicolor*)—using individually marked birds at feeding stations and at a site 0.8 km away where no supplemental food was available. No birds traveled between the two sites. Differences in fat were small, especially in the tree-foraging chickadees and tits, suggesting food supplies were sufficient in natural areas. There was no significant difference in predation by *Accipiter* hawks (Rogers and Heath-Coss 2003).

Of these species, all but the cardinal and titmouse are essentially northerly birds that regularly or irregularly (Black-capped Chickadee) migrate south every autumn. The usually sedentary cardinal and titmouse as well as the Mourning Dove (*Zenaida macroura*), in contrast, all began expanding their ranges northward in the mid-twentieth century at a time when backyard bird feeding became widespread but global average temperatures had not yet risen as significantly as in more recent decades (Root and Weckstein 1995). Their overwinter survival may initially have been aided or enabled by these supplementary food sources, while today some of their descendants live through the winter in areas where there are no bird feeders.

The Carolina Wren (*Thryothorus ludovicianus*) is another bird that has expanded its range northward substantially in the last several decades. A study of wren density and overwinter survival near Ann Arbor, Michigan, compared populations where birds had access to feeders, in residential areas and city parks, with those in rural areas where they did not. Wren densities in the city parks were twice what they were in residential areas, while these in turn were four times higher than in rural areas with no feeders. The first two habitats were also slightly warmer, but over the course of the winter wrens disappeared from both parks and residential areas that lacked feeders. By the end of winter, survival was higher where birds had access to feeders. The Carolina Wren requires 14 g of insects to meet its daily metabolic needs; in winter it forages mainly on the ground, where cold and snow can make insects inaccessible for long periods. At the northern edge of its range—where a single peanut can supply one-third of a Carolina Wren's daily metabolic needs—visiting a bird feeder is therefore much more efficient than foraging for insects (Job and Bednekoff 2011). In contrast, seed-eating ground feeders like doves, cardinals, and sparrows can generally find wind-blown food over the snow, and arboreal feeders such as chickadees are not usually prevented by snow from securing food they have previously stored or come across while foraging. Thus, the importance of supplemental food sources to overwinter survival depends substantially on each species' feeding ecology.

Survival through the winter depends not only on food supply but also on avoiding predation. Feeders have an impact on both elements. When food is easily available at a feeder, birds have more time as well as more energy to scan for predators. A study of Willow (*Parus montanus*) and Crested (*Lophophanes cristatus*) Tits in western Sweden at 57° N, where midwinter daylight is five hours and temperatures may descend to −25° C, found that less than half the tits far from feeders in late October were still alive in early April, while nearly twice as many survived among those that used feeders. Part of the higher survival rate among tits near feeders was due to changes in behavior that may have reduced predation. While tits feeding on natural sources are vocal and highly visible as they move in groups through trees searching for food, those using feeders were found to spend most of their time motionless and silent, concealed by branches near tree trunks, flying quickly and directly to feeders to pick up a seed and then returning to their perch. Northern Pygmy Owls (*Glaucidium passerinum*), the primary predators of tits in winter, were able to capture more of the foraging tits than the concealed ones (Jansson et al. 1981).

Supplemental feeding may also influence the subsequent spring's breeding schedule and success. When half the Song Sparrows on Mandarte Island, British Columbia, were provided supplementary food through one winter, there was no difference in survival by adults and only inconclusive evidence that more young with access to the feeders survived. The following spring, pairs with territories where there had been feeders began laying eggs 25 days sooner than the unfed pairs, but their broods were the same size and the later-nesting, unfed pairs fledged more young (Smith et al. 1980). A similar experiment with Blue Tits (*Cyanistes caeruleus*) in County Down, Ireland, found that even when winter supplemental food was cut off six weeks before normal egg-laying dates, the birds that had had the benefit of additional food began laying eggs 2.5 days sooner than the control population. Their clutch and subsequent brood sizes were no greater than those of tits that had foraged entirely in the wild that winter, but they fledged an average of almost one more chick per nest (Robb et al. 2008).

In Britain, where recent changes in agricultural practices have reduced the winter food available in fields to farmland birds, the Winter Food for Birds and Bird Aid programs go beyond backyard feeding stations to supply additional seeds on open land for declining species that do not readily come to feeders. Three years of controlled experiments comparing abundance of Yellowhammers (*Emberiza citronella*), Corn Buntings (*E. calandra*), and Eurasian Tree Sparrows (*Passer montanus*) in fields with additional provisions and in others without found no significant positive trends coming from supplementary feeding. More birds were in some of the fields with added food, but there was no increase in local breeding the following springs. Other factors, such as the different needs each species has during the breeding season and how far individuals may disperse, as well as the question of whether *enough* food was provided during the winter to stabilize a declining population, need to be examined to determine the potential long-term benefits of supplementary feeding (Siriwardena et al. 2007).

PREDATION

Predation may take a significant toll on birds in winter. In temperate regions, there are more raptors at this time of year. At midlatitudes, the absence of foliage on most trees and shrubs makes small birds more conspicuous than during other seasons. Similarly, waterfowl and shorebirds that winter in open settings and in aggregations are easier for predators to find than when they are dispersed on territories. Finally, longer nights and fewer well-concealed roosting sites increase the opportunities for owls and other nocturnal hunters. In the tropics, there are more bird-eating snakes and mammals, as well as predatory birds that bear no resemblance to what wintering birds have previously learned to recognize as dangerous. In Colombia, the bright red Andean Cock-of-the-rock (*Rupicola peruvianus*), primarily a frugivore, has been observed chasing and eating a Canada Warbler (*Cardellina canadensis*) and chasing a Swainson's Thrush (*Catharus ustulatus*) (Mahecha et al. 2018).

Birds feeding on the ground may be conscious of their conspicuousness. In an experiment performed near Terre Haute, Indiana, Dark-eyed Juncos, American Tree Sparrows, Northern Cardinals, Song Sparrows, White-throated Sparrows (*Zonotrichia albicollis*), Swamp Sparrows (*Z. georgiana*), and Field Sparrows (*Spizella pusilla*), in order of abundance, were kept in an enclosure with alternate strips of shade and sun with equal food in each. The birds consistently preferred the shaded areas, even when the midday temperature was 30° C below their thermoregulatory zone. Their preference for foraging in the shade was highest when temperatures were warmest and the thermal benefit of being in the sun was reduced. Sometimes 75% of the flock was in the shaded area, which covered only 37% of the enclosure. The birds faced southward least often on sunny days, when predators may use solar glare to spot them (Carr and Lima 2014).

A three-winter project monitoring shorebirds in a small estuary of the Firth of Forth, Scotland, where at least 12 species wintered, found that Redshanks (*Tringa totanus*) and Dunlins (*Calidris alpina*) were the species most taken by Eurasian Sparrowhawks, Merlins (*Falco columbarius*), and Peregrine Falcons (*F. peregrinus*). The three raptors' success rates per pursuit were, respectively, 11.6%, 8.8%, and 6.8%. The Redshanks lost 50% of their population each year to raptor predation, and 90% of the juveniles were taken, while the total loss to Dunlins was 5%, 14%, and 34% during the three winters from an annual local raptor population of 5–10 Sparrowhawks, 1–2 Merlins, and 3 Peregrines. Predation rates were higher than the raptors themselves actually required, because 24.5% of their catches were stolen by Carrion Crows (*Corvus corone*), forcing each raptor to hunt again for itself. For the Redshanks and Dunlins, raptors were responsible for 86% of winter mortality and starvation for only 1.1%. The significance of winter predation is further accentuated by the fact that among the study's 47 individually marked surviving Redshanks, at least 36 returned the following winter, indicating that mortality in other seasons was low (Cresswell and Whitfield 1994).

At Bolinas Lagoon, California, six falcon and hawk species and three owl species prey on wintering shorebirds. Merlins and Short-eared Owls (*Asio flammeus*) were found to be the two most significant predators, but American Kestrels (*Falco sparverius*) and Northern Harriers (*Circus hudsonius*) regularly resorted to shorebirds, especially during colder weather, when Dunlin flocks spread out to feed and Least Sandpipers (*Calidris minutilla*) were in salt marshes rather than on mudflats. Total raptor predation took 21% of the local Dunlins, 11.9% of Least Sandpipers, 7.5% of Western Sandpipers (*C. mauri*), 13% of Sanderlings (*C. alba*), and 16% of the dowitchers (*Limnodromus* spp.) (Page and Whitacre 1975). Further research there found that juvenile Dunlins were far more likely to be caught than adults; 73% of the Dunlins taken by Short-eared Owls were juveniles and, depending on the month during the winter, approximately 62–92% of those taken by Merlins were juveniles. The young birds may have been less experienced at evading predators and more likely to be on the periphery of flocks, where they were more vulnerable (Kus et al. 1984).

At Boundary Bay, on the southern edge of the Fraser River estuary in British Columbia, where 50,000 Dunlins winter, Peregrines were the major predator, with a capture rate of 14%. To avoid roosting at high tide, when shorebirds are most vulnerable because they are forced to concentrate most densely near places where Peregrines may be concealed, Dunlin flocks flew out to sea one to two hours before high tide, returning two to four hours later; "aerial roosting" is evidently worth the energetic expense to avoid predation.

To avoid predation by Peregrine Falcons, flocks of Dunlins on the Pacific coast sometimes spend the hours when the tide is highest flying over the sea. They then become most vulnerable when they return to land and are most hungry.

Peregrines were most successful when the Dunlins returned and the tide was still sufficiently high that they were feeding near wetland edge vegetation and could be approached with a surprise attack. At that time, the Dunlins were probably most hungry and least vigilant (Dekker and Ydenberg 2004). Predation rates were increased when many Gyrfalcons (*Falco rusticolus*) or Bald Eagles (*Haliaeetus leucocephalus*) were present, because Peregrines then ceased hunting ducks and focused only on Dunlins, and 36% of the Dunlins they caught were stolen by the larger raptors (Dekker et al. 2012).

The recovery and population expansion of these predators since the banning of DDT in 1973 has affected the wintering distribution of the shorebirds on which they feed. Between 1975 and 2010, falcon numbers in Christmas Bird Counts increased by a factor of 6.5 on the Pacific coast and 3.1 on the Atlantic coast. As falcons occupied more sites used by wintering Dunlins on the Pacific coast, the Dunlins responded by concentrating at the larger and safer of these. But since Dunlins usually return to the same wintering site in their second and subsequent winters, while falcons move opportunistically along the coast, the more concentrated Dunlins may not have significantly reduced their risk of predation. And the danger they face has been further increased by the simultaneous growth in Bald Eagle populations, by a factor of 7.7 on the Pacific coast and 10.8 on the Atlantic; as more eagles steal catches from falcons, the falcons need to redouble their pursuit of the Dunlins (Ydenberg et al. 2017).

Raptors themselves are also vulnerable to winter predation. A five-year study using radiotelemetry to track individual wintering Sharp-shinned (*Accipiter striatus*) and Cooper's (*A. cooperii*) Hawks in Indiana and adjacent Illinois found that 75.4% of adults and 9.4% of juveniles survived each winter. Predation accounted for 50% of known Cooper's Hawk mortality and 80% of Sharp-shins. Owls were the major identified predator, and one Sharp-shin was killed by a Cooper's Hawk. Sharp-shins wintering in woodlands left their roosts well after dawn and returned long before dusk to avoid encounters with owls (Roth et al. 2005).

ACCLIMATION TO COLD

At high latitudes, where winter is most extreme, birds have the coldest temperatures to withstand, the least daylight in which to find the food that will fuel their metabolism, and the longest nights through which these energy reserves must carry them until they can feed again. North of the Arctic Circle, the sun never rises above the horizon during some of the winter. The duration of this sunless period depends on the latitude, but there are several hours of twilight during which birds search for food. Temperatures can descend to −50° C. Extremes have been recorded in Siberia of −68° C and in Canada of −65° C (Irving 1960:332).

Arctic birds have evolved a wide range of physical and behavioral mechanisms to withstand cold. Rock Ptarmigan (*Lagopus mutus*) on Svalbard, an island at 77–80° N in the Greenland Sea, live through a winter with 10 weeks lacking all daylight. They are active throughout that period, thanks to a highly insulating plumage, a digestive system that adapts to the less nutritious food available in winter, and a crop that can store what they find, to be consumed later during the long hours they spend in snow burrows that protect them from wind and insulate them against heat loss. When necessary, they can fast for days, drawing on the fat reserves they deposited in autumn (Skokkan 1992).

Still farther north, Snowy Owls (*Bubo scandiaca*) winter at 82° N on Ellesmere Island. The insulation of their plumage equals the highest value of any Arctic mammal; their standard metabolic rate is lower than that of any other bird of comparable weight; their thermoneutral zone, in which they need no additional physical or behavioral methods for controlling heat production, descends to 2.5° C; and they can survive ambient temperatures below the lowest recorded in the Northern Hemisphere (Boxall and Lein 1989).

Even a few much smaller birds that lack the insulation to protect themselves from the cold have found a narrow ecological zone in which they can survive the Arctic winter. Of the 130 species of birds that breed in mountains and plateaus of northeastern Sakha Republic, Siberia, at 63° N, only 20 remain in winter, when the average daily temperature may remain below −50° C for three to five months and may occasionally go below −65° C. Most of these birds inhabit forest. Brown Dippers (*Cinclus pallasii*) have recently been found wintering there along rivers that remain unfrozen because warm springs keep the water near 10° C. The birds may be able to survive in and directly above the water in the thin layer of relatively warm air, but not beyond, where the air is much colder (Dinets and Sanchez 2017).

Few birds encounter such extreme challenges to survival, but maintaining body temperature is the overriding concern for all birds living in cold places. Heat escapes from a warm body to cooler surroundings at a rate proportional to the difference in temperature between the body and its surroundings. Thus, to maintain a constant body temperature, a bird's heat production must equal its rate of heat loss. Birds use a combination of insulation and metabolic heat production to stay within survival range; for large birds, with a lower surface-to-body-size ratio, the insulation of feathers may be sufficient. Small birds must deploy additional physiological means and may also change their behavior to reduce energy expenditure and heat loss.

The normal body temperature of a resting bird is 40 ± 1.5° C, so the temperature difference with the winter environment may be 30–90° C (Reinertsen 1983). When birds are active, their body temperature may rise 2–3° C above the lower range of the temperature at rest, but for some Arctic birds their standard body temperature is the same in winter and summer, and in the same range as that of birds in temperate latitudes (Irving 1960:332–333, 336). An experimental study of Greenfinches (*Carduelis chloris*), however, found that at −20° C their body temperature was 0.8° C higher than in summer and at −40° C it was 3.2° C higher; in winter, the birds could maintain their body temperature at −40° C but in summer their body temperature dropped if the birds were exposed to 0° C (Saarela et al. 1989).

CHANGES IN BEHAVIOR

For small birds, the first adjustment to temperatures colder than the normal winter span may be a change in behavior that reduces energy loss. In the familiar mixed-species flocks composed of Downy Woodpeckers, Tufted Titmice, chickadees, and White-breasted Nuthatches (*Sitta carolinensis*) moving through leafless deciduous woods, increased wind speed causes the birds to forage lower down, where they are less exposed. They also reduce their progress through the woods, by both making fewer moves and moving more slowly. At temperatures from 0.1° C to 10° C, foraging Tufted Titmice advance approximately 28.7 m per minute when the wind is 0.32–3.52 km per hour, but they stop more often and advance only 4.4 m when the wind is 10.8–14 km per hour (Grubb 1978). On extremely cold days in Fairbanks, Alaska, with mean temperatures of −40° C, Black-capped Chickadees travel less far in search of food and retire to roost earlier than on warmer days; in Ontario, on days below −29° C, they feed less and seek places exposed to the sun and sheltered from wind (Kessel 1976).

Kinglets, smaller than any of these birds, cannot reduce the rate at which they feed. They need to consume at least the weight of their own body mass each day; in severe conditions they require twice this quantity. In winter, they feed during 99% of the daylight hours. Goldcrests, the smallest member of European mixed-species flocks, avoid competition and aggressive behavior toward conspecifics when the temperature drops. They limit unnecessary movements such as wing flicking, rapid flights, and hovering. Their rate moving through the trees also decreases—at 15° C, a Goldcrest progresses 8 m per minute, at −18° C, 0.71 m per minute. While their winter diet is mainly springtails (Collembola), abundant on conifers and rich in fat, they will also take larger items they come across that they would ignore in summer, such as flies, moths, and hymenopterans. These they swallow whole, not removing the legs, wings, and other parts as they normally do from insects having large but nonnutritious appendages. The Goldcrests also take less notice of predators; at −20° C they ignore Eurasian Sparrowhawks and react less to their principal predator, the Northern Pygmy Owl, even if either of these attacks another Goldcrest nearby. But the hawk and the owl have also adjusted their behavior and have less interest in small prey like kinglets than they do during other seasons (Thaler 1991).

INSULATION BY PLUMAGE AND CIRCULATION

Feathers are a bird's outermost barrier between the cold and its body. Fluffed feathers can increase insulation three- or fourfold over sleeked feathers, at no energetic cost (Rautenberg 1980). On winter days, it is easy to see chickadees and sparrows erect their outer feathers to retain more air against their skin. Some small Arctic birds at rest do this to the point that they appear nearly spherical. The contour feathers of many Arctic birds have less rigid terminal barbs, as well as softer

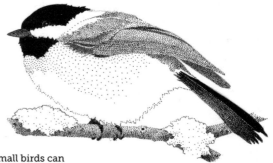

Black-capped Chickadees and other small birds can substantially increase their insulation against cold by fluffing up their feathers.

barbules containing extended fine processes, than do those of temperate latitudes, so they interlock where the feathers of birds from warm climates separate (Irving 1960:345). The unpigmented white winter feathers of Rock and Willow Ptarmigan are unique in having air-filled cavities in the barbules, further increasing insulation (Skokkan 1992). Unlike their more southerly relatives, many Arctic birds, including grouse and Snowy Owls, have feathers covering much of their feet to reduce heat loss from these extremities.

Waterbirds have an additional challenge. Water temperatures may be as low as 1.8° C before freezing. While seals and whales have evolved a thick layer of blubber, birds depend on plumage that is both water repellent and insulating. Loons, alcids, and sea ducks winter as far north as 71° N. Like the Snowy Owl, these birds have a low metabolic rate; insulation derived from plumage and physiological mechanisms essentially eliminates the need to put energy into heat production. Common Eiders (*Somateria mollissima*), the largest of the sea ducks, are well known for their highly insulating plumage. They further minimize heat loss by constricting blood vessels supplying peripheral tissues, thereby lowering skin temperature. The Long-tailed Duck (*Clangula hyemalis*), the smallest of the sea ducks and therefore more subject to heat loss, also depends primarily on its insulating plumage and may deploy vasoconstriction as well (Jenssen and Ekker 1989).

On birds that do not have feathers extending to the toes, venation on the legs with arteries and veins close together spreads heat between the two. The arterial blood gives up its heat to the incoming venous blood. In some Arctic birds, there is closer contact between the arteries and veins of the feet and legs than in their relatives inhabiting warmer areas. Meanwhile, blood destined for the brain is cooled by passing near naked and poorly insulated areas like the eyes, nasal cavity, beak, and palate; this prevents heatstroke when the body may be producing extra heat. Finally, because exposure to cold increases the viscosity of blood, this slows its circulation (Midtgard 1989).

FAT RESERVES

Fats, or lipids, stored in the body provide the fuel for all activities as well as for maintaining metabolism and insulation against cold, both by day and through the winter night when most birds can no longer feed. The amount of fat a bird puts on is a balance between its needs and the potential handicaps of extra weight, such as decreased maneuverability when escaping predators. The first fat to be deposited is beneath the skin; this is also the last fat to be used, because it is so important for insulation. The first fat used is around the abdomen; larger birds also deposit fat in bone marrow. There may also be small amounts in muscle fibers (Blem 1990).

Total capacity to add fat seems not to be the limiting factor, since many birds increase their fat reserves far more before long migratory flights than they do in winter. In summer, these same birds, like tropical species, have almost no fat reserves that would carry them through any food scarcity the following day (King 1972). Larger birds and, within species, more northerly populations that have larger bodies—and must get through longer winter nights—can store more fat. Dark-eyed Juncos in Michigan are 2.7 g heavier than juncos wintering in Alabama (Nolan and Ketterson 1983).

On a daily level, fat deposits are greatest at the end of the day, when birds are about to enter the long night without possibility of replenishment. By delaying fat deposition until the end of the day, birds also reduce the period they may be less able to escape predators (Macleod et al. 2005). On a seasonal level, deposition increases toward midwinter, when nights are longest and average temperatures may be coldest. Temperature seems to be the most significant factor stimulating acquisition of fat reserves, but its influence varies in different birds from daily to long term. For Gray Jays (*Perisoreus canadensis*) living at 66° N in interior Alaska, where temperatures often fall below −40° C and there are four to six hours of daylight in midwinter, temperature, snow cover, and wind all influence the daily variation in fat deposition (Waite 1992). Great Tits (*Parus major*) in experiments in Germany consumed the same amount of food on days with constant and variable temperatures but gained more weight by the end of cold days and lost less during the following night, because of more efficient digestion (Bednekoff et al. 1994).

Ground foragers such as sparrows may have to endure longer periods when food is inaccessible because of snow cover than do the jays and tits that forage in trees and store food. The mechanisms stimulating their fat deposition may therefore respond to longer cycles than daily variation in weather. Experiments with White-throated Sparrows showed that deposition was better correlated with 20-day averages than with any shorter period (Blem 1990). This is more adaptive to sudden fluctuations in food availability. In Dark-eyed Juncos, day length, snow cover, and time of day all play a role, but these may all be linked to temperature as the ultimate cause, since winter day length and snow cover correlate with latitude, which usually determines temperature (Nolan and Ketterson 1983).

Physically and ecologically similar species may store different amounts of fat. Experiments with Dark-eyed Juncos and American Tree Sparrows wintering in Indiana found that juncos could survive considerably longer without food than the sparrows—a mean of 43 hours (range 34–67) compared with the sparrow's mean of 30 (range 21–40)—even though the juncos lost a greater percentage of body mass while deprived of food. Since the two species have similar diets and roosting habits, a possible reason for the difference in fat storage may be that Tree Sparrows typically feed in more open situations, farther from cover than do juncos, and may benefit from being lighter when they have to evade predators (Steube and Ketterson 1982).

Similarly, different populations of the same species may vary in fat deposition, depending on the local environment. Resident House Finches (*Haemorhous mexicanus*) in Colorado fatten while those in southern California do not. American Goldfinches (*Spinus tristis*) in southern California do not fatten, but those in eastern Texas, where the winter climate is similar, fatten as much as do goldfinches in Michigan; this may be because eastern goldfinches are irregular migrants, wintering each year at different latitudes, and the disposition to fattening is genetically encoded (Dawson and Marsh 1989).

SHIVERING

During the day, when birds must be mobile to find food and avoid danger, they must maintain their body temperature despite the loss of heat when the ambient temperature is significantly lower. Beyond the energetically low-cost fluffing up—which, however, conflicts with movement—and reduction of circulation to extremities, birds can resort to shivering, or "thermogenesis." Shivering is a tremor of the skeletal muscles that produces heat but not coordinated movements. It may be stimulated by skin temperature, because birds have thermoreceptors on the upper surface of their wings, on the back of the neck, and on the spinal cord. The maximal heat gain from shivering depends on the external insulation by feathers; this can be four to five times what a bird produces at its basal metabolic rate when resting (Rautenberg 1980).

The energetic cost of shivering is not high, compared with other regular activities. The flight muscles, 15–25% of the body mass of most birds, are significantly responsible for shivering, but actual flight can increase a bird's metabolic rate 14-fold (Dawson and Marsh 1989). It has not been demonstrated experimentally, but the increased heart rate associated with shivering suggests a redistribution of circulating blood during cold exposure, perhaps to nourish the muscles that are shivering (Midtgard 1989).

ROOSTING

At high latitudes, the long winter night, when temperatures are usually even lower than by day, is the most challenging aspect to survival. Birds must maintain a body temperature sufficiently high to ensure that basic metabolic functions continue until they can refuel in the morning. The first resort to reducing energy expenditure is to remain motionless and to fluff up to increase insulation. The next is to find or create a site with protection from wind, precipitation, and exposure to low temperatures.

Where it is available, snow is more insulating than cavities in trees; birds large and small will burrow in it, sometimes using the burrow for a single night or for many. The temperature in a snow burrow, before any heat is generated by the occupant, depends on the temperature of the snow and the external air temperature. The lower the air temperature, the greater the difference compared with the burrow's temperature.

All species of grouse are known to roost in snow. In Nevada, Greater Sage-Grouse (*Centrocercus urophasianus*) begin using snow within a week of its first fall. They dig into open, level snow or, if that is not available, use unpacked, soft drifts on the lee side of shrubs. When they dig into drifts, the opening they create closes as the roof collapses behind them; in the morning, the grouse dig their way out on the other side of the drift. In level snow, the burrows may be as long as 152 cm and as deep as 35 cm. The temperature within the burrow may be between −3° C and −12° C while outside it is −35° C (Back et al. 1987). Capercaillie (*Tetrao urogallus*), a Eurasian grouse up to 87 cm long, dig a burrow and begin shivering immediately to raise the temperature in it; during the night, however, the birds do not shiver, indicating that they create a microclimate from which they lose no heat. In the morning,

Capercaillie shiver again to raise their body temperature to the level required for activity (Reinertsen 1988).

Small birds, including Common Redpolls (*Acanthis flammea*) and Siberian Tits (*Parus cinctus*), also make burrows for themselves; they cannot heat them to the same degree as a large bird, but one redpoll's burrow was found to be −14° C when the outside temperature was −35° C. The burrow was warmer than a tree cavity would be, but there are evidently hazards to snow burrows, at

Common Redpolls may protect themselves from cold and wind by roosting in snow burrows or small crevices.

least for small birds, since few boreal species use this widely available resource. Weather conditions during the night may change the consistency of the snow, making it difficult or impossible for small birds to dig out of their burrow. Burrows may be detected more easily by nocturnal mammals on the ground that cannot find or reach elevated roost sites in trees (Reinertsen 1988).

The Tibetan Ground Tit (*Pseudopodoces humilis*), endemic to treeless alpine meadows on the Tibetan Plateau at 2500–5500 m above sea level, excavates one nest burrow in the ground in spring and another for use during the winter. The winter burrows maximize the available nocturnal thermal benefits by orienting toward the afternoon sun and away from prevailing winds. The burrows range in length from 100 to 160 cm, with chambers 20–40 cm deep to avoid colder surface soil temperatures. Longer tunnels may reduce wind disturbance, while shorter ones could allow chambers to be warmed sooner (Ke and Lu 2009).

Far more small birds roost in tree cavities or their artificial equivalents like birdhouses. The benefits include some insulation as well as protection from wind and concealment from predators. The cavities used by Mountain Chickadees (*Poecile gambeli*) and Juniper Titmice (*Baeolophus griseus*) in Utah were found to reduce wind speed enough so that temperatures were 12.1–13.7° C higher inside the cavities; part of the increased temperature may have been generated by the birds themselves. The energy savings resulted in an additional 5.7–7.3 hours the birds could have gone without food (Cooper 1999). Several woodpecker species excavate new cavities in autumn or winter for use until the next breeding season; Downy Woodpeckers orient these away from the prevailing wind, while cavities excavated at other seasons are randomly oriented (Jackson and Ouellet 2002). Among other birds that nest in cavities, some use them in winter while others do not. Great and Blue (*Cyanistes caeruleus*) Tits retreat to cavities; the Willow Tit roosts elsewhere in trees (Reinertsen 1983).

Why do not more birds use available cavities? The supply is limited, and adult residents of the species that nest in cavities have usually claimed them before young of the year can gain access or migrant winterers have arrived. Trying to drive out another bird that has already used the cavity is dangerous just before nightfall, as well as energetically costly at an hour when birds seek to conserve their reserves (von Haartman 1968).

Even when cavities are numerous, some birds roost communally. As many as 14 Eastern Bluebirds (*Sialia sialis*) were found roosting in one nest box in Indiana during a period of especially cold weather when other boxes nearby were unoccupied. The birds arranged themselves like spokes of a wheel, all with heads inward, in two or three layers. This configuration may have reduced the risk of suffocation as well as generating warmth from the air exhaled by each bird (Frazier and Nolan 1959). In Colorado, as many as 100 Pygmy Nuthatches (*Sitta pygmaea*) roosted in a hollow stub of a dead yellow pine, with another 50 using other cavities in the same tree. From other communal nuthatch roosts, dead birds have been collected that probably suffocated underneath many others (Knorr 1957).

Some birds that, unlike the bluebird and nuthatch, neither flock nor nest in cavities will also gather in sheltered sites. Black-tailed Gnatcatchers (*Polioptila melanura*) in Arizona use the empty globular nests of Verdins (*Auriparus flaviceps*), which Verdins themselves build specially for winter and use as individual roosts (Webster 1999). Sixteen gnatcatchers have been seen using a single Verdin nest for the night. They may by day have occupied 56–59 ha of linear habitat along a stream wash in individual territories. The birds entered singly during 15–25 minutes between sunset and the end of dusk; the next day, most emerged in 1–3 minutes at dawn, a few as much as 20 minutes later. Within the nest, the temperature was 8.6–30.2° C above the outside temperature on different nights (Walsberg 1990).

Rosy finches (*Leucosticte* spp.) are known to use a variety of structures as winter roost sites, including buildings, wells, and mine shafts, which all share features with the shallow caves they use in natural landscapes. In Montana, a flock of 60 Gray-crowned Rosy Finches (*L. tephrocotis*) was found roosting for at least several weeks in a mine shaft with an entrance measuring 2 × 2 m. They perched at least 8–10 m below the entrance; the ambient temperature at that level was a steady 9–10° C, while outside the shaft it dropped to as low as −28.6° C. The birds left in the morning shortly before first light and returned around 3:00 p.m., occupying the shaft for 15–16 hours. Presumably, they had already fed enough without having to continue as long as light persisted, and it was more efficient for them to spend more hours not moving (Hendricks 1981).

Golden-crowned Kinglets (*Regulus satrapa*), in contrast, are at 5–6 g the smallest birds to remain in the North American boreal forest during winter, and they continue feeding until darkness. They move in groups of two to four as part of mixed-species flocks and settle separately from the other species wherever night overtakes them. While the Downy or Hairy Woodpeckers in the group may return to the hole they excavated for winter use, the kinglets find a sheltered evergreen twig, huddle together, tuck their heads underneath their feathers, and fluff themselves up, staying immobile through the night (Heinrich 2003).

Despite the thermal advantages of cavities in snow, wood, or rock, and the additional benefits of body contact, most landbirds roost in a place where vegetation is the only

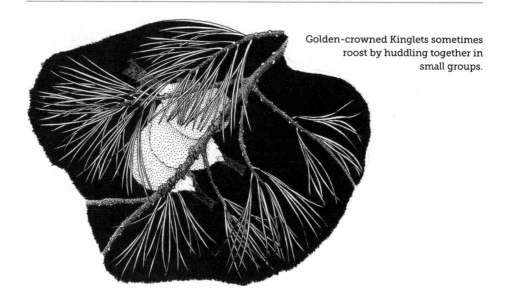

Golden-crowned Kinglets sometimes roost by huddling together in small groups.

shelter from wind and concealment from predators. Birds may be solitary or in flocks, depending in part on the availability and extent of suitable sites.

Foliage, clusters of twigs, and tangles of vines do not, however, provide any significant insulation. For Dark-eyed Juncos roosting in junipers and yews in Indiana when temperatures ranged from −17° C to 20° C, it was only 0.2° C warmer within the trees, but the wind speed was 81% less than the speed beyond the trees (Webb and Rogers 1988). American Goldfinches roosting in dense vegetation reduce their energy expenditure by one-third compared with those in more open sites (Reinertsen 1988).

HYPOTHERMIA

Roosting sites may not, however, themselves provide sufficient protection from wind and insulation from cold, especially for small birds. An additional measure deployed by some small birds is nocturnal hypothermia—lowering their body temperature several degrees to reduce their metabolic rate and therefore expend less of their energy reserves before dawn, which may be as much as 20 hours away.

Birds seem to make the determination at dusk, setting the level of hypothermic metabolism within 30 minutes of roosting. Their breathing rate slows quickly while their body temperature descends only gradually for the first few hours until leveling off. Based on their fat reserves, the ambient temperature, and the length of the night, birds adopt the metabolic rate they can afford to maintain until dawn. In the morning, hypothermic birds rewarm by raising their oxygen consumption through vigorous shivering. They need to have retained enough energy reserve to do this. The rate of warming when arousing from hypothermia is inversely related to body weight, so hypothermia is too costly to large birds at low temperatures (Reinertsen 1983).

The efficiencies of hypothermia vary depending on ambient temperature. A Willow Tit lowering its body temperature to 32° C from the normal sleeping level of 38° C would save 35% of its energy when the ambient temperature is 20° C, 15% at 0° C, and 10% at −20° C (Reinertsen and Haftorn 1986). A Black-capped Chickadee in temperate regions where the night temperature is 10° C can save as much as 75% of its nightly energy expenditure by lowering its body temperature by 10° C, but less in the much colder Arctic night. In Alaska, Black-capped Chickadees do not enter hypothermia even at temperatures of −50° C; there, they are at least 1 g heavier than those of temperate latitudes and have more insulating feathers (Reinertsen 1983). Alaskan chickadees also adjust their behavior to reduce energy loss; they roost 70% of the 24-hour day and even while foraging spend 10% of their time at rest (Grossman and West 1977).

Some species use hypothermia regularly, others only under extreme conditions. In central Norway, Willow Tits routinely reduce their body temperature at night, while Great Tits and Common Redpolls normally do not. Each species deploys different behavioral and ecological means before nightfall to ensure it has enough energy to survive until morning. Territorial Willow Tits may resort to the food they cached earlier in the year or in the day. Great Tits roost in cavities, which Willow Tits never do; they also have a body mass 50% greater than Willow Tits, so they deplete proportionally less of their energy reserves overnight. The nomadic redpoll, with a body size midway between the two tits, leaves areas when birch seeds are declining. It also stores food in its crop before roosting; tits have no crop, so at night they can draw upon only the food they have already digested. Finally, redpolls begin feeding earlier in the morning than do tits and will also forage on moonlit nights. Experiments with all three found that when the Great Tit and redpoll were deprived of food for two to four hours before roosting, they entered profound hypothermia (Reinertsen and Haftorn 1986).

Some birds may also reduce their body temperature during the day. A study of Black-capped Chickadees in Quebec at 48° N found that among individuals within the same flock, daytime body temperature among actively foraging birds varied from the normal 41° C down to 35.5° C, a few degrees higher than the nocturnal 33.8° C when chickadees are sleeping, usually sheltered in tree cavities. Individual variation occurred at all external temperatures: at −15° to −20° C some birds in the flock had body temperatures close to 41° C while others were close to 36° C, and on days when the external temperature ranged between −5° and +5° C, some birds had body temperatures below 39° C. There was no correlation between body temperature and extent of body fat or social rank within the flock. What factors prompt a chickadee to lower its body temperature during the day are unknown, but this does not reduce its ability to forage efficiently and may contribute in some other way to survival on at least a daily level (Lewden et al. 2014).

Hypothermia has been found not only in birds of high and middle latitudes but also in birds inhabiting deserts and even subtropical and tropical rainforests where temperatures may fluctuate widely between day and night. Pigeons, cuckoos, swifts, hummingbirds, and goatsuckers as well as passerines in several families deploy it, so hypothermia is likely to have evolved independently in different groups of birds (Wang 1988). It is known for the Golden-crowned Kinglet, at 5–6 g, and the Gray Jay, which weighs 65–85 g. Because of its larger size, the jay can remain in north-central Alaska at 66° N, where temperatures are as low as −40° C; at the winter solstice there are two hours of daylight and jays persist in feeding during twilight. They roost at 4:00–5:00 p.m. and their body temperature drops for the next 12–14 hours, stabilizing between 5:00 and 7:00 a.m. before rising sharply to daytime levels when they become active at 10:00 a.m. (Waite 1991). The kinglet winters much farther south but may also occasionally be exposed to temperatures of −40° C. Finding sheltered roost sites and huddling together are its first resorts, but at 0° C these alone are not enough to maintain sufficient energy reserves for a night of 14 hours (Blem and Pagels 1984).

Hypothermia has risks as well as benefits. A hypothermic bird is less responsive to external stimuli and reacts more slowly, if at all, to a predator. If nocturnal predation is a serious risk, birds are likely to use hypothermia only when low temperature is combined with weather conditions that reduce foraging success and when fat reserves—and caches, for birds that store food—are low. But to avoid the costs of hypothermia, a bird must forage more during the day, gaining more weight to sustain its body temperature overnight at normal levels; this additional activity depends first on the availability of sufficient food and comes with the risks of being more conspicuous to diurnal predators as well as less agile in escaping them (Pravosudov and Lucas 2000). Thus, the level to which birds lower their body temperature may be a balance between maximal energy efficiency and predator avoidance.

TORPOR

A few species of birds have the capacity to lower their body temperature even further and remain in a dormant state for several days, more like mammalian hibernation than nightly hypothermia. The best-known case is the Common Poorwill (*Phalaenoptilus nuttallii*), a partial migrant in western North America, where some birds live year round in the southwestern United States and in northern Mexico. In December 1946, a sleeping poorwill was found in a rock hollow in the Chuckwalla Mountains, part of the Colorado Desert, in Riverside County, California. When picked up, it remained immobile. Later, the bird was banded and thus identified as the individual that spent most of that winter and the two following ones in the same crevice (Jaeger 1949).

That bird was not monitored daily (or nightly, when poorwills are normally active), so the duration of its dormant periods is not known, but experiments on other poorwills have found wild birds torpid for four days and captive ones for nine days (Ligon 1970). Nor has it been determined whether torpid poorwills in the Southwest are wintering birds from farther north or residents, but the region is one where cold periods reducing the availability of the poorwill's prey, large nocturnal flying insects such as beetles and moths, generally last only a few days. Further experiments have shown that poorwills can lower their body temperature to 5° C, the lowest naturally recorded for any bird, and reduce their oxygen consumption by over 90%; they can fly when their body temperature is only 27° C (Csada and Brigham 1992).

Arousal from torpor is slower than in hypothermic birds, which are capable of enduring much colder temperatures; at 20° C, the poorwill requires 50 minutes, at 10° C, 140 minutes, and at 6° C, 12 hours (Withers 1977). Captive experiments with a few other goatsuckers, including the European Nightjar (*Caprimulgus europaeus*), have found similar capacities for torpor, but this has not been documented in the wild (Ligon 1970). The ability to survive for a few days when weather conditions make foraging impossible is potentially beneficial to goatsuckers at all seasons and all latitudes.

Hummingbirds, with a very different ecology, have also evolved the capacity for torpor. Being so small, they have little ability to store energy, which they consume at a tremendous rate on a diet high in sugars but low in fats—effectively living at a subsistence level. Their metabolism is the highest of any birds, equal to mammals of the same size, like shrews— but shrews can feed by night as well as by day, while hummingbirds are limited to foraging during daylight. With the usual ways to conserve energy not enough, hummingbirds enter a torpid state at night even when ambient temperatures are high and food is abundant. Their body temperature sinks from the daytime 40–42° C to 18–20° C. When ambient temperatures are 15–20° C, hummingbirds can conserve 85% of their energy by becoming torpid; at lower temperatures they save less (Kruger et al 1982).

The Black-chinned Hummingbird (*Archilochus alexandri*) breeds from southern British Columbia south through the Rocky Mountains to the Southwest and into Mexico, wintering in Mexico, coastal Texas, and increasingly around the Gulf Coast to Florida, so it encounters relatively cold nightly temperatures over much of the year through most of its range. When torpid in ambient temperatures of 13–25° C, it maintains a body temperature within 1.2° C of that level, but below that range it increases its metabolism to maintain a body temperature of 18–20° C. Changes in heart and breathing rates demonstrate the efficiency of torpor: at 32° C, an active Black-chinned Hummingbird's heart beats 480 times per minute and at 1° C, 1,020 times per minute, but when torpid only 45–180 times per minute, and least at low ambient temperatures. Similarly, its waking breathing rate is 245 times per minute at 33° C and 420 times per minute at 13° C, but when torpid, hummingbirds breathe only sporadically; some species may have periods of apnea (Baltosser and Russell 2000).

SUMMARY

For most birds, their sole winter focus is survival, with the two challenges of securing enough food to meet what, at least in high and middle latitudes, are likely to be greater energy demands than during the breeding season, and avoiding predation, often in a very different environment from that season and sometimes in a place where there are more predators. Even the normal range of winter weather phenomena affects survival, with some species more vulnerable to cold, snow cover, extended periods of rain, and other conditions, as each of these factors may affect their ability to forage. Brief but extreme weather may cause still greater mortality than long persistence of normal winter conditions.

Many birds must shift their diet, both residents when their summer food dies, departs, or becomes inaccessible, and migrants that move to different regions and habitats, often during the time of year when food there is scarcer than in the spring and summer of higher latitudes. This season of greatest competition seems likely to be responsible for some of the behavioral and anatomical specializations of many birds that feed in a more generalized manner on more abundant food during the breeding season.

The widespread practice of providing food for birds in winter has varied impacts. Comparative studies of winter survival of birds with supplemental food available and others without show that except in extreme weather, most birds do as well on their own. Birds that have expanded northward in recent decades may be particularly dependent until populations have evolved in these local conditions.

Predation is a significant factor for smaller birds during winter. Many must feed in more open environments than during summer, and the urgency to feed during short winter days makes them more exposed and perhaps less vigilant. Their danger is compounded by the greater concentration of wintering raptors and, in warmer latitudes, more bird-eating mammals and snakes. Young birds and those least experienced in their winter habitat are most vulnerable.

Birds wintering in cold places have evolved a wide range of behavioral and physical adaptations to maintain their necessary body heat despite the challenges of low temperatures, potentially less available food, and long nights through which they must live on the energy reserves they accumulated during a short day. Their plumage is highly insulating, and many birds can reduce the blood circulating to their peripheral tissues, thereby further limiting heat loss. Fat deposits are also highly insulating, but birds must balance their need for insulation and the fuel reserve that fat provides with their need for maneuverability when escaping predators, or, if predators themselves, their own need for agility. Most birds therefore put on more fat at the end of the day when they complete feeding and must get through the long night. The fattening patterns in different species respond to the various weather factors that affect their ability to feed. Shivering, the tremor of skeletal muscles, can also generate heat that is then trapped by insulating plumage.

Many landbirds seek or create sheltered roosting sites where they are protected from wind and the lowest temperatures. Some birds, from redpolls to grouse, dig burrows in snow. Where trees are present, more birds seek out cavities where they may roost singly or communally, which generates more heat. But since these are scarce and well defended, most birds roost in the open, where some vegetation may at least reduce exposure to wind.

Since shelter alone may not create a sufficiently warm environment, some small birds deploy nocturnal hypothermia, lowering their body temperature several degrees to reduce their metabolic rate and expend less energy. They seem able to make this determination at dusk. Soon after roosting, their breathing rate slows, while body temperature descends gradually; both increase again before dawn. The risks of hypothermia include reduced responsiveness to danger and slower reactions.

Finally, some birds can reduce body temperature even further, entering a state of torpor. This is best known in the Common Poorwill, which can remain torpid for several days, and in hummingbirds, which do it nightly because their small body size and diet high in sugars and low in fats give them little capacity to store the energy required to get through many hours without feeding.

CHAPTER 6

THE WINTER DAY

AMONG THE MOST SIGNIFICANT seasonal changes birds face in winter is the amount of daylight. Only a few birds, such as Arctic Terns (*Sterna paradisaea*) and the austral seabirds that migrate over oceans to spend their summer and winter at similar latitudes, enjoy days of equal length most of the year. For birds remaining at high latitudes, the hours of light are vastly reduced for several months, and even the birds wintering near the equator, with a constant maximum of 12 hours of light, see less of the sun than in their breeding range. Birds must adjust their time budget to ensure that they secure enough food, while the longer nights often prompt a shift in roosting locale and behavior.

TIME AND ACTIVITY BUDGETS

Most of the research on winter budgets has been done in northern latitudes, and most of the study subjects have been birds that are easy to monitor through the day because they live in open areas and do not move very far. Waterfowl and some raptors lend themselves well to this kind of observation. Results from research based on a few hours of the day and even sunrise-to-sunset projects, though they may be conducted over several winters, still account for only a portion of the bird's 24-hour day and potential activity patterns. Tactile feeders such as dabbling ducks may in fact be feeding during the hours they cannot be observed.

A comparative study of Mallards (*Anas platyrhynchos*) and American Wigeon (*Mareca americana*) at the Eufaula National Wildlife Refuge in Alabama found that Mallards usually spent less than 30% of daylight hours feeding while wigeon fed 45–71% of the day. The wigeons' diet of algae and leafy aquatic vegetation of low quality required them to spend more time foraging, while in the evenings the Mallards flew to flooded cornfields where they found more nutritious food that reduced pressure to feed during the daylight hours (Turnbull and Baldassarre 1987). Eurasian Wigeon (*M. penelope*) wintering on the Solway estuary in Great Britain, 23° north of Eufaula, Alabama, and with far fewer hours of light, feed on grass shoots for 14 hours each day, including many hours in darkness (Mayhew and Houston 1999). The American Wigeon in Alabama may therefore also continue feeding at night.

Diving birds pursuing fish cannot find food at night, so daytime observations may give a more complete picture of their activity budget. On the coast of Rhode Island, a multiyear study found that Common Loons (*Gavia immer*) spent 23–38% of the winter day feeding, 5–18% in maintenance activities, and the balance of the day drifting in the water; their activities do not differ by time of day or tidal cycle, but the wide variation in feeding time may reflect the changing abundance of fish relatively near shore in different years (Ford and Gieg 1995). In a tidal cove off Assateague Island, Virginia, loons feed 55% of the daylight hours and give 25% of their time to maintenance. Feeding bouts there are most intense after the rising tide has slowed but before tide in the cove is highest. In

Common Loons, which feed separately by day, may roost in loose aggregations in coastal coves at night.

late afternoon, loons stop feeding to preen and sleep for a half hour, and then, about 20 minutes after sunset, drift from their individual territories to form a loose aggregation in the deepest part of the cove, where they will spend the night (McIntyre 1978).

Common Mergansers (*Mergus merganser*) wintering in southern New Mexico on an artificial reservoir stocked with fish spend 76% of the day loafing or sleeping, 5.7% flying, 8% preening and stretching, 3.8% swimming, and less than 4% fishing. Their major prey in the reservoir are two species of shad (*Dorosoma* spp.), which may be more abundant in the reservoir than in natural waters (McCaw et al. 1996). In contrast, Red-breasted Mergansers (*M. serrator*) near the Charlestown Navy Yard in Boston Harbor spend 20–30% of the day foraging and similar amounts of time preening and swimming or being alert. Daytime sleeping and flying are less frequent. Here, in a polluted water body where fish are probably at less than natural numbers, the mergansers surface with a fish on 6.4% of their dives (Miller 1996).

Some birds are highly mobile in a regular pattern over each winter day. In Sant Jordi Gulf, Spain, approximately 41,000 Mediterranean Gulls (*Larus melanocephalus*), almost half the species' total population, including some from breeding colonies on the north shore of the Black Sea, begin their day by following fishing trawlers up to 65 km from the coast; they feed on fish parts discarded from the vessels. When the fishery day ends each afternoon at 4:30 p.m., the gulls fly inland as much as 10 km to feed in olive groves, where they pick up fruit that has fallen to the ground. From there, they move to a nearby reservoir to drink and bathe, and at dusk they return to the sea to roost offshore (Cama et al. 2011). Mediterranean Gulls in areas where they depend less on human effort and infrastructure may have a more flexible daily regime.

Other birds shift their range over the winter when daylight diminishes below the number of hours the birds require for feeding. Above the Arctic Circle, in northern Norway at 70° N, there are less than 4.5 hours of twilight in midwinter; from November 20 to January 20 the sun is below the horizon. Three sea ducks winter on the coast, feeding in the sounds between islands. Common Eiders (*Somateria mollissima*) are present year round, King Eiders (*S. spectabilis*) only in winter, and most Long-tailed Ducks (*Clangula hyemalis*) move out in midwinter. All three feed on mollusks, crustaceans, and other invertebrates that have low energy density and must be consumed in bulk for a bird to maintain energy balance. Long-tailed Ducks, much smaller than the eiders and therefore less able to retain body heat, feed mainly on snails and mussels and may not be able to consume enough during the darkest parts of winter; some leave in midwinter. In December and January, all the remaining ducks extend their feeding into the afternoon darkness, and Common Eider males stop displaying before females until the spring. As the days lengthen—growing to 20 hours of sunlight in April—Common Eiders that spent 73 minutes diving during the shortest winter days increase to 144 on the longest, King Eiders go from 57 to 161 minutes, and Long-tailed Ducks from 148 to 382 minutes. While all the ducks increase their absolute feeding time, the proportion of the daylight hours they use decreases for Common Eiders and Long-tailed Ducks. King Eiders continue to feed for the same proportion of each day, perhaps because they are preparing for their March flight north (Systad et al. 2000).

Birds of prey that feed on birds and mammals spend less of the day actively pursuing them than do species consuming smaller items. Raptors that are territorial and sedentary and live in open country have been best studied. Snowy Owls (*Bubo scandiaca*) near Calgary, Alberta, where winter day length averages 10 hours, spend 98% of those hours perched, most often alert, but sometimes either resting or sleeping. Most of their diurnal hunting occurs in the last two hours of light, when the owls often shift to higher perches from which they have a better vantage point. With a local diet of small rodents, the number required exceeds what Snowy Owls are seen to capture during daylight; they also hunt at dusk and at night, when their prey is most active (Boxall and Lein 1989).

In the Lower Peninsula of Michigan, Snowy Owls spend more time hunting in daylight than in Canada, where days are shorter; here, they search for and pursue rodents 31% of the day, with the balance of the day divided equally between loafing and observing. Most diurnal hunting is in the first hour after dawn and the last occurs one or two hours before sunset, with 92.5% of hunting time spent observing and pursuing from perches, 6.8% walking or hopping on snow, and 0.7% searching while flying. Hunting from an observation perch averages 99 minutes in duration, with a success rate of 0.7 per hour of effort. Episodes of ground hunting average 14 minutes, with a success rate of 0.4 per hour. The morning success rate is 59.2% and the afternoon 47.6%. As in Alberta, Snowy Owls continue hunting into the night (Chamberlin 1980).

Ferruginous Hawks (*Buteo regalis*) in eastern Colorado often establish winter territories at colonies of black-tailed prairie dogs (*Cynomys ludovicianus*) and therefore do not need to spend time searching for prey over wide areas. Eighty-four percent of their day is spent perching, mostly on poles and deciduous trees that give them a vantage point over a colony, and 16% flying, including pursuing prey. Ninety-five percent of their successful hunts are launched from a perch—but only 5% of their kills are consumed undisturbed. Bald (*Haliaeetus leucocephalus*) and Golden (*Aquila chrysaetos*) Eagles steal 75% of their kills. More food (36%) comes from scavenging at carcasses, but 18% of this is also stolen by other raptors (Plumpton and Andersen 1997).

For hawks that feed on more thinly dispersed prey and survey large areas from the air, winter weather conditions have a greater impact on how the day is spent. Red-tailed Hawks (*Buteo jamaicensis*) soar more as wind velocity and sunlight increase and cloud cover and humidity diminish. Thermals of warm air rising from the ground are less available than in other seasons, so in winter soaring birds take advantage of wind that has been deflected upward from hills and other obstacles. The greater the wind velocity, the more and the higher Red-tails will soar (Preston 1981).

Like the raptors that are inactive most of the day because they require only a few captures of large prey to meet their needs, birds that can gorge intensely at concentrated food sources need not spend much time exposed to predators and to cold. This strategy is deployed by grouse, which have large crops that store food for later digestion. In High Arctic Greenland (74° N) and in Svalbard (77° N), Rock Ptarmigan (*Lagopus mutus*) maximize foraging efficiency by feeding in areas where musk ox (*Ovibos moschatus*) and reindeer (*Rangifer tarandus*), respectively, have dug through the snow to expose vegetation for themselves (Schmidt et al. 2018).

Similarly, during the brief winter daylight or twilight, Willow Ptarmigan (*L. lagopus*) in Alaska as far north as 73° N browse on shrubs that are accessible over the snow; they store food in their crop that will last through the long night they spend in their snow burrows, and they have enough to carry them through the following day should that be too stormy to emerge and feed (Skokkan 1992). Even much farther south, in British Columbia, Willow Ptarmigan are active an average of only 52 minutes per day. Other northerly grouse also have a concentrated feeding period and a much longer inactive one: Black Grouse (*Tetrao tetrix*) in Finland are active 90 minutes per day, Capercaillie (*T. urogallus*) in Norway three hours, and White-tailed Ptarmigan (*L. leucurus*) in Colorado three hours (Hewitt and Kirkpatrick 1997).

In the Appalachians of Virginia, however, the winter food of Ruffed Grouse (*Bonasa umbellus*)—herbaceous leaves, buds, catkins, and fruits—is more widely scattered and the birds need five to six hours to find enough food, twice the time used by Ruffed Grouse farther north. Virginia grouse may be active any hour of the day, but least often in the final hour of daylight. Ruffed Grouse here are less numerous than in the central part of their range, where they typically feed intensely at dawn and dusk and are sometimes also active in the middle of the day. There, greater concentrations of food may sustain larger grouse populations as well as reduce the time they require to feed (Hewitt and Kirkpatrick 1997).

For most small birds at middle and high latitudes, the factors influencing time and activity budgets are more complex. Their size prevents them from carrying energy stores that will last them many hours or days without refueling, and their food sources are usually evenly dispersed, so that the birds require more time and energy to locate them and to travel between sources. Some foods may have more value but may also require more energy expenditure to secure; shifting to a less nutritive source may consume less energy, although it may require more time and associated risk. Temperature variations affect small birds more than larger ones, and small birds may well be less resistant to wind, rain, or snow, all of which may also have a direct impact on the quantity, availability, and quality of their food. Finally, most small birds are visual feeders, so they have only a fraction of the 24-hour cycle to secure what they need to carry them through until the following day.

To deal with all these factors, two organizing theories propose strategies for how small birds can optimize the daylight time available to them. Intensive feeding early in the day replenishes what has been lost over the long night, but the added weight may make them less maneuverable in evading predators. The alternative is to concentrate feeding at the end of the day, when the risk of diurnal predation is nearly over. If food is reliably available, waiting until late in the day may be more useful; when food is unpredictable because of its inherent nature or the effects of weather, then consuming as much as possible early in the day, or whenever food is found, is more strategic. The analysts who create models for how each approach would play out with different variables acknowledge the difficulty of testing the theories on wild birds, concluding that most small birds in fact deploy elements of both alternatives: they feed most intensively both early and late in the day—even when food may be most available in the middle of the day (McNamara et al. 1994).

A study of Common Cranes (*Grus grus*) in an area of central Spain comprising wetlands and farmland on which barley, wheat, and corn were the major crops found that the factors of food predictability and predation risk also influenced a large bird for which nocturnal energy depletion was less a concern. Some of the 30,000–50,000 cranes wintering there remained in family units that each had its own territory; the majority of the cranes were in roving flocks. The families that defended a territory—selected because it had reliable food resources—were isolated from the neighboring flocks and therefore more vulnerable to predators; these birds increased their feeding rate through the day, peaking at the end when the risk of predation was least. Cranes in the flocks ranged widely each day in search of food, which varied from place to place over the course of the winter; their numbers reduced the chance of a predator going undetected, as well as the odds of any one bird being pursued, so these cranes maximized their intake early in the day. By the end of the day, the total intake for both groups of cranes was the same (Bautista and Alonso 2013).

At lower latitudes where weather may have less impact on survival, time may be used differently. Brown Thrashers (*Toxostoma rufum*) at the Rob and Bessie Welder Wildlife Foundation along the Aransas River in south Texas consistently spend 42–51% of their time foraging between 5:30 a.m. and 5:30 p.m. They loaf 40–49% of the day and devote the remaining time to preening, flying, bathing, drinking, and scratching. Territorial defense occupies less than 1% of the day's first and last hours. On cooler days, foraging may increase slightly in time and/or rate at the expense of loafing (Fischer 1981). In the mangrove forests of Jamaica, the preferred habitat for wintering American Redstarts (*Setophaga ruticilla*), males spend 85% of the day foraging, while in summer in New Hampshire, where the days are longer and redstarts are foraging for both themselves and their young, they spend only 43–65% of their time pursuing prey. The redstart's winter activity budget comes close to the high levels most resident birds of tropical rainforests devote to foraging (Lovette and Holmes 1995).

Shorebirds that winter on coasts have yet another factor in their time budgets—the tides. Sanderlings (*Calidris alba*) at Bodega Bay, central California, use both outer beaches and tidal sand flats. They begin arriving for the winter in August. From then through October, they spend more time roosting than during November–March. In the autumn, Sanderlings spend an average of 9.4 hours feeding (75% of daylight) and in winter 10.4 hours (95% of daylight), reducing their diurnal roosting. No Sanderlings feed on the beaches at night, and they are only occasionally active on the tidal flats at night. The tides affect both roosting and feeding at each site: on the beaches, more birds cease feeding and roost during midafternoon high tides than during morning ones, regardless of the tide's height, while on the sand flats the proportion of birds roosting depends on the height of the tide, with more birds continuing to feed when more of the flats remain exposed; here, too, more birds roost during afternoon high tides. This suggests that feeding is a greater priority early in the day and, in the tidal flats, perhaps at all times (Maron and Myers 1985).

NOCTURNAL FEEDING

For some birds specializing in low-energy food that must be consumed in large quantities, short winter days do not provide enough hours for feeding, and they shift their time budget to become active at night, supplementing what they find during the day. Some foods may be more available or harvested more efficiently at night, and the risk of predation may be lower. The importance of each of these factors varies in different species over the course of the winter, influenced by season, lunar cycles, temperature, and impacts of snow or ice that may cover feeding sites.

As with other anatomical features that seem most adapted to winter activities, such as bill length in shorebirds, the extent and method of nocturnal feeding have influenced the relevant organs. A comparison of the rod-to-cone ratio in the retina of three shorebirds that in the breeding season live in almost perpetual daylight and in winter each have distinctive nocturnal feeding methods showed that each had a different proportion of rods, which gather more light in dark conditions. The Black-bellied Plover (*Pluvialis squatarola*), a visual feeder picking food off the surface of sand or mud both by day and by night, had the greatest density of rods. Short-billed Dowitchers (*Limnodromus griseus*), a tactile feeder probing into mud at all hours, had the second-greatest density of rods, and Greater Yellowlegs (*Tringa melanoleuca*), which switches from visual to tactile feeding at night, had the greatest density of cones (Rojas de Azuaje et al. 1993). Cones function in bright light to form sharp images and to distinguish shades of color, which may be most useful to the yellowlegs, because it looks for prey in shallow water, where there may be glare or reflections.

In shorebirds, the shift to nocturnal feeding may begin while on migration, when at the end of summer the birds leave high latitudes of very long days and reach areas where darkness fills more hours. For Dunlins (*Calidris alpina*) from northern Europe and

Where winter days are short, Dunlins and other shorebirds also feed during the night, especially when the tide exposes flats that were inaccessible during the day.

Russia, the Danish Wadden Sea is the first major stopover site for southbound migrants. They feed on tidal flats both by day and by night. Birds foraging at dusk continue as darkness falls, but most shift location to places where prey is densest in upper sediment, especially where this is soft. During the day, Dunlins feed most by pecking at the surface, on moonless nights they spend most time probing deeply, and on moonlit nights their behavior is intermediate. They take different polychaetes, snails, bivalves, and crustaceans over the course of the 24-hour cycle as their availability changes (Mouritsen 1994).

Nocturnal activity in shorebirds is not limited to species where tides may prevent diurnal feeding. Eurasian Golden-Plovers (*Pluvialis apricaria*) and the Northern Lapwings (*Vanellus vanellus*) from Iceland and northern Europe have taken to wintering on arable farmland in eastern Britain. In southeastern Norfolk, where much of the land is devoted to grains, oilseed rape, and sugar beets, both species feed and roost on these fields and avoid grasslands and pasture. The birds feed primarily on worms near the surface in this cultivated ground, and they could feed all day but rarely do; they are often inactive on the days around the full moon, suggesting that they feed most successfully when nights are well lit. Golden-plovers forage consistently at night in all phases of the moon, irrespective of cloud cover, rain, or temperature. Lapwings feed less on nights with more cloud cover and when lower temperatures make the ground harder. Both species then pursue larger, deeper-burrowing earthworms; they use both acoustical and visual cues to find them. On dark nights, the flocks are smaller, perhaps because large flocks would attract predators that would be hard to detect; when the moon is full, the birds are in larger flocks that may be able to spot any approaching danger (Gillings et al. 2005).

Sixteen shorebird species use the tidal flats at North Humboldt Bay on the California coast. Nocturnal feeding varies by species and season, influenced by the extent and hours of the tides, the amount of moonlight, and weather conditions. The American Avocet (*Recurvirostra americana*) and Black-bellied and Semipalmated (*Charadrius semipalmatus*) Plovers feed equally by day and by night, while among the sandpipers, the smaller species are more nocturnal than the larger ones. During winter, nocturnal feeding occurs on 42% of nights, with more individuals of all species feeding during the day (Dodd and Colwell 1996).

Several species there, from the Marbled Godwit (*Limosa fedoa*) and Willet (*Catoptrophorus semipalmatus*) to the two dowitchers (*Limnodromus* spp.) and the Dunlin, feed more often on nights with a visible moon, because much of their invertebrate prey is most active near the surface when the moon is full. These birds increase their daylight activity during days closest to the new moon, when the night is dark and foraging is less productive. Many shorebirds feed longer on days with lower temperatures and wind speeds, and longer during nights that are warmer. They also feed more at all hours during rain, which brings more prey closer to the surface. Finally, more birds feed during the day when tides expose the mudflats most briefly, and more of even the nocturnal feeders are active during the days when the daytime tidal exposure is longest, suggesting that diurnal feeding is preferred (Dodd and Colwell 1998).

On the northeast coast of Venezuela at 10° N, where temperatures are higher and winter day length longer than at temperate latitudes, a dozen shorebirds winter on a lagoon with little tidal shift. Here, Whimbrels (*Numenius phaeopus*) feed only in daylight, specializing on fiddler crabs (*Uca cumulanta*), which are 3–10 times less active at night, while most smaller shorebirds feed in soft mud at lower tidal levels and are active at all hours; the invertebrates and fish in this zone are 3–30 times more abundant at night (McNeil and Rompré 1995). The smallest sandpipers and plovers forage mainly in the intertidal zone exposed only at low and intermediate tides and therefore need to take advantage of the nocturnal low tides more than do larger species that can feed when water is higher (Robert et al. 1989). Plovers continue to be visual feeders at all hours and small sandpipers remain tactile feeders, but the Greater and Lesser (*Tringa flavipes*) Yellowlegs, which feed equally by day and by night—in shallow water rather than on exposed flats—shift from diurnal visual feeding to side-sweeping their bill through the water, finding prey by touch (McNeil and Robert 1988).

At 40° S on the coast of Argentina in Río Negro Province, Red Knots (*Calidris canutus*) spend September–November and February–April feeding by day in dense flocks at low tides on mussels (*Brachidontes rodriguezi*) attached to rocklike sediments called restingas on the shoreline below cliffs and sand dunes. An hour before dusk, the knots leave the restingas to scatter widely on sand flats, where there are also mussel beds, although at lower densities. Here, they will feed at night, returning at dawn to the restingas. Even on nights with a full moon, no knots feed on the restingas. They may avoid them because the local Great Horned (*Bubo virginianus*) and Short-eared (*Asio flammeus*) Owls are then a danger (Sitters et al. 2001). In Tierra del Fuego, at 54° S, where the days are longer still, knots feed at restingas during the day and roost at night in shallow lagoons, where they do not feed (Harrington 1996:40).

Waterfowl are the other major group that feeds both by day and by night. The balance between diurnal and nocturnal activity varies among species by food source and site. Geese and swans that graze on fields and pastures away from water are only rarely active after dark, when they would be most vulnerable to mammalian predators. At night, they roost in open water far from shore. At a refuge in southern Illinois where as many as 60,000 Canada Geese (*Branta canadensis*) disperse over thousands of hectares of farmland to feed every day, only when snow covers most of the ground and the moon is bright do they leave their lake roost to feed after dark (Raveling et al. 1972). But Brant (*B. bernicla*), Snow Geese (*Chen caerulescens*), and ducks such as American Wigeon that feed on tidal mudflats adjacent to their roosting site take advantage of the nightly exposure of intertidal zones (Owen 1991).

On the coast of Norfolk, England, Brant feed most at night during the three hours surrounding high tide. Then, they can often swim into areas that are dry at low tide, thereby using far less energy to reach a food source than during the day, when they fly to pastures and grain fields to graze. Neither the phase of the moon nor the time after dusk when the tide is highest affects their feeding pattern, but Brant feed longer at night following warm days, when they put more energy into aggressive encounters, preening, and wing flapping than on colder days. Among these flocks, some individually marked birds were found to feed every night, others only infrequently. Adult males, which enjoy dominance during the day, fed least at night (Lane and Hassall 1996).

Ducks like these Northern Pintails that feed on land often sleep during the day and disperse over fields at night, when they are safer from predators.

Some diving ducks are strictly diurnal, others active through the 24-hour cycle. Harlequin Ducks (*Histrionicus histrionicus*) in Resurrection Bay, Alaska, where daylight is as little as 6.4 hours in midwinter, feed on moving prey and do not dive after dark, when they roost farther offshore (Rizzolo et al. 2005). Similarly, Surf (*Melanitta perspicillata*) and White-winged (*M. fusca*) Scoters on the east coast of Vancouver Island, British Columbia, feeding on bivalves near the shoreline, stop at the end of the day and spend the night in deeper water (Lewis et al. 2005). Radio-tagging of all these birds found them to be inactive at night. But many bay ducks in the genus *Aythya* feed at night. Lesser Scaup (*A. affinis*) wintering in the Indiana Harbor Canal dive for oligochaete worms, equally available at all hours, in feeding bouts averaging 11 minutes, followed by resting periods. They feed during 23.7% of the 24-hour day, 29% of the night, and 16% of the daylight hours (Custer et al. 1996).

Human hunting pressure may also have influenced the nocturnal activity as well as the density of some waterfowl. At the Lacassine National Wildlife Refuge in southwestern Louisiana, where ducks cannot be shot, as many as 285,000 Northern Pintail (*Anas acuta*) spend up to eight hours sleeping on a lake each winter day and then disperse between 30 minutes before and one hour after sunset to flooded rice fields as far as 24.4 km away. Legal hunting ends at sunset, but here, even before the hunting season commences, pintails leave the lake at dusk; at some refuges in the Sacramento Valley of California, however, pintails adjust their departure time for less safe areas to postsunset only when the hunting season begins (Cox and Afton 1996).

ROOSTING AND SLEEPING

Of all the activities of a bird's winter day, roosting and sleeping are the ones most available to choice—as to time, location, and social situation. Amount and schedule of sleep vary by species and by season, depending on other demands. For birds that sleep through the hours of darkness, roosting occupies more hours than any other activity. Whether a bird sleeps during the day or at night, sleep is believed to restore the physiological properties of some body tissues, especially in parts of the central nervous system. In winter, sleeping is the most energy-efficient way to pass the time, provided the bird is adequately fed, because birds then typically lower their body temperature. In poor weather, sleeping or resting may consume less energy than foraging requires, and birds delay or reduce their activity. Birds like grouse that remain in snow burrows not only through the long night but also during daylight are further conserving energy.

The terms "sleeping" and "roosting" are not totally interchangeable, because birds may sleep briefly in the middle of other activities rather than retreat, alone or in a flock, to rest in a different place, what is typically called roosting. Birds in roosts are often awake, and some highly aerial species, including certain terns, swifts, and swallows, stay aloft continuously except when nesting; whether they actually sleep is unknown. For birds roosting in exposed places, sleep is normally of short duration, with breaks in which the bird opens one or both eyes and scans its surroundings. Experiments monitoring birds sleeping alone and in groups have found that their sleeping bouts are longer when birds are in a group. Longer or deeper sleep may therefore justify the energetic demand of traveling long distances to a communal roost, where more birds awake at any moment increase the collective vigilance against predators (Ydenberg and Prins 1984).

Whether birds that roost in cavities sleep longer or more intensely is not known, but many of these, like tits and chickadees that forage all day, are believed to sleep through the night. On winter mornings, they typically begin moving at civil twilight, when the sun is 6° below the horizon; roosting time is less well correlated with sunset or civil twilight. Birds that are more intermittently active by day sleep at irregular intervals between other activities (Amlaner and Ball 1983).

Many ducks spend part of the winter daylight hours highly visible in groups sleeping on open water. Ducks can sleep with one eye open and one hemisphere of the brain awake. They open one eye several times each minute when in a state between what electrophysiological studies define as quiet sleep and relaxed wakefulness. In the Camargue, in southern France, wintering Gadwalls (*Mareca strepera*) are awake 80% of daylight hours at the beginning and end of winter, and 50% during midwinter. In early winter, when they are recovering from the energetic expenses of molt and migration, they spend more time feeding, as they do also in late winter, when less food is available, the climate is colder, and body mass decreases; they are then preparing again for migration and breeding (Gautier-Leclerc et al. 2000).

ROOSTING IN FLOCKS

Many birds that are territorial when breeding find it necessary or advantageous to roost in flocks during winter. In some habitats, the sites with least risk may be limited—as for shorebirds needing a place above the high tide line that is also distant from concealed predators—and birds are obliged to roost communally in these scarce locales. Within these sites, there may be a hierarchy of preferred places dominated by higher-ranking birds. In flocks of Dunlins wintering in Bodega Bay, California, more juveniles are found at the center of roosting groups and more of the birds in the center are females, which are larger than males (Ruiz et al. 1989). The benefits of this configuration are unclear, but in Great Britain adult Redshanks (*Tringa totanus*) force juveniles to the windward side of roosts, where they lose whatever microclimate benefits there may be in collective roosting. Even for adults, the benefits of buffering against wind may be less than the energetic cost of flying long distances to reach a more sheltered roost site. This suggests that predator avoidance and detection may be the more significant benefits (Ydenberg and Prins 1984).

At the Cape Romain National Wildlife Refuge on the coast of South Carolina, 11 shorebird species ranging in size from peeps (*Calidris* spp.) to Whimbrels and American Oystercatchers (*Haematopus palliatus*) use a variety of roost sites at high tide, whenever during the day that occurs. A study over four winters found that the key factors affecting daily and long-term choice of roosting sites included size, region within the refuge, substrate, and directional aspect, especially as this was affected by wind. Roosting birds may reduce risk of predation when in larger flocks, but predator avoidance seemed less a determining factor of roost site than other attributes of the site. Of 32 roost sites surveyed, a few were used more than 80% of the time, while most were used less than 50%. Different species at the roosts also had variable sensitivity to disturbance by passing boats, with some species much more tolerant than others (Peters and Otis 2007).

Some of the largest winter bird roosts are found in the southern United States, where several million blackbirds may gather nightly in wetlands, rice fields, canebrakes (*Arundinaria*), deciduous thickets in swamps, or stands of conifers. The roosts are used all through the year, but at greatest density during winter. Red-winged Blackbirds (*Agelaius phoeniceus*) are the most numerous species, joined by Common Grackles (*Quiscalus quiscula*), Brown-headed Cowbirds (*Molothrus ater*), and Rusty Blackbirds (*Euphagus carolinus*), as well as European Starlings (*Sturnus vulgaris*). At coastal sites, there may also be Boat-tailed Grackles (*Q. major*). Some roost sites are used for several years; in other places, birds shift sites over the course of a winter. Most birds arrive before dusk, earlier on cloudy days, and disperse at dawn—at one roost in a Texas coastal marsh, flying as much as 83 km to feed in rice fields. The birds are often vocal all through most nights, but quiet on stormy ones. There is a hierarchy among species and between sexes: at an Arkansas roost in deciduous thickets, starlings were highest in the trees, Common Grackles and male red-wings together slightly lower, then cowbirds and female red-wings, and Rusty Blackbirds and more female red-wings closest to the ground. American Robins (*Turdus migratorius*) may join these roosts, with most staying at the periphery. In marshes of reeds (*Phragmites communis*) 3 m high and wild rice (*Zizania aquatica*) 2.5 m, where there can be less vertical stratification, birds usually perch less than 1 m over the water; the base of the plants may provide a securer perch, more shelter from wind, and concealment or less access for aerial predators (Meanley 1965).

Cities have also become attractive roost sites for starlings and blackbirds. In Europe and North America, starlings roost by the thousands on city buildings that are well lit and have no natural predators; birds begin going there in June or July, when they have finished nesting, suggesting that safety, not thermal benefit, is the attraction; they remain through late winter. At Grand Army Plaza, in New York City at the southern entrance to Central Park, starlings are joined by Common Grackles. Similarly, in Panama and elsewhere from Mexico through Central America, Baltimore (*Icterus galbula*) and Orchard (*I. spurius*) Orioles, which arrive in August and leave in mid-March, gather at dusk to roost in flocks of hundreds in urban sites that may be used in successive years (Wetmore et al. 1984:366, 376). In Brazil, flocks of Purple Martins (*Progne subis*) began shifting from marshes to roosting in cities during the 1960s; one roost in northern Brazil is within a petrochemical compound, where thousands of martins sleep on pipes and catwalks of the refinery (Brown 1997).

Many landbirds that roost in flocks share certain ecological features. Birds that disperse widely by day to feed in open country often roost in very large flocks at permanent locations, while species that feed on patchy resources wander in small flocks and roost wherever they find concentrations of food. In North America, the several blackbird species that flock in winter, sometimes with starlings, fit this pattern, as do the Purple Martins in Brazil. In Europe, Rooks (*Corvus frugilegus*) and Jackdaws (*C. monedula*), both grain feeders, also form roosts of thousands. Their relatives, the Common Raven (*C. corax*) and Carrion Crow (*C. corone*), more opportunistic predators and scavengers, roost in smaller groups at shifting locations (Ydenberg and Prins 1984). Similarly, Cedar Waxwings (*Bombycilla cedrorum*) and crossbills (*Loxia* spp.), which roam great distances in small flocks searching, respectively, for fruit and ripe conifer seeds, roost in these small flocks in temporary locations. Resident birds that are territorial in winter usually roost solitarily or in family units within their territories even as migrant conspecifics are traveling to communal roosts. In Britain, territorial resident Chaffinches (*Fringilla coelebs*) roost and feed alone while migrant Chaffinches disperse over large areas and roost in flocks at permanent locations (Ydenberg and Prins 1984).

Another benefit of communal roosting is the potential for information gathering. Many kinds of birds that nest in colonies as well as those that roost in flocks during other seasons are known to observe others returning at the end of the day and departing in the morning, looking for signs that these have found a productive feeding site. The birds that leave a roost earliest are likely to be going back to where they fed the day before; other birds that had less success will follow them. This is not deliberate information *sharing*, because the knowledgeable birds have nothing to gain by having more birds feeding with them, unless there is greater safety in increasing the number over what was there the day before. But the birds may have nothing to lose if more join them at a feeding site that is ephemeral, such as a school of fish, and where the original knowledgeable birds could not consume everything available.

This is often the winter feeding situation for Common Ravens, when they feed at large carcasses that might become covered by snow or moved by mammalian scavengers before the birds have picked them clean. In Maine, some ravens have traditional roosts near

Common Ravens roost in flocks, sometimes changing their site every few days or weeks to be near the carcasses they scavenge.

stable sources of food, like town dumps, while others shift roosting sites every several days or few weeks, usually to within 1 km of a newly discovered carcass of a deer, moose, or other large animal. Roosts are always in dense stands of white pine (*Pinus strobus*), where the birds arrive shortly before sunset, coming singly, in pairs, or in small groups for the next hour. Some birds move to another roost sometime well after dark, and in the morning those that have remained all leave together at dawn, with all departing within seven minutes of the first birds, and all usually flying in one direction. Experiments with individually marked birds revealed that the ones that first discover a new carcass are likely to be followed on successive mornings by birds unaware of it. An experimental naive bird released at a roost the previous evening will the next morning follow the birds that are going directly to a carcass they have already located. And when an experimental bird is the first introduced to a carcass and then joins a roost of naive birds, the knowledgeable bird is followed the next morning or within a few days to the carcass it alone knew of (Marzluff et al. 1996).

Unusual among communal roosting birds, Barnacle Geese (*Branta leucopsis*) do not go directly to a feeding site in the morning; instead, they gather in smaller flocks near their roost before dispersing. On the Dutch coast, 3,000–4,000 Barnacle Geese roost at night in salt marshes on the island of Schiermonnikoog, feeding during the day in pastures on the island and nearby mainland. Each morning, they fly in small groups to varying places on the island's nearby mudflats, spending almost an hour there—10% of the available daylight—when the pastures are only a five-minute flight away. Birds coming to the flats from elsewhere are an indication of poor conditions, as are birds arriving at the roost late at the end of the prior day, because this means they needed more time to feed wherever they were. Birds spend longest in the morning at the mudflats on days following the evenings with late returns to the roost sites. Presumably, the geese that ultimately depart first are those most confident of finding or returning to a promising feeding area. The extent to which this gathering is an actual information center remains to be tested (Ydenberg et al. 1983).

In Ithaca, New York, where House Finches (*Haemorhous mexicanus*) depend substantially on feeders for winter survival, they roost in small groups of up to 11 birds, but more often averaging 3.5, fewer than are together in daytime feeding flocks. In autumn, they roost first in deciduous trees close to feeding sites, and then, after the leaves have dropped, in conifers, which may be farther away. In winter, when roost sites are scarcer, House Finches continue using the same tree, or another one nearby, longer than in autumn; a few birds may move more than 3 km. They enter the tree 25 minutes before sunset and depart in the morning an average of 9 minutes after sunrise, having first spent several minutes preening, sunbathing, and vocalizing (Dhondt et al. 2007).

Several warblers wintering in Central America and on Caribbean islands roost in groups, irrespective of their diurnal dispersal and social habits. By day, Prothonotary Warblers (*Protonotaria citrea*) in Costa Rica form single-species flocks or join local mixed-species flocks feeding in wooded areas; at dusk, they move in small single-species flocks to adjacent mangrove forests to roost. In central Panama, away from coastal mangroves, Prothonotaries gather in isolated trees at the end of the day, arriving singly and in small groups near dusk, and then depart as a flock 20 minutes after sunset for a more distant roost site. One such preroost gathering site used in 1973 was still used in 1991 (Warkentin and Morton 1995).

In the Dominican Republic on Hispaniola, Prairie (*Setophaga discolor*) and Palm (*S. palmarum*) Warblers that spend the day singly in desert thorn scrub move to communal roosts in adjacent strips of mangroves, while individuals of these species in other habitats on the island do not move at night (Latta et al. 2003). On the east coast of Puerto Rico, 87% of Northern Waterthrushes (*Parkesia noveboracensis*) that maintain individual territories in several types of mangrove forest and in dry forest all leave their territory each day to roost in stands of red mangrove (*Rhizophora mangle*), some birds traveling as much as 2 km, alone or in loose aggregations. They depart their territory between an hour before sunset and 30 minutes after sunset, returning the next morning within an hour of sunrise. Some marked birds were found at the same roost two years later (Smith et al. 2008).

For all these species, the benefits of mangrove forests may include greater protection from wind and rain than other vegetation provides; the standing water below may reduce the presence of snakes and mammalian predators.

SOLITARY ROOSTING

In cold winter climes, small birds that roost solitarily typically choose sites that provide some thermal benefits or shelter from wind, rain, and snow. Tree cavities are often used, mainly by birds that also nest in them. These may be scarce, especially for the birds that, unlike woodpeckers, cannot excavate cavities for themselves. Resident birds have the advantage over winter arrivals. In Europe and North America, where most European Starlings roost in large flocks, some continue, singly or in pairs, to use their nest hole (Clergeau 1981) through the winter. A few birds that feed in small flocks, such as tits and chickadees, disperse to roost singly, also probably to maintain ownership of a protective cavity that may be used in the following nesting season. Some birds that feed solitarily will roost together only on extremely cold nights, returning to their territories the next morning, like the Black-tailed Gnatcatchers (*Polioptila melanura*) that crowd into empty Verdin (*Auriparus flaviceps*) nests (Walsberg 1990). Larger birds, less likely to lose heat, may roost in more exposed places.

The roosting timetable of the Great Tit (*Parus major*) has been studied in detail in the Netherlands. In winter, adult males and females remain on their breeding territory but roost in separate cavities. They begin their day an average of 26 minutes before sunrise, starting at a higher level of light intensity than during summer. Males emerge first and go to the entrance of their mate's cavity, waiting a minute or two for her to emerge. On mornings with rain or snow, tits may look out of their hole but only come out 10 minutes later. Heavy clouds and strong wind delay them less, and they leave their roost site earlier on cold mornings. The pair spends the entire day foraging, and the female returns to her cavity an average of 12 minutes sooner than the male; this is usually before sunset, except in December, when it may be 5 minutes after sunset. In contrast, at 67° N in Lapland, Great Tits start foraging two hours before sunrise, continuing through the one hour of daylight and two more hours after sunset (Kluijver 1950).

Owls often roost by day in conifers where these are available, because evergreens provide more darkness, shelter from weather, and concealment from both predators and the many smaller birds that may harass them. On Assateague Island, Maryland, in an area of pinewoods and shrub swamp, radio-tagged Northern Saw-whet Owls (*Aegolius acadicus*) were found to have home ranges—not strictly defended like territories, and sometimes overlapping—of 38.5 to 248 ha. Each bird roosted in its home range, using the same or a nearby site in a conifer for days or weeks, then moving hundreds of meters away and, after several days, frequently returning to the prior area (Churchill et al. 2002). Saw-whets nest in the abandoned holes excavated by woodpeckers, but on Assateague did not use them for winter roost sites.

Radio-tagging various warblers in Jamaica has revealed that some species are solitary roosters. Ovenbirds (*Seiurus aurocapillus*) wintering in coastal second-growth scrub each maintain a territory and roost in it, generally within the most heavily used portion. They usually settle at a different roost site every evening, averaging 34 m from the previous one used (Brown and Sherry 2008). Slightly higher, in the hills with coffee farms, Black-throated Blue Warblers (*Setophaga caerulescens*) each have a diurnal foraging range within a coffee plantation, but two-thirds of them depart shortly after sunset to roost solitarily, sometimes as far away as 1 km in adjacent forest where the trees are taller, the foliage thicker, and the canopy closed. Birds that moved to the forest changed their roost site as often as nightly (Jirinec et al. 2011).

SUMMARY

Only the birds that go far beyond the equator in winter have day length comparable to what they enjoyed during the spring and summer. From the highest latitudes, where birds may live through weeks of darkness or only brief light to the equator, both diurnal permanent residents and migrants have fewer hours in which to feed than during their breeding season. In winter, many birds may in fact need more time than daylight provides to secure the food required to sustain them through 24 hours. Some tactile feeders, especially shorebirds and waterfowl, shift at least part of their feeding to night, which may be safer if they feed in exposed situations. In addition, for shorebirds some prey items become more available at night, when these are closer to the surface in sediments. Raptors and other birds that catch large prey items spend less time actively hunting; much of their day is spent perching, expending little energy. Other birds that can store substantial amounts of food in their crop for later digestion similarly gorge and then rest. Most small birds, however, use all available daylight to search for food; this requires balancing their need to refuel with the risk of losing maneuverability if they feed heavily early in the day.

Roosting and sleeping are the two activities most available to choice for wintering birds. Actual sleep varies by species in total amount and duration of each bout, with cavity roosters likely to sleep through all the hours they are there, while birds in exposed situations may sleep only for very short intervals between scans for predators. In addition to its restorative benefits, sleep is the most energy-efficient way to pass the time in cold regions, provided the bird is adequately fed, because birds then typically lower their body temperature. Some highly aerial species that never land during the winter sleep while on the wing, if at all.

Roosting—for the night or parts of the day—is a form of rest during which birds may be awake. Many birds that were territorial in the breeding season roost in flocks afterward. This is often safer because, although they are more conspicuous to predators, some birds are always alert to danger. Shorebirds that lose their foraging sites when the tide is high also roost communally in the remaining places that are safest from predators. The limited number of good roosting sites for large flocks of birds means these may host many species; in roosts of both single and multiple species, there is often a dominance hierarchy, with birds of certain ages, sexes, or species occupying the more sheltered spots.

For birds that disperse widely during the day in search of food, communal roosts may also serve as information centers; in the morning, birds that were less successful watch and follow those first to depart, because these are most likely to be returning to a place they fed the day before. But even some species in which each individual defends a territory by day may travel considerable distance to roost with others in what may be a safer site; each bird returns to its territory in the morning.

Birds that roost solitarily seek sites that provide the best available concealment and protection from wind, rain, or snow. Such places may be distant from their actual feeding area. In mixed-species flocks that forage together by day, each bird usually roosts singly, although in some species individuals may share a tree cavity. A roosting place may be used for days or weeks or changed nightly.

CHAPTER 7

ANTICIPATING SPRING

JUST AS SOME BIRDS begin storing food in late summer and autumn for retrieval in winter, during winter itself many birds are also anticipating the following season. Among nonmigrants, territorial defense, pair formation, and nest site selection may all be part of a winter day's activities. Some migratory species such as waterfowl form pairs wherever they winter. There is much less evidence of pairing in other migrants—males, from shorebirds to songbirds, typically arrive at their intended breeding site several days before females.

The behavioral and physiological shifts toward breeding activity are controlled by chemotransmitters—hormones and factors (the term for those with an unknown chemical structure)—released by the hypothalamus in the brain to the pituitary gland, which in turn releases into the bloodstream hormones that stimulate annual growth of the testes and ovaries. In birds, the gonads atrophy after use in each breeding season, reducing unnecessary weight and bulk, and must be regrown. The hypothalamus and pituitary gland themselves are influenced by changes in day length, and to varying degrees in different species by temperature, food availability, and social factors.

Some of the anticipatory behavior is easy to observe—and may in fact occur while days are shortening, not lengthening, long before any gonadal recrudescence. In North America, European Starlings (*Sturnus vulgaris*) become active at nest sites in autumn and early winter, entering cavities or boxes frequently, sometimes bringing new material. They sing and display near their chosen site. A study in New Jersey found that 95% of the individually marked starlings that roosted in pairs in nest boxes during winter used them for nesting the following spring. Some of the birds had nested in the same box the previous year. Except on the very coldest nights, no additional starlings chose to roost in the boxes, so shelter from cold, wind, or storms was not likely a factor for starlings seeking cavities rather than traveling to communal roosts. A further indication of the value of early occupancy was the occasional dead male starling found in a box, with wounds that would have come from a rival; no females were found dead (Lombardo et al. 1989).

In Great Britain, where winters are milder than in most of North America, a 1943 review of resident birds found anticipatory reproductive behavior between October and January, ranging from courtship, copulation, and nest building to actual egg laying, in 66 species. In today's usually warmer winters, the number is likely to be higher. Rooks (*Corvus frugilegus*) form communal roosts used by members of several colonies, but from autumn on, many adults return to their rookery in early morning, where for a half hour or more pairs settle on their nest from the prior season, call noisily, and circle around the trees before going off to feed. From late November into December, males may feed females, and in mild weather pairs may copulate. By February, which may be the coldest part of winter, some remain for the night at their rookery. Jackdaws (*C. monedula*), which

Rooks from several colonies may roost together in winter, but many adults return to their rookery on autumn and winter mornings to spend time at their nest before dispersing to feed.

often share roosts with Rooks, similarly begin returning to previously used nest sites in November. English House Sparrows (*Passer domesticus*) occasionally build nests and lay eggs through autumn and winter, sometimes raising young successfully. During mild winters in Ireland, Grey Herons (*Ardea cinerea*) at large heronries start repairing their nests in January, but isolated small colonies are not visited until late February or March. At Bass Rock on the east coast of Scotland, Northern Gannets (*Morus bassanus*) begin carrying nesting material in late December (Morley 1943).

Singing is often the first sign of prebreeding behavior, but it must be distinguished from the songs some birds use to establish territories in autumn and early winter. Northern Shrikes (*Lanius borealis*) wintering in southwestern Idaho arrive there in November and sing at the edges of their expansive territories until the end of the month; they do not resume singing until early February, when some are seen in pairs (Atkinson 1993). Among North American birds that are not territorial or do not sing to defend a winter territory, some, such as Mourning Doves (*Zenaida macroura*), begin singing within a few days after the winter solstice.

A survey of 51 European species that winter in or migrate through Egypt found that the three species that sang most copiously long before departure—Serins (*Serinus serinus*), Common Linnets (*Linaria cannabina*), and Corn Buntings (*Emberiza calandra*)—were in flocks rather than territorial. Chiffchaffs (*Phylloscopus collybita*), in smaller, looser parties, were the next most frequent singer, followed by a few species found singly or in pairs. Immediately before leaving Egypt, many more species sang at least occasionally, while others closely related to the frequent singers never sang at all (Moreau and Moreau 1928).

Northern Gannets in Scottish colonies start bringing new material to their nest in December.

WINTER PAIRING

Winter pairing is most widespread and has been best studied in waterfowl. While in autumn swans and geese usually migrate with their mate and any surviving young and return as pairs the following spring to their breeding site, in most duck species pair bonds have ended long before summer and the birds travel unattached to their winter destination. Among dabbling ducks, the timing of pair formation seems correlated with the completion of the male's molt out of the female-like "eclipse" plumage into the alternate plumage it wears all winter and into the following summer. Female ducks, which retain a cryptic plumage all through the year, molt on a different schedule, replacing body, head, and tail feathers in late winter prior to migration; some continue their molt into spring migration (Bluhm 1988).

Females of several *Anas* species have such similar plumage that males will court any of them—hence the occasional hybrids between Mallards (*A. platyrhynchos*) and Northern Pintails (*A. acuta*) and American Black Ducks (*A. rubripes*), among others. Thus, it is up to the female to recognize an appropriate mate, based on the unique male plumage of each species as well as the distinctive elements of its courtship display. Females do in fact prefer males that have attained their full alternate plumage to those still molting (Bluhm 1988).

Female ducks benefit in several ways from securing a mate during the winter. The male they accept protects them from harassment by other males, increases their access to food, and provides additional predator vigilance. Males outnumber females in winter, so females can be selective in choosing a mate. The benefits carry over to spring: ducks lay big clutches of large eggs that develop into young strong enough to feed themselves at hatching, and for this egg production females depend on their lipid and protein reserves acquired during winter, because they arrive at their breeding site before much food is available. A study of American Black Ducks wintering in Missouri found that paired females had greater lipid masses than unpaired ones throughout autumn and winter (Heitmeyer 1988). For males of all dabbling ducks, guarding a female during winter incurs increased energy costs as well as greater risk of mortality; these males are likely to be in better condition and therefore a better choice for the female. The benefit to the male, if the pair bond survives into the breeding season, is the increase in the fitness of his potential mate, so she will lay more eggs of superior quality that he has fertilized (Rohwer and Anderson 1988).

Pairing begins at different times in each species, corresponding with the completion of the male's molt. The stages of molt, courtship behavior, and pairing are in plain view wherever these ducks winter. Gadwall (*Mareca strepera*) begin on the breeding ground, continue on migration, and complete pairing in October–November. Most Mallards, American Black Ducks, and Northern Pintails pair in November–December; Northern Shovelers (*Spatula clypeata*) and American Wigeon (*M. americana*) both begin in November. A study of shovelers at Cape Hatteras, North Carolina, found that 43% of females were paired in December, 66% in January, and 97% in February (Hepp and Hair 1983). Finally, Green-winged (*S. carolinensis*) and Blue-winged Teal (*S. discors*), the last to complete their prealternate molt, both pair from late December or early January to March (Bluhm 1988).

There are also ecological factors influencing the timing of courtship in dabbling ducks, and since molt schedule varies widely in the group, ecology could in fact have had an evolutionary impact on the schedule. Gadwalls, the first to form pairs, feed on leafy vegetation and algae, which are less nutritious than the foods of other dabbling ducks. Pairs are dominant over unpaired birds and probably have greater access to preferred food resources. Unpaired Gadwalls are usually at the periphery of flocks, more exposed to both predators and wind. Thus, there is an incentive to both males and females to form a pair quickly. In southwest Louisiana, where Gadwalls begin arriving in late September, 45% of females are paired by mid-October and 81% by late November. Mallards, American Black Ducks, and Northern Pintails, which all consume more seeds, do not reach this level until December or January (Paulus 1983).

At the northern limit of wintering dabbling ducks, wherever fresh water usually remains unfrozen, local conditions also influence the onset of courtship. A study comparing American Black Ducks at three locations near Ottawa, Canada, found that the quality of available food correlated with courtship schedule. Unlike in the other widespread North American *Anas* species, male black ducks do not have an alternate plumage significantly different from the female's, so the male's completion of molt may be less relevant. At the site with the most nutritious and easily available food, the ducks began courtship in January and devoted more of the day to it. Where food was intermediate in quality, courtship did not begin until mid-February, when temperatures rose above the level where ducks needed to spend all their time feeding. At the least favorable site, where ducks had to dive to reach aquatic vegetation, expending more energy for less reward, they began courting in late February and devoted the least time to it. For black ducks as well as other waterfowl, the northernmost wintering sites may thus be determined by the birds' ability not merely to survive but rather to initiate courtship early so they can return soonest to their breeding sites (Brodsky and Weatherhead 1985).

In diving ducks, pairing begins even later: early January to late February for Redheads (*Aythya americana*), while for the Canvasback (*A. valisineria*) a few form pairs in February, but most do so on migration in March and April (Rohwer and Anderson 1988). In the best-studied merganser species, the Smew (*Mergus albellus*), courtship begins in late January in southern Sweden, intensifying in late February and early March, with one-third of the birds paired by March and April (Nilsson 1974). Diving ducks may form pairs later than dabbling ducks because males are less able to defend underwater food resources; in addition, while the male is diving, he is unable to defend a female. Since diving ducks often forage in flocks in open water, predation may be less of a risk, but should the flock take off while a male is underwater, his chance of reuniting with a mate is reduced (Rohwer and Anderson 1988).

By midwinter, long before the ducks have departed, courtship displays may be followed by actual copulation (Hepp and Hair 1983), but this takes place before the female's ovaries have developed, so copulatory behavior may simply be reinforcing the pair bond (Spurr and Milne 1976). For none of these species is there definitive proof that the winter pair bond is the same as the one on the breeding territory, but northbound migrating ducks usually associate in pairs. Female ducks return most regularly to where they have been born or have bred before, together with a male, which is typically less loyal to a prior breeding site (Bluhm 1988).

A few projects that marked waterfowl individually have found pairs reuniting in at least a second winter. On the Southern High Plains of northwestern Texas, a pair of Gadwalls trapped and marked in January of one winter was retrapped together in December of the following one; because Gadwalls are scarce there, the likelihood of a pair re-forming is greater than where they are more numerous (Fedynich and Godfrey 1989). On the coast of British Columbia, pairs of Barrow's Goldeneyes (*Bucephala islandica*) that breed together sometimes reunite in winter and defend a joint territory (Savard 1988). Some marked Common Eiders (*Somateria mollissima*) wintering at the Sands of Forvie, in eastern Scotland, that disperse locally to nest were found in the same pairs in subsequent winters. While there may have been benefits to the eiders in reestablishing a winter pair bond, these females did not nest any earlier the following spring than did females with a new mate (Spurr and Milne 1976).

Winter courtship behavior has been observed in a few other birds sharing typical dabbling duck habitat. In southern California, where Western Grebes (*Aechmophorus occidentalis*) rarely breed but often associate in pairs, females may vocalize while holding their head near the water with mandibles open as they approach a male, repeating this behavior before the male dives. When the male resurfaces with a fish, he gives it to the female. In one case, a male dove 29 times, catching a fish on 9 of the dives and giving it to the female (James 1989).

Some other birds that migrate to nest at high latitudes where the breeding season is short may also begin pair formation in advance. Snowy Owls (*Bubo scandiaca*) lay clutches of

11–13 eggs; this alone consumes several weeks of the brief Arctic summer, since eggs are usually laid only every other day, sometimes at intervals of as long as five days (Parmelee 1992). In Calgary, Alberta, male Snowy Owls have been seen in February hooting and displaying with wings partly raised while facing females several hundred meters away—a distance similar to that in some displays given within the breeding range. In Calgary, these displays elicited no response from the females, but on other occasions at the end of winter there, pairs of owls have been seen together on the ground or flying north side by side at a great height (Boxall and Lein 1982).

Similarly, Northern Shrikes, which breed in the northern taiga and taiga-tundra ecotone, arrive there in pairs; in winter in Idaho, males have been seen following females, and pairs may perch near each other (Atkinson 1993). With species such as these two, very thinly dispersed and little observed both in winter and when nesting, the evidence of lasting pair formation in winter is even more circumstantial than in waterfowl. Whether or not the pairs stay together during their spring migration, however, the impulse to begin courtship begins at least for some individuals long before breeding commences.

For long-distance migrants, proof that any paired activities are more than temporary associations of convenience is still more challenging to demonstrate without individually marked and tracked birds. In the survey of common Egyptian winterers, three species— Black Redstarts (*Phoenicurus ochruros*), European Stonechats (*Saxicola rubicola*), and Sardinian Warblers (*Sylvia melanocephala*)—were paired through most of the winter, but it is not known whether these relationships continued when the birds departed or were simply females joining males to receive the benefits of his territorial defense (Moreau and Moreau 1928). Male Black Redstarts arrive at their breeding sites several days or weeks before females, while 50% of stonechats return to Germany in pairs (Cramp 1988:689, 743). Sardinian Warblers returning from Egypt overlap the breeding range of others that did not migrate, so distinctions between the sexes in arrival times are not known (Brooks 1992:374).

On Barro Colorado Island in Panama, 60% of Canada Warblers (*Cardellina canadensis*) are in pairs in the mixed-species flocks following various antwrens in October–November, and 79% in April–May. Since the antwrens have permanent territories, it is likely the warblers use these birds as their own territory; no flock has two Canada Warblers of the same sex. Were the warbler pairs in fact permanent associations, one would expect them to arrive at their breeding sites together, but males typically arrive well ahead of females (Greenberg and Gradwohl 1980). At fruiting trees on Barro Colorado, Summer Tanagers (*Piranga rubra*), Rose-breasted Grosbeaks (*Pheucticus ludovicianus*), and Baltimore Orioles (*Icterus galbula*) are also seen in pairs; these too are likely to be temporary associations (Leck 1972). In Amazonian Peru, in the city of Pucallpa at 8° S, a pair of Northern Hemisphere wintering Peregrine Flacons (*Falco peregrinus*) was regularly seen perching 2–5 m apart on building structures throughout the day; at dusk, they began hunting small bats synchronously, once both pursuing the same bat. From Brazil, there are similar observations, including a migrating male and female that stayed together in Recife for a week (Kéry 2007).

In contrast to the uncertainties about winter pair purpose and duration in long-distance migrants, nonmigratory Black-capped Chickadees (*Poecile atricapillus*) have been studied extensively, and the benefits of winter pairing are strikingly similar to those proposed for waterfowl. Resident chickadees normally pair in autumn at the onset of flock formation, and usually these pairs breed together the following spring. If both survive, the pair remains together in successive years. Females in winter flocks acquire the hierarchical rank of their mate, so females paired with a high-ranking male receive less frequent and less intense aggression from others in the flock as well as a higher feeding rate when accompanied by their mate. Even the bottom pair has more success than any single bird. Since female chickadees generally have a lower winter survival rate, the benefits to a male paired with one is a greater likelihood of her survival and superior condition in spring (Lemmon et al. 1997).

Among other common, easily observed nonmigrant songbirds, there is a wide range of pairing schedules. British European Robins (*Erithacus rubecula*) each form a separate territory in September after they have molted at the end of the breeding season, but females may join males on their territory as early as December; they will not begin nesting until March (Cramp 1988:606). Pairs of Northern Mockingbirds (*Mimus polyglottos*) usually maintain joint territories—although some females, especially in the northern part of their range, have a separate winter territory, often contiguous—and males begin singing and displaying in February (Derrickson and Breitwisch 1992). Northern Cardinals (*Cardinalis cardinalis*) may remain paired all year on their breeding territory or may join loose winter flocks. Males may leave these flocks in late winter to defend a territory, to be joined later by a female, but sometimes a pair leaves a flock together; the pair may or may not have been mated the previous year. Even at the northern end of the cardinal's range, in Ontario, courtship begins in February; nest building starts weeks later (Halkin and Linville 1999).

WINTER BREEDING

Unusually warm weather, unnatural light effects, or an unseasonal superabundance of food may occasionally stimulate birds to nest in autumn and winter. A cold front, however, may discourage parents from continuing incubation or make it impossible for them to find enough food for their young, or for the poorly insulated young to withstand the temperature. A pair of Killdeers (*Charadrius vociferus*) was found incubating a clutch of four eggs in North Charleston, South Carolina, December 5–15, 1998; they abandoned the nest December 16 after the arrival of a cold front (Smith et al. 1999). In North America, Barn Owls (*Tyto alba*) usually nest from March into the summer, but in southern Illinois a pair used a nest box continuously over 23 months beginning in September 1993 and produced five clutches of eggs. The only failure came in the last week of December 1993, when the three chicks in the nest died of cold exposure; the adults did not begin another clutch until early March (Walk et al. 1999).

Nonmigratory thrushes seem more easily stimulated to nest in winter than many birds. In the unusually mild British winter of 1953–1954, when November temperatures averaged 1.8° C higher than normal and December temperatures 2.5° C higher, several common birds were reported nesting, thrushes most frequently. European Robins,

European Blackbirds (*Turdus merula*), and Song Thrushes (*T. philomelos*) nested in England, Wales, and Scotland. January reverted to normal weather, but some nests of each species fledged young (Snow 1955). In more recent years, Song Thrushes in Britain and western Europe have been found nesting in every month of the year (Cramp 1988:998).

Artificial light may also stimulate breeding. In downtown Columbus, Ohio, an American Robin (*Turdus migratorius*) began building a nest outside an office building on December 12, 1965; by December 26, it was incubating three eggs. Two of the eggs hatched January 6, 1966, and the chicks were fed worms, but on January 9, the temperature dropped to −11° C and the ground froze; the nest was abandoned that day. At this site, nest building had begun with the installation of a large, brightly lit Christmas tree across the street; the additional light may have stimulated nesting behavior (Kress 1967). Similarly, in December 1959 a pair of European Blackbirds in Berlin built a nest in the middle bar of a neon light forming the letter *A* and fledged young the following February (Grummt 1962). In southwestern Pennsylvania, an American Robin incubated a clutch of three eggs in a nest in a maple tree January 6–20, 1965, during a period when the temperature fell to −55° C. As nearly 13 cm of snow fell on January 20, some piling on the nest rim, the robin continued incubating but deserted the nest the next day. Incorporated in the nest were Christmas lights that had decorated the maple; the tree had probably been illuminated when the robins began (Berger 1966).

A few birds that feed on conifer seeds routinely nest in winter. They may in fact nest in any month of the year, whether daylight is increasing or decreasing; their irregular breeding cycles are triggered by the erratic abundance of their preferred food. Pinyon Jays (*Gymnorhinus cyanocephalus*) feed primarily on the seeds of the dominant pine or juniper species in different parts of their range and give insects to their young. In southwestern New Mexico, where they depend on pinyon pines (*Pinus edulis*) for their immediate food needs and also store the seeds for retrieval days or months later, the jays' nesting season is correlated with the three-year cycle of cone development, which in turn depends on a specific sequence of moisture and warmth that does not always occur; flocks of jays therefore move over large areas to find substantial crops of cones in their green stage and do not breed every year (Ligon 1978).

Where Pinyon Jays encounter abundant maturing green pine seeds, they may initiate nesting in late January or early February. As with the dabbling ducks, they will have begun courtship in November, so pair bonds are well established and nesting can commence quickly. In autumns with good cone crops, male testes begin growing in December, even before the solstice. Unusual among jays, Pinyon Jays build extremely well-insulated nests, which give additional protection to the eggs and young of winter broods. Females remain on the nest during incubation and the early nestling stage, fed by their mate. Experiments with captive birds have demonstrated that the presence of green pine cones, not just abundant supplies of other food or increasing day length, is necessary to stimulate gonadal growth (Ligon 1978).

The other conifer specialists, crossbills, travel far over the boreal forests of the Northern Hemisphere in search of specific cone crops suited to the bill of each species. White-winged Crossbills (*Loxia leucoptera*) will nest in every month of the year, but least frequently in late November and December. In January and February, they require a large crop of white spruce (*Picea glauca*) or red spruce (*P. rubens*) cones that are holding their seeds through the winter; a period of unseasonable warmth will make the seeds fall out of the cones, and the crossbills will desert their nests (Benkman 1992). North American Red Crossbills (*L. curvirostra*) have been found nesting from mid-December to early September; they need no more than 10.5 hours of daylight to begin breeding (Adkisson 1996) and do not require any insect food for their young (Tordoff and Dawson 1965).

BREEDING IN WINTER RANGE

Some migrants have established new populations by remaining to nest where they winter rather than returning to their normal breeding range. The many known examples, from both evolutionary and recent times, are all birds of Northern Hemisphere origin. Island populations far removed in space and time from their ancestral source show how this can happen. On the Hawaiian Islands, never connected to any landmass, all birds are descended from ones that originated elsewhere. Several of the now indigenous birds are derived from northern migrants blown far off their normal migration route that then never left. The Hawaiian Common Gallinules (*Gallinula chloropus sandvicensis*) and Short-eared Owls (*Asio flammeus sandwichensis*) are now sufficiently distinct from their North American mainland ancestors to be treated as different subspecies. Canada Geese (*Branta canadensis*) that once reached the islands have evolved into the endemic Nene or Hawaiian Goose (*B. sandvicensis*), and two colonizations by Mallards have produced the endemic Hawaiian Duck (*Anas wyvilliana*) and the Laysan Duck (*A. laysanensis*). Similarly, a roving flock of White-winged Crossbills that at least once wandered far south of the bird's current range—perhaps during a period of glaciation when spruces were common in the Southeast—came to Hispaniola, adapted to the island's pine forests, and remained; the descendants are now a distinct race, *Loxia leucoptera megaplaga*.

The phenomenon is not unique to islands. The nonmigratory Sandhill Cranes (*Grus canadensis pratensis*) of Florida and Georgia (and another race in Cuba) and the Killdeers (*Charadrius vociferus*) of South America (and various Caribbean islands) probably originated from migrant birds that remained. Africa has far more examples of Palearctic birds that may have shifted from wintering to resident status. On that continent, there are 32 migratory Palearctic species with isolated African breeding populations south of the Sahara, another 27 species with nonoverlapping populations in both continents separated by the Sahara, and 12 Palearctic species that occasionally nest in Africa but do not have significant permanent populations there (Leck 1980).

Of the species with permanent African populations, it may be impossible in some to determine which actually originated north or south of the Sahara. But the incentive to cease migration within the winter range must be stronger in Africa than in other regions, because these 71 species come from a total of about 160 European birds that today winter in Africa. Of the species that currently breed only sporadically in Africa, nearly all have settled 20 or more degrees south of the equator, where summer day length as well as temperature range is closest to the conditions at the southern edges of their Palearctic range (Moreau 1966:123). Unlike the North American wintering species that have established themselves from Hawaii to the Caribbean, all north of the equator, these birds have had to adjust their breeding cycle to the Southern Hemisphere spring.

Some of the most recent African colonizers demonstrate how the process evolves. The European population of Black Storks (*Ciconia nigra*) normally winters north of the equator; sometime after 1900 (Voous 1960:20) Black Storks began breeding farther south in Africa and are now found from 16° S to the southern end of the continent. In southern Africa, they nest from April to September, the same months they would in Europe; here the local winter months are most opportune, because rivers and other water bodies are drying, concentrating fish, frogs, and other food (Harrison et al. 1997:86).

The White Stork (*C. ciconia*) winters in much of southern Africa, and since about 1933 a few pairs have nested in southern Cape Province, South Africa (Harrison et al. 1997:82). White Storks do not begin breeding until they are three to five years old and some birds remain in Africa for at least their first post-hatching summer, so the shift to breeding in the Southern Hemisphere spring may not be difficult. White Storks in Cape Province lay eggs in September–November (Brown et al. 1982) and young are on the nest in December. One banded then at Bredasdorp was recovered 3300 km north near the southern border of Tanzania in March; it may have joined Eurasian storks migrating north, but its parents did not leave South Africa (Moreau 1966:123).

European Bee-eaters (*Merops apiaster*) commonly winter in southern Africa, arriving mainly in October and departing by mid-April. In 1886, a colony was found in southwest Cape Province, South Africa; this species now breeds in at least 30 locations in South Africa and Namibia, with a population estimated at 20,000 in 1988. The birds have retained a migratory habit, although not traveling nearly as far as their Palearctic ancestors; most winter in central Africa north to about 10° S, with a range that matches that of six other South African aerial insectivores, all swallows. The African population has not evolved any subspecific differences, but it has shifted its molt schedule to the nonbreeding season, July–August, matching the sequence but not the actual time of Palearctic wintering bee-eaters, which molt in the species' original nonbreeding season, October–March. Two geographically distinct populations may reflect separate colonizations: one breeding in southwest Cape Province during September–January and departing in February, the other north of there, from southern Namibia east to the Orange Free State, South Africa, during October–February and departing in March–April (Brooke and Herroelen 1988).

An estimated 90 million Eurasian House Martins (*Delichon urbicum*) enter Africa each autumn, wintering from West Africa across the continent south to Cape Province. Since 1892, there have been sporadic nestings by House Martins singly or in small colonies in Namibia and South Africa, but no establishment of a permanent local population. Some observations in Zambia suggest how the birds may be stimulated to nest. There, a flock of about 120 House Martins was seen fluttering around the nests of an active colony of Red-throated Cliff Swallows (*Hirundo rufigula*) in October; some entered some of the broken mud nests, which are close in shape to their own, rearranged the grass lining, and pecked at the mud edges. There was no evidence of roosting (Keith et al. 1992:195–196).

Elsewhere in Zambia, House Martins in immature plumage have been seen in October flying around the nests in a colony of Lesser Striped Swallows (*H. abyssinica*); some may have roosted there (Fellowes 1971).

In the Western Hemisphere, the two known recent shifts to breeding within the winter range are both swallows. In 1980, six pairs of Barn Swallows (*H. rustica*) were found nesting in Mar Chiquita, in southeast coastal Buenos Aires Province, Argentina. Since the swallows use artificial structures and the area is well populated with birdwatchers, it is not likely they were previously overlooked. Barn Swallows have since expanded along the coast as well as inland and northward in Buenos Aires and La Pampa Provinces, nesting primarily under bridges and in culverts under roads in agricultural areas, with a few also in urban sites. The twentieth-century development of roads and railways created nesting sites where none had previously existed. Most of the monitored colonies started with a few pairs that grew to tens of pairs. As with the African colonists, the Barn Swallows have rapidly shifted their molt schedule to their new nonbreeding season. As of 2015, Barn Swallows were found breeding over 244,000 sq km, with an additional 730,000 sq km of suitable potential habitat to the north. None winter in Argentina; it is believed these birds migrate to the north coast of South America, from northern Brazil to Venezuela. With so much more habitat available in Argentina, the limit of this population may be determined by the carrying capacity of its new wintering area (Grande et al. 2015). The rapid numerical and geographical expansion makes it unlikely that all the Barn Swallows nesting in Argentina are descended from the six pairs in the first discovered colony. Genetic analysis found the Argentinean birds to be no different from those in New York and California. Continued recruitment from wintering birds has probably fueled the expansion (Billerman et al. 2011).

During surveys of Barn Swallow colonies in Buenos Aires Province between 1994 and 2002, a few distinctive nests of Cliff Swallows (*Petrochelidon pyrrhonota*) were found at four sites under concrete bridges. Most of the nests were incomplete, perhaps abandoned, and while adult Cliff Swallows were present at some, no nests had any sign of occupation by chicks. All were in colonies of Barn Swallows (Petracci and Delhey 2004). Cliff Swallows typically nest in larger colonies than do Barn Swallows; to complete nest building and lay eggs, they may require the social stimulation of more birds than were present (Brown and Brown 1995). More recently, a larger, successful colony was discovered in Villa María, Córdoba Province, northwest of the first known attempts. Here, Cliff Swallows nested under a concrete bridge over the Río Tercero in November 2015. Of 33 finished nests, at least 8 fledged a minimum of 15 young during December. Finally, on the other side of the Andes, attempted nests have been found near Santiago, Chile (Salvador et al. 2016).

Highly migratory seabirds may also occasionally shift their breeding timetable to their wintering area. Leach's Storm-Petrels (*Oceanodroma leucorhoa*), which breed in the North Atlantic and North Pacific, have been found in nest burrows on islands off the South African coast and east of New Zealand. On Dyer Island, east of Cape Town, as many as 19 were located in February 1997 and at least four nests held chicks. Other islands farther east on that coast have also had Leach's Storm-Petrels, but no active nests have been found (Whittington et al. 1999). In the Pacific, this species normally winters south to 16° S.

In November 1980, ornithologists on the rarely visited 5-ha Rabbit Island, at 44° S in the Chatham Islands, found two Leach's Storm-Petrels, one in a burrow that could have been excavated by one of the six seabird species that regularly nest on the island. The two petrels were in fresh plumage, while this species typically begins its molt in November (Imber and Lovegrove 1982). The inaccessibility of most islands in the Southern Ocean and the infrequency of their inspection by ornithologists may mean that other, similar colonizations by Northern Hemisphere seabirds are unobserved.

CARRYOVER EFFECTS

The general winter conditions and individual experiences of each bird can have an impact on the following breeding season. For resident birds where the transition from winter to spring is gradual, these carryover effects, or seasonal interactions, are most reflected in the mating opportunities and initiation of nesting. For migrants, how well they survive the winter may affect their departure schedule, entire journey, and first days and weeks where they will breed. Birds that end the winter in good condition are likely to have the advantages of establishing a territory and mating sooner, giving them more time to raise one or more clutches, and perhaps even setting them ahead for the molt and other preparations they will make in due course for the following winter.

As spring approaches, both resident and migrant birds increase their feeding rate. Resident males will need to expend more energy to defend a territory and attract a mate, if they have not already done so, while females use the added protein to form eggs, and, in some species, both may replace some body feathers. Most migrants usually complete whatever degree of molt they undertake into their alternate plumage before they depart. On arrival, their priorities will be the same as those of resident birds, but before that some will have had to travel long distances over which they may not have been able to feed extensively or at all. Birds nesting at high latitudes where spring comes late will find little food available and must continue to depend on what fat reserves they accumulated at their wintering sites or en route.

The key factors influencing carryover effects are weather conditions wherever a bird has wintered, population pressure from conspecifics that affects competition for food and other resources, and, for birds that have an individual winter territory, its quality. The impact of carryover effects, positive or negative, is much stronger on resident birds and those with strong migratory connectivity; for both, potential mates have experienced the same winter conditions. Birds with populations that disperse over a wider region each winter or do not return to the same breeding area the following spring are less affected, since in any pair the male and female may have endured or benefited from different conditions. Similarly, for bird populations that winter in the same area but generally have individual territories in different habitats, the effects of local conditions on future breeding success may be diluted. Finally, the impacts of winter can be vitiated or ameliorated by conditions encountered on migration, which variably affect populations or sexes that are on different routes and schedules. The recent expansion of technology that enables geographic pinpointing of individual birds over the course of their annual cycle has contributed substantially to current understanding of carryover effects.

Among some resident birds, there is another carryover factor: rank in a winter hierarchy. Paired chickadees are in better condition through the winter and are more successful at establishing a breeding territory the following spring than single birds that were in the same flock. Similarly, among nonmigratory Song Sparrows (*Zonotrichia melodia*), age and rank affect future success. On Reifel Island, British Columbia, Song Sparrows remain all year and territories are limited; adults winter on their territories of the previous breeding season and young birds form loose aggregations. If there is an adult male territorial vacancy in March, it is nearly always filled by the dominant male from the local flock of birds born the previous year. Young females replacing adults are, however, not necessarily the dominant ones in the winter flock (Knapton and Krebs 1976).

In addition to the carryover effects reflected in physical condition, birds may return to their breeding site with other benefits of their winter experience. First-year Marsh Warblers (*Acrocephalus palustris*) wintering in Africa acquire there much of the vocal repertory they will use when singing the following spring and the rest of their lives. A study of 26 birds from Belgium that wintered in southeast Africa found that they imitated the songs and alarm calls of at least 113 African species. In Belgium, each of the Marsh Warblers imitated an average of 31.2 European species and 45.0 African ones; some of the European species were ones they encountered only in Africa. Other Marsh Warblers recorded in Finland, France, and Sweden also included many African bird sounds. During their winter in Africa, the Marsh Warblers incorporate more and more imitations as the austral summer progresses, until they depart in April (Dowsett-Lemaire 1979). While this study did not examine whether individuals with the greatest repertory, or largest African component, had any competitive advantage during the breeding season, a large repertory is thought to be beneficial.

Finally, events during the breeding season may also have carryover effects both on the subsequent winter and beyond to the following spring. Experiments manipulating the length of the breeding season for Manx Shearwaters (*Puffinus puffinus*) tagged with geolocators on Skomer Island, Wales, found that birds that completed chick rearing later, because they had been given a younger chick to raise, had less time in the wintering area, spent less time each day resting, and in the following year delayed their start to breeding, resulting in eggs and chicks of lighter weight and less likelihood of fledging. Meanwhile, shearwaters that finished chick rearing early, because the experiment gave them an older chick, did not leave the colony any sooner, but in winter at sea they rested more, foraged less, and had a subsequent breeding season similar to that of birds on a normal schedule (Fayet et al. 2016).

EFFECTS OF WEATHER

Weather variables in the winter range that affect the condition of birds over the course of the season and when they depart can also influence the start of nesting, its overall success, and whether some birds attempt to breed at all. For large birds like Whooper Swans (*Cygnus cygnus*) that nest at very high latitudes and have young with a slow growth rate, ensuring that the young are able to migrate in autumn before local water bodies freeze may be a challenge. Whooper Swans wintering in southern Sweden breed in Russia, where they must initiate nesting as soon as they arrive. Females cannot delay egg laying to attain better body condition. Winters in Sweden, at the northern end of the swan's winter range, are variable; when low temperatures freeze important feeding areas, the swans shift to croplands, an available but inferior food source. In the years following mild winters, when the swans depart in good condition, 17% of the swans returning the following autumn are young born that year, while after severe winters only 10.4% of the next winter's population is young (Nilsson 1979).

Barn Swallows nesting in northern Italy winter in Africa primarily between Ghana and the Central African Republic, where rainfall is variable, resulting in more flying insects in some years than in others. Amount of rainfall correlates directly with several key elements of swallow reproduction the following season. Females find male Barn Swallows with long outermost tail feathers more attractive, and in winters with more rain these feathers, grown while in Africa, are in fact longer than during drier winters. In spring in Italy, the males with the longest outer tail feathers secure mates and begin nesting sooner than those with shorter tails. Their clutches are larger and fledge more young; pairs with clutches begun 11 days sooner are 11% more likely to lay a second clutch that will also fledge more young. At these Italian colonies, the postbreeding population is 15% larger following rainy winters in Africa than following dry ones (Saino et al. 2004). In winters with low rainfall, some swallows may leave Africa early, while others stay longer even in areas with poor conditions. The females that remain longest arrive last in Italy, breed later, and have lower fecundity. Staying longer in Africa in dry years seems to affect females more than males, perhaps because for late returnees to Italy the time required to produce eggs is greater than for males to begin mating (Saino et al. 2017).

In the much shorter Finnish summers, Pied Flycatchers (*Ficedula hypoleuca*) lay only one clutch, with late ones having fewer eggs than those produced by the first birds to arrive. The flycatchers winter in treed savanna and gallery forest in the Sahel, where rain falls during the African summer; in different years, the rains generate varying amounts of foliage that over the winter support the insects on which the flycatchers will feed. In years following good rainfall, more Pied Flycatchers return to Finland, some nesting later in the season than after years with less African rainfall. These late nesters are likely birds that in drier African winters would not have survived to return at all (Laaksonen et al. 2006).

Of North American long-distance migrants, the Kirtland's Warbler (*Setophaga kirtlandii*) is an ideal subject for research on carryover effects of rainfall, because both its breeding and winter ranges are very small. The birds winter in second growth in the Bahamas archipelago and breed in jack pines (*Pinus banksiana*), primarily in a few counties of the northern Lower Peninsula of Michigan. The extent of rainfall in March, just before Kirtland's Warblers depart at the end of the normally dry winter, correlates with the arrival of males in Michigan and with their breeding success. They arrive later in years with less March rainfall in the Bahamas, and first-time breeders are in poorer condition, perhaps because—

The breeding success of Kirtland's Warblers is determined in part by the extent of rainfall in the Bahamas during March. Rainier years there put the males in better condition for migration and establishing territories.

while precipitation may be equal over any island—the younger males had inferior winter territories or were less experienced at finding patchy food resources. All males raised more fledglings following rainier Marches—at the rate of 0.23 fledgling for every additional 1 cm of rainfall; this can translate into one more fledgling per male. Conversely, a 10-day delay in arrival resulted in an average 0.74 fewer fledglings per nest (Rockwell et al. 2012).

Studies on a broader scale have shown similar impacts. American Redstarts (*Setophaga ruticilla*) that breed in eastern North America winter in the Caribbean; breeding bird surveys from the Gulf Coast to eastern Canada have found them most numerous following winters with higher rainfall in the Caribbean, while populations of western North American redstarts, which winter in Mexico and northern Central America, seem not to vary with amounts of rainfall in their winter range. In Jamaica, where annual precipitation can range from less than 40 mm per year to over 200 mm, there is a corresponding fivefold variation in vegetation and abundance of arthropods. Redstarts wintering in Mexico and Central America may use wetter habitats like humid forest and mangroves, where effects of drier winters are less severe (Wilson et al. 2011).

In addition to higher redstart breeding populations in eastern North America following wetter winters in the Caribbean, the varying amounts of moisture in individual winter territories can be linked to subsequent breeding success, just as with the Kirtland's Warblers. Analysis of stable carbon isotopes from claw samples of redstarts breeding in Maryland revealed the quality of their prior winter territory in terms of water availability; for adult males, quality influenced the number of young produced, while first-year males from higher-quality winter territories had greater success in both mating and nesting, even though this was lower than for adult males. Winter habitat quality did not, however, influence reproductive success for females (Rushing et al. 2016).

The variation in annual rainfall in much of the Neotropics is due to the El Niño–Southern Oscillation (ENSO), which generates hemispheric impacts. Yellow Warblers

(*S. petechia*) breeding in southern Manitoba winter primarily in southern Central America and northwestern South America, a region that receives below-average rainfall during El Niño years. This has the greatest effect on winter habitat used by female Yellow Warblers, which tends to be drier in any year; surveys in Manitoba found that breeding productivity was lower following El Niño years, and highest in La Niña years, when more rain falls in this population's winter range. Productivity in Manitoba may be further compromised by local effects of El Niño—warmer springs and earlier flowering and leafing out of plants that may put the arriving Yellow Warblers out of sync with their insect prey. In the Mexican part of the Yellow Warbler's winter range, El Niño causes above-average rainfall, and La Niña, which brings more rain farther south, makes for drier years here, reversing the potential impacts on different warbler populations (Mazerolle et al. 2005).

IMPACTS ON POPULATIONS

Carryover effects can reflect the quality of winter habitat used by entire populations. Brant (*Branta bernicla*) that nest near the Tutakoke River in the Yukon-Kuskokwim Delta in Alaska winter on the western coast of Baja California, Mexico, concentrating in three estuaries, where they feed primarily on common eelgrass (*Zostera marina*). Both quality and quantity of the eelgrass decline with latitude. Brant wintering at the northernmost site return to Alaska and begin nesting an average of 2.2 days sooner than birds from the most southerly estuary. In the short Arctic summer, even two days may make a difference in the nutrition and growth rate of goslings; from individually marked birds with known wintering sites, it was found that Brant from the southern end of the winter range had a 10% lower survival rate of first-year birds (Schamber et al. 2012).

The Icelandic race of the Black-tailed Godwit (*Limosa limosa islandica*) has expanded dramatically over the last century, for reasons that are not yet understood. On the coasts of Britain, Ireland, France, and the Netherlands, the increasing population has been forced to settled at sites little or never before used where there is less prey and survival rates are lower. In Britain, the wintering population has increased fourfold since 1970; godwits at estuaries on the south coast, the traditional wintering area, have a survival rate of 94%, while on the newly settled east coast the survival rate is 87%. When prey is depleted on the south coast, godwits there shift inland to flooded meadows where they feed on earthworms; east coast birds seem not to have taken up this habit. From individually marked birds, it has been shown that the south coast birds arrive in Iceland up to 10 days sooner than east coast birds and therefore have the opportunity to establish territories in superior habitat (Gill et al. 2001). In the original Icelandic habitat, marshes, 53.7% of godwit pairs fledge young, while only 29.3% of pairs fledge young from nests in bogs of dwarf birch, colonized in the last few decades by birds displaced from the marshes. Finally, since godwits tend to return to their prior wintering sites—with the earliest nesters likely to return first and fill the superior estuaries—the birds that were at poorer estuaries their first winter are likely to continue in less competitive sites at both ends of their annual cycle, for a lifetime of lower survival and fecundity (Gunnarsson et al. 2005).

IMPACTS ON INDIVIDUALS

Stable carbon isotopes can register the extent of moisture in a bird's territory, as transferred to feathers from vegetation through the leaf-eating insects a bird consumes. Differences in isotope level will reveal whether a bird spent the winter in moist or dry habitat, if that is where it grew the feathers being analyzed. This has opened new opportunities to assess the impact of one season on entire life histories, showing that birds with individual winter territories that may vary in quality between one neighbor and the next take aspects of their experience with them when they depart.

The most extensive work has been done on American Redstarts that winter in the Caribbean. Prior research had already determined that redstarts with territories in mangroves or moist lowland forest are in better condition, depart earlier, and have higher annual survival rates than those wintering in drier scrub habitats. Males from the superior winter territories also arrive in their breeding range sooner, mate with females that lay their first egg sooner, and together fledge young earlier. Female arrival date is not influenced by their winter territory, but the first females get to mate with the earlier-arriving males defending better breeding territories, since males from poor winter habitat often arrive several days after most females. For redstarts, which raise only one brood each year, nests initiated later in the season generally produce fewer young—but early nesters may still have the chance to lay a replacement clutch if the first one is lost. Males that begin nesting earlier may fledge one more offspring per nest and the entire brood a week sooner than later-arriving males; for the first females, the gain may be two more offspring, fledged a month earlier than from nests of the last females to return (Norris et al. 2003).

The very last male redstarts to arrive, usually birds born the previous year and therefore less likely to have had a superior winter territory, may not find a female available. In contrast, the first males to return will already have mated and begun nesting; they also have the opportunity for extrapair copulations with other females that have not yet laid eggs. Tests using blood samples of adult and nestling redstarts have shown that 59% of broods contain at least one offspring not sired by the male attending the nest. In these nests, as many as 40% of the young come from extrapair copulations, produced by only 5–16% of local males (Reudink et al. 2009).

Finally, body condition at the end of winter affects the distance first-year redstarts will migrate north, and where they will breed the rest of their lives. Using individually marked Jamaican redstarts from various habitats that have returned the following autumn, when isotopes in their new feathers can reveal the latitude at which they nested and molted, correlations can be made between their initial wintering site and their breeding latitude. Immatures in Jamaica that had mangrove territories are among the first of their age class to depart; they travel relatively short distances within North America and settle to nest within the southern portion of the species' range. Most immatures winter in scrub; they leave later and migrate farther north. This dispersal pattern is not linked to the latitude at which the redstarts were born, or to their sex or size. The first immature birds to reach North America from the Caribbean will find spring more advanced in the south, at the stage of leaf and insect emergence they need, and will remain. In future years, they will return to this latitude. When immatures departing the Caribbean later pass through here, the region will already be saturated and the birds will continue north. Even when immatures that wintered in mangroves and scrub depart Jamaica on the same day, the scrub residents will travel farther. The conditions on the first winter territory thus determine the breeding range for each bird's lifetime, even if when returning to Jamaica as an adult it then gains a superior territory, since adults in moist territories also depart sooner, whatever their spring destination (Studds et al. 2008).

CAPITAL BREEDERS AND INCOME BREEDERS

Beyond the differences in winter habitat quality that enable some birds to reach their spring destinations sooner and in better condition than others, certain birds that breed at high latitudes where little food may be available when they arrive depend for their own survival and for production of eggs on nutrients they have acquired wherever they wintered. Other species going to the same or similar destinations spend a few days or weeks feeding intensively at staging areas between their wintering area range and the final flights to their breeding site. The species that draw on reserves accumulated before they reach their nesting area are called "capital breeders." Those that fuel their egg-laying and incubation phases with food consumed on-site are "income breeders." There is a continuum between the two that varies among species and even within species, depending on the quality of the winter and the amount of food available when they reach their breeding site. Most passerines are income breeders, because they can rest and refuel several times on migration after they leave their winter range.

At one extreme is the Adelie Penguin (*Pygoscelis adeliae*), which winters at sea off Antarctica, feeding on fish and krill, and must fast while on land. The females arrive at their colony 10 days before laying the first egg, having begun egg formation while at sea; each egg requires 19–24 days to produce and females normally lay two, with a three-day interval in between. The yolks come from a mix of food taken at sea and body reserves. When the clutch is complete, the male takes over incubation and the female returns to the sea after her fast; she has lost an estimated 70 g of body mass each day (Meijer and Drent 1999).

The Brant that breed in Taymyr, in Arctic Siberia, winter on the west coast of Europe. Many of them spend the entire winter in the estuary of the Wadden Sea in the Netherlands. At the beginning of spring, they are joined by others from farther southwest on the coast, most remaining until late May before undertaking their nonstop flight of 4000 km to Taymyr. Before departing they increase their body weight to 1600 g from 1250 g, which was its lowest point at the end of winter. Individually weighed and marked females showed that the heaviest birds returned the next autumn with the most young, suggesting that those with the greatest "capital" were most successful (Ebbinge et al. 1982).

Giant Canada Geese (*Branta canadensis maxima*), the largest race of this species, use both capital and income at the start of the breeding season. They have a much shorter spring migration than the Brant; one population wintering at Silver Lake, near Rochester,

Adelie Penguins cannot feed on land, so the eggs females lay are formed from food they have ingested many days earlier while at sea during the winter.

Minnesota, and nesting at Lakes Winnepeg and Winnipegosis, Manitoba, travels only 885 km, with stops, over a maximum of seven days. In early February, as soon as snow begins melting in Minnesota, the geese begin to shift their diet from corn in farm fields to include new shoots of bluegrass (*Poa pratensis*), which provides protein to supplement corn's carbohydrates. By the time they depart in early April, the weight of mated females has increased an average of 36%, males 26%. Unpaired birds consume and gain far less. The additional lipids, protein, and water are estimated to cover the paired geese's nutritional needs on arrival for basic maintenance, territorial defense, and the time the female spends without feeding on the nest during egg laying and incubation (McLandress and Raveling 1981a). While still in Minnesota, paired geese often copulate, especially in late March; egg yolk development, which takes 12–13 days, begins before the birds depart, but actual egg formation requires some additional protein and minerals that females will ingest after they arrive (McLandress and Raveling 1981b).

Red Knots (*Calidris canutus*) and Ruddy Turnstones (*Arenaria interpres*) that winter on European coasts breed on Ellesmere Island and in Greenland, between 70° and 84° N. Some stop to refuel in Iceland on their way north, while others fly directly from British coasts (Cramp 1983:274, 616). At Alert, on the north coast of Ellesmere Island, stable isotopes of carbon and nitrogen were measured in eggs and in the blood of adults; the levels derived from marine food taken in estuaries before departure differed from those in the terrestrial arthropods the two sandpipers consumed at Alert. The eggs in the earliest-laid turnstone clutches contained some marine-derived components, while their later clutches and those of knots did not. Knots lost 15–20 g in their first week after arrival, their weight falling from approximately 150 g to 135 g, and it is likely that none of their capital remained when females began laying eggs. For both these species, the bulk of capital gained at wintering sites or staging areas may have been expended on the northward flight over the Atlantic and the Greenland ice cap (Morrison and Hobson 2004).

SUMMARY

Just as some birds are preparing for winter in late summer and autumn, during winter many birds are also anticipating the following breeding season. Permanent residents may be defending territories they will use, pairing, and even building nests. Many waterfowl begin courtship and pairing in autumn and winter. Other migrants, including long-distance ones, begin singing before they depart their wintering site, even some that have not been territorial all winter. The physiological stimulants to various forms of prebreeding behavior are controlled by the release of chemotransmitters by the hypothalamus in the brain to the pituitary gland, which in turn releases hormones into the bloodstream that prompt annual growth of the testes and ovaries. The hypothalamus and pituitary gland are influenced by changes in day length, temperature, food availability, and social factors.

Both migrants and residents can benefit from winter pairing. In waterfowl, females receive less harassment and have more time to feed when a mate defends them; for the male, enabling his mate to attain better condition before departure increases the chances she will lay more eggs of superior quality that he has fertilized. Among resident birds like chickadees that pair off in autumn at the onset of flock formation, the benefits are similar. Mated females outrank single ones, whether they are part of the dominant pair or lower down, and gain more access to food.

Occasionally, birds are stimulated to nest in autumn or winter, months ahead of their normal schedule. Unusually warm weather, unnatural light effects, or a superabundance of food may prompt this. Later, however, a cold front may discourage parents from incubating or eliminate whatever they were feeding their young. A few species that feed their young ripe conifer seeds routinely nest at any time of year, including winter.

In both evolutionary time and in very recent historical time, some Northern Hemisphere migrants have remained in their winter range and begun breeding there, establishing new, usually sedentary, populations. This phenomenon is most frequent in Africa, where many of the recent European avian colonists breed at similar latitudes south of the equator with climatic conditions comparable to those in their original home. In the Western Hemisphere, the only transequatorial incidences are of wintering Barn and Cliff Swallows establishing breeding colonies in Argentina since the 1980s.

As spring approaches, both resident and migrant birds increase their feeding rate, which enables residents to expend more energy on territorial defense and egg formation, and migrants to initiate their flights. Temperature and amount of rain, as these influence food availability in the winter range, directly affect birds' condition at departure and may carry over to their subsequent breeding success. The impact of winter conditions may affect entire populations, especially of waterfowl and shorebirds that are concentrated in a few locales, or may vary individually for birds that each have a territory of different quality. Carryover effects are especially important for birds of high latitudes, where there may be little food available when they arrive, so that initiating the breeding cycle depends on energy gained during late winter or at stopover sites on migration. These are "capital breeders," while "income breeders" find all the resources they need at their breeding site.

CHAPTER 8

DEPARTURE

E VERY BIRD THAT HAS traveled any distance from its nesting area to avoid the effects of winter there eventually decides it is time to begin another reproductive cycle and return to the region if not the same place where it was born or bred the year before. The precise stimulants to departure vary by species and within species, depending on sex, age, condition, and wintering latitude, but are the result of a long evolutionary process that has created a finely tuned set of internal mechanisms that coordinate with external factors. Departure schedules and subsequent migration rates reflect the optimal time for birds to arrive at their nesting areas in order to maximize production of young in that breeding season. Writers as early as Aristotle and those of the Bible, as well as poets from Chaucer onward, have noted the consistent returns of familiar birds. Today, however, we are still learning how they do it.

STIMULANTS TO DEPARTURE

The first birds of tropical origin that extended their breeding range to higher latitudes may have sought to avoid the intense local competition with conspecifics for food, mates, and nesting sites. Birds that originated at higher latitudes and withdrew in winter to warmer areas, especially in eras of glacial expansion, may have felt a similar sense of winter crowding and a need to disperse more thinly during the breeding season when they required additional space to find enough food to feed young as well as themselves. Colonial species like seabirds with limited nesting sites left these to scatter over the sea and return in spring to their colonies. Whatever the origin of their migration, all these birds needed to develop means to time and to prepare for their departure from their wintering area.

The greater the distance between the wintering and breeding ranges, the stronger the need for internal mechanisms to set the spring departure timetable, because wintering birds far from their summer destination have no direct cues regarding conditions there as spring advances. Similarly, the greater the number of generations any species has been migrating, the farther from its original nonmigratory life cycle and timetable it will have evolved; the original promptings to avoid competition will have less impact than the need to synchronize arrival with optimal conditions in the breeding range. This further promotes the need for endogenous mechanisms governing the annual departure of migrants from their wintering area (Rappole 2013:173).

The nervous system is the mediator between the environment and the physiological functions that prompt and enable departure and migration. Experiments on birds kept in natural conditions, as well as on those given varying amounts of artificial light and feeding regimes, have consistently shown that the photoperiod, the increase or decrease in day length, is usually the key stimulant to these changes. For birds wintering near the equator, however, where day length is constant, strictly endogenous factors control the timing.

Within each bird, a variety of hormones first stimulate the change in metabolism that lets increased food consumption (*Zugdisposition*) result in the substantial gain of lipids and proteins before departure that migrants require to meet the energy demands of their flights. This is followed by the phase of migratory restlessness (*Zugunruhe*), as has been measured in many passerines using cages that register their nocturnal activity, and in some experiments, also their orientation. These terms for weight gain and restlessness coined by some of the first investigators in Germany have become part of the English ornithological vocabulary.

Many of the same hormones control the growth of gonads in both resident and migratory birds prior to reproduction, although there is substantial variation in timing even among closely related species and within different populations of some species. Experimentally castrated birds may or may not show zugdisposition or zugunruhe, depending on when in the year this is done, the light regime to which the birds are exposed, and other factors. Castrated White-crowned Sparrows (*Zonotrichia leucophrys*) later given testosterone implants gain fat and became restless (Newton 2008:362).

Comparative experiments on populations of Dark-eyed Juncos (*Junco hyemalis*) in Berkeley, California, where both resident and several migratory subspecies are present in winter, demonstrate the range of responses to different hormones and to natural and artificial photoperiods. When all subspecies were exposed to the same rates of artificial light increase, migrant juncos, some of which breed as far north as Yakutat Bay, Alaska, at

60° N, could be brought to a state of zugunruhe two months earlier than normal; in these birds the testes were still undeveloped, so gonadal hormones were not the stimulus, nor were they responding to changes in photoperiod. The resident subspecies had a growth in gonads but not of fat, while migrants increased fat levels. When treated with extra gonadal hormones, residents and migrants responded equally with gonadal growth but did not gain weight, indicating again that these hormones did not themselves stimulate migration. Growth hormones from the anterior pituitary gland, however, prompted both weight gain and gonadal growth. Activity in the pituitary itself was stimulated by artificial increases in day length, although this had less effect on the resident than the migratory junco subspecies; migrant races gained more fat in both natural and artificial light increases. Males became heavier sooner than females, matching the difference in their usual departure times (Wolfson 1945).

Relatively few Western Hemisphere birds winter near the equator, where day length varies by no more than 10 minutes over the course of the year; most research on how migrants at that latitude are prompted to depart has been done in Africa. There, as with the Berkeley juncos, field observations of species in which several races from different breeding areas winter together have also shown how the stimulants prompting weight gain, molt, and departure vary in response to the distance and timing of each population's migration. In eastern Democratic Republic of the Congo, several subspecies of Yellow Wagtail (*Motacilla flava*) winter in mixed flocks for five to seven months. Neither photoperiod nor weather conditions change over the season. For these wagtails, the cues have become completely internalized. Their collective breeding range runs from the British Isles and western Europe to Arctic Russia and east to Central Asia. The birds of each region begin their preparations for departure at different times during the winter, based on how far they will travel. Thus, while they all begin adding weight three months before departure, the southern European breeding wagtails (*M. f. flava*) start in December and the northernmost

Yellow Wagtails of several subspecies winter together in Africa, with each population departing at a different time for its breeding destination. Here, from left to right, birds from northern Scandinavia and Russia, from Iberia and southern France, and from Britain.

(*M. f. thunbergi*) in January. Molt is on a similarly sliding timetable, in February for *M. f. flava*, while some *thunbergi* have not completed their molt when they arrive north of the Arctic Circle. In *M. f. flava*, the first gonadal development is in mid-February and is close to completion by departure time in March–April; in *thunbergi*, the first development is in late March and does not advance very much by departure in April–May. The very first wagtails depart in February, the race *M. f. feldegg*, which breeds in southeastern Europe, Asia Minor, and Central Asia (Curry-Lindahl 1963).

Farther west, in Nigeria slightly north of the equator, wintering Yellow Wagtails span a climatic range running from dry to wet by the end of winter. Here, rather than mixing, the northernmost breeding birds winter farthest south. As in the Congo, wagtails with the least distance to travel depart first, but here it is at the end of a long dry season when insect numbers are much diminished; birds then spend at least 75% of daylight hours feeding. When sufficiently fattened in February and early March, they gradually move north to the southern end of the Sahara, from which most will fly nonstop to North Africa before crossing the Mediterranean. Meanwhile, the wagtails from northern Europe that have wintered in southern Nigeria remain there longer, benefiting from the flush of insects that comes with the start of the rainy season. They begin fattening about 25 days later than the birds that were farther north, moving in due course to the latitude those birds have already vacated, by which time rains have also come to this region. These northernmost wagtails do not reach Scandinavia until 50 days after the first birds to depart this latitude reach Italy. For both the wagtails that leave Nigeria when insect food is scarce and those that remain until more emerges, the departure time seems linked to the optimal arrival time where they breed rather than to any local conditions (Wood 1992).

Experiments with other equatorial winterers have demonstrated the strength of the endogenous system for birds wintering where the photoperiod is constant—as well as the dependence on photoperiod for those wintering at other latitudes, even within 10 degrees. Some Willow Warblers (*Phylloscopus trochilus*), which winter in Africa at the equator, were held for 15 months under 12-hour light regimes. They maintained their schedule of molt, weight gain, and zugunruhe. Closely related Chiffchaffs (*P. collybita*) winter several degrees north of the equator; when held under the same conditions as the Willow Warblers, their body weight did not change, molt was reduced, and zugunruhe was irregular or ceased by 12 months (Gwinner 1972).

Similarly, groups of Pied Flycatchers (*Ficedula hypoleuca*) and Collared Flycatchers (*F. albicollis*) kept under light conditions of 10° N, 0°, and 20° S responded differently, based on their normal African winter range. Pied Flycatchers winter north of the equator and went through their normal cycle of molt, gonadal growth, and zugunruhe in the 10° N experimental conditions, but not in those of 0° and 20° S, while Collared Flycatchers, which winter primarily near the equator and farther south, responded normally to all three latitude simulations (Gwinner 1996). In Central and South America, Dickcissels (*Spiza americana*) winter at 10° N; in another set of experiments they were similarly kept in light conditions of 12 hours. They maintained their normal schedule of molt, zugdisposition, and zugunruhe for only one year, whether they were first introduced to the steady photoperiod in January–February while in Panama or in August after breeding in Illinois. After one year, molting and fattening then went out of phase in different ways among different individual birds, indicating that the sequence of these activities is not necessarily linked in a single cycle and each may require different stimuli (Zimmerman 1966).

In autumn in every hemisphere, it is the decrease in day length that stimulates migration. For birds that go far beyond the equator in winter to spend that season in the opposite hemisphere's spring and summer, it is again the decrease in day length that prompts return to their breeding range. Of North American birds, few go farther in winter to the opposite hemisphere's summer than the Red Knot (*Calidris canutus*) population that travels 15,000 km from Arctic Canada to Tierra del Fuego, Argentina. Other knot populations winter near the equator in Brazil and on the southern coast of the United States. Across this range of latitudes, from 54° S to at least 30° N, knots are exposed to entirely different photoperiods; the population at each latitude begins its spring migration at a different time, in response to the local photoperiod. In Tierra del Fuego, where days shorten rapidly, most knots complete their prebasic molt by mid-February, begin fattening, and depart by March. As they proceed north with long flights from one major staging area to the next, they encounter other knot populations. On the central coast of Argentina, most birds leave between mid-March and mid-April, and on the equatorial northern coast of Brazil, where day length does not change, by the end of April. The small population wintering in Florida, where increase in day length must be the stimulant, also leaves in April for Delaware Bay, the final staging area used by all Atlantic coast knots in the Western Hemisphere. From there, the departure among the several populations that were each on a distinct timetable is highly synchronized, with nearly all birds leaving by late May on their nonstop flight to the Canadian Arctic (Harrington 2001).

PROCESSES

The extent of the key processes that take place before actual departure—molt, fattening, and gonadal development—varies tremendously among migrants. Some species have little or no prealternate molt, while others, such as the Scarlet Tanager (*Piranga olivacea*), replace all their body and tail feathers. Some birds continue molting feathers not essential for flight itself even after they have begun migrating—as can be seen vividly in North America in the occasional male Scarlet Tanager that in spring is still a splotchy mix of green and red. Similarly, the period devoted to premigratory fattening and the actual gain in body mass depend in part on the distances ahead to be traveled and whether the route includes frequent stops where birds can refuel. While gonads may begin to develop during winter, this is usually last in the sequence of preparatory body changes, because migrating birds that have already increased their weight with lipids and proteins do not benefit from any additional expansion of organs not assisting their flight.

Scientists studying the influence of photoperiod on the entire annual avian cycle divide the bird's year into a long photorefractory period, when the reproductive system shuts down, and a shorter period of photosensitivity necessary for gonadal development.

Experiments with various passerines consistently show that for some species the shutdown begins when days lengthen beyond a certain point, varying by species and by any population's normal latitude. After that, the decrease in day length that follows is what returns the system to "readiness to function." Finally, a specific level of increase in day length in winter stimulates the pituitary to secrete hormones, starting the recrudescence of the gonads. Thyroid hormones are believed to control the bird's entire time-measuring system. In nonmigratory European Starlings (*Sturnus vulgaris*) at 52° N in England, the gonads grow and mature rapidly in March, function during April, and regress in May until the following spring; photorefractoriness begins in May and ends between mid-October and mid-November (Dawson et al. 1988). For the many birds of middle and high latitudes that are capable of laying first or subsequent clutches after the summer solstice, the refractory period begins later.

Birds wintering on the opposite side of the equator from their breeding range experience a sequence first of shortening day length before they depart their nesting area in autumn, then increasing day length in the austral spring after they cross the equator, and finally the beginning there of another autumn. Their gonadal response to changing photoperiods must therefore differ from that of birds that do not cross the equator for winter. In the Garden Warbler (*Sylvia borin*), which winters in Africa at 20° S, gonadal development begins only slightly in the lengthening austral spring days after it arrives, accelerating in late December and January when the days are diminishing; as day length in subequatorial Africa continues declining in February and March, development pauses until the warblers return to the Northern Hemisphere. The hormonal changes that prompt gonadal recrudescence also stimulate zugunruhe and prevent the gonads from fully maturing before migration. In the Western Hemisphere, similar patterns have been found for the Bobolink (*Dolichonyx oryzivorus*), which in South America likewise winters at 20° S (Gwinner 1988).

Understanding the influence of molt on departure time requires marking and being able to track individual birds; to date, less research has focused on this than on the other aspects of migratory readiness. A study in Arkansas of Mallard (*Anas platyrhynchos*) females, which initiate their prealternate molt long after males, found that in January some females were halfway to completing their molt while others had not yet begun. Female Mallard molt can begin any time between late November and early March—indicating that it is not photosensitive but controlled either endogenously or perhaps by availability of food. In the marked birds, molt status was not correlated with age, and departure dates, beginning in mid-February, were not correlated with body condition in January. Forty percent of the early molters and 62% of those midway through their molt in January departed Arkansas by March 17, while only 13% of those that had not begun molt in January had migrated by March. There was no evidence that the stage of molt prohibited or impeded the start of premigratory fattening. In this case, it may be that the readiness to depart was determined by the females' destination, with birds that were traveling less far, irrespective of molt condition, leaving sooner than those ultimately going farther north (Dugger 1997).

There has been much more research on the extent and rate of fat deposition before spring migration. Sedentary birds may not gain anything at the onset of spring, and short-distance migrants that have no significant barriers of water or desert to traverse may gain only a small fraction, while Northern Wheatears (*Oenanthe oenanthe*) and Yellow Wagtails in northeast Nigeria about to fly at least 1000 km over the Sahara put on 30–40% of their body weight in fat before departing (Ward 1963). The Red Knots wintering in South America that use only a few staging areas separated by long flights can increase their weight by as much as 84% at each stop (Harrington 2001). In the same manner, "capital breeders" gain enough before departure or at their final staging place to sustain them through the first days at their nesting site.

Among passerines, the group most amenable to experimental monitoring, fat is usually laid down in the 10 or 12 days before departure, sometimes more rapidly—although in some near-equatorial winterers like Yellow Wagtails the process is slow and begins much earlier. Away from the equator, changes in day length prompt the weight gain. Bramblings (*Fringilla montifringilla*) wintering in Britain that were kept at the local 8-hour winter day length increased weight rapidly within eight days of exposure to 14.5-hour days typical of April. Zugunruhe and gonadal development began with weight gain. But Bramblings exposed to the same changes in light regime while kept on a minimal diet that prevented them from adding mass had equal zugunruhe and gonadal development (Lofts et al. 1963). Thus, the urge to migrate may be prompted independently of actual physical readiness, although in a natural situation it is likely that the two will go together; Bramblings wintering in Britain fly over the North Sea to Scandinavia (Perrins 1994:482) and might not be able to complete that crossing on the experimentally reduced diet.

Wild White-crowned Sparrows of the race *Zonotrichia leucophrys gambellii*, which breeds in Alaska and northern Canada, were weighed regularly over several years near Pullman, Washington, between January and their departure in late April and early May. Their highest midwinter weight was in early January. From then, it declined gradually during the prealternate molt to its lowest level, between March 20 and April 10. Finally, as molt tapered off, the birds began a very intensive buildup of fat, more than doubling their lipid reserves within 10 days at the end of April. In the five years of the study, the start of weight gain was always within a period of eight days on the calendar, and the inception each year had no correlation with local weather conditions. Known individuals were consistent in their start dates over more than one year. The absence of variability suggests that changes in photoperiod are the likely stimulant (King and Farner 1959).

Other studies have focused on changes in diet during this period of zugdisposition when birds increase their body mass. In southeastern Louisiana, wintering Hermit Thrushes (*Catharus guttatus*) feed on fruit in winter but shift to an exclusively arthropod diet just before their departure. Thrushes in captivity given a purely arthropod diet gained more body mass and subcutaneous fat than did those given a mixed diet, but body mass did not correlate with onset or intensity of zugunruhe. Because Hermit Thrushes are short-distance migrants that can resume feeding every day on their route, the elements of their diet may be less important than the timing of their departure (Long and Stouffer 2003).

In contrast to the rapid weight gain of the passerines and of other shorebirds, Bristle-thighed Curlews (*Numenius tahitiensis*) wintering on Laysan Island in the Pacific take on fat stores slowly. They arrive from Alaska in July and August after a 4000-km flight with a mean body mass of 450 g. By November, it is 500 g, and at the time of their departure in late April and early May body mass averages 656 g, with the occasional bird weighing almost 800 g. At this time, the percentage of body fat relative to total body mass ranges from 36% to 47%. The slow deposition of lipids may be because Pacific islands lack the rich intertidal feeding zones that most shorebirds wintering on continental coasts can exploit. The frequent winter storms and high winds that continue for days on Laysan also limit feeding. There, Bristle-thighed Curlews take exclusively terrestrial foods including arthropods, land crabs, seabird eggs, lizards, and carrion, while in spring other shorebirds wintering on the island, including Pacific Golden-Plovers (*Pluvialis fulva*), Ruddy Turnstones (*Arenaria interpres*), and Wandering Tattlers (*Heteroscelus incanus*), shift to feed in the island's lagoon on abundant brine flies (*Neoscatella sexnotata*) and brine shrimp (*Artemia salina*), which may be too small to supply the needs of the much larger curlew (Marks 1993).

DIFFERENTIAL TIMETABLES

The departure timetable has several variables. As we have seen, among the individuals of any species wintering in the same region, these may include sex, age, physical condition, and destination. For populations of a species spread over different latitudes, additional features distinct to each locale such as photoperiod and precipitation levels may be important. Where several closely related or ecologically similar species or subspecies winter together in the same habitat, the distance to their breeding range may be the determining factor for each. All these elements, however, are ultimately linked to the optimal time for each bird to reach its breeding destination. Thus, winter departure schedules are closely associated with feeding ecology in the breeding range, where diet may differ from what a bird consumes in winter. Among landbirds, those that feed on seeds are likely to migrate less far—if they migrate at all—than are insectivores and are able to return to their breeding range sooner. Specialists in arthropods on tree trunks and branches or on the ground will find food earlier in spring than birds that eat flying insects or ones living on seasonal foliage. These ecological constraints create the schedule of returning migrants familiar to birdwatchers.

The migrants that travel least far from their breeding range are also most responsive to local weather conditions at the end of winter, since these are likely to be similar at their not-too-distant destination. Their departure will therefore vary more from one year to the next based on local annual fluctuations in temperature, snowfall, and other factors. The farther birds winter from their breeding range, the more fixed their departure timetable will be, even when this varies by sex or age. The very last migrant species to pass through on migration probably winter most distantly but, given how far they have to travel, may not be the last to leave their wintering area, as demonstrated by the Red Knots that depart Tierra del Fuego long before knots have left Brazil or Florida.

A few birds have more individual variation in their departures; these are species likely to be most opportunistically wide ranging during winter. Individually marked and tracked White Storks (*Ciconia ciconia*) in Africa have shown how far each bird may wander searching for good feeding sites. One adult male from a colony in Radosze, in northeast Poland, that over four different winters moved over 12° of latitude and 1200 km, from Sudan to Kenya, had departure dates between February 21 and March 22. In three of those years, the departure date fell between February 21 and March 10, and the bird returned to its nest April 10–15. In the fourth year, however, conditions were much poorer in Africa and the stork did not depart until March 22, arriving in Radosze on May 16 (Berthold et al. 2002).

The sex-based difference in departure dates for birds wintering at the same location is generally attributed to selection pressure on males to establish territories in the best possible locales before females return, in order to be most competitive and begin nesting sooner. In the Hermit Thrush, a short-distance migrant, there is relatively little winter latitudinal separation by sex, but males tend to leave 20 days earlier than females (Stouffer and Dwyer 2003). Unique among birds, female Spotted Sandpipers (*Actitis macularia*) migrate earlier in spring than males to reach and claim breeding territories in which they will mate with a succession of as many as four males, providing each with a clutch of eggs to incubate. In a population on a small island in north-central Minnesota where birds were marked individually, the mean arrival date for experienced females was May 15, for experienced males May 20, for inexperienced females May 22, and for inexperienced males May 29 (Oring and Lank 1982). Female Spotted Sandpipers generally winter north of males, and a male from this island in Minnesota was seen in September in Guyana, South America (Oring et al. 1997), but it is not known whether the birds of each sex and class departed from their wintering sites in the sequence that they later arrived in Minnesota.

For closely related and ecologically similar species wintering together, such as shorebirds, travel distance and the time when the advance of spring makes high-latitude environments suitable for nesting are key factors. On the Banc d'Arguin in Mauritania, one of the major wintering sites for Eurasian shorebirds, 15 species are abundant. From there, the most northerly breeders will travel more than 10,000 km, twice the distance of the more southerly breeders. Of the 13 larger species that could easily be monitored, the average departure date for each species was found to correlate almost perfectly with the midpoint of its breeding latitude, at a rate of about two days for each additional degree of latitude. Eurasian Oystercatchers (*Haematopus ostralegus*), with a midpoint at 55° N, have an

At coastal estuaries where many species of shorebirds winter together, each species, like these Sanderlings, departs on a schedule closely correlated with the distance it must travel to its breeding grounds.

average departure date of March 29, while Sanderlings (*Calidris alba*), at 70° N, leave May 5 on average. For Red Knots, the most northerly with a midpoint at 75° N, the average departure date is slightly earlier, May 2; this may reflect more days spent later at European staging areas to prepare for a longer final flight (Piersma et al. 1990a).

Bar-tailed Godwits (*Limosa lapponica*) wintering on the Banc d'Arguin have an average departure date of April 25 (Piersma et al. 1990a). This population, estimated in 1990 at 540,000, with another 160,000 wintering farther south on the African coast in Guinea-Bissau, breeds in western Siberia, some 8300 km away from the Banc d'Arguin. The African birds fly to the Dutch Wadden Sea, where, like so many other shorebirds, they feed for a few weeks before making the final flight to their breeding range. They arrive at the Wadden Sea just when the estimated 115,000 Bar-tailed Godwits that wintered there and elsewhere on the coasts of western Europe are completing their own preflight zugdisposition; these birds, however, will fly only another 2500 km to northern Scandinavia and western Russia. Even though for the European population the average daily weight gain in April is 1.5 g, this will enable them to reach their destination, where spring comes earlier than in Siberia. The godwits from Africa that replace them in May, when the tidal flats of the Wadden Sea are more productive, may gain 5.6 g per day. For each of these populations, winter departure seems clearly linked to the productivity of their final staging area and the advent of suitable breeding conditions at their different destinations (Drent and Piersma 1990).

Where feeding conditions at the end of winter are variable, this may influence departure date. A study from 1972 to 2008 of Whooper Swans (*Cygnus cygnus*) wintering in Ireland at the Kilcolman Reserve, County Cork, found that in years when February temperatures were higher and grass began growing earlier, birds were sooner able to add the 25% of fat to their body mass necessary for their nonstop flight to Iceland. Whooper Swans are one of the heaviest birds to use sustained flapping flight; for them, the migration to Iceland is energetically costly. In years when they were able to begin feeding on new grass earlier, many departed sooner than in years with colder Februaries. Swans leave Ireland in March–May and lay eggs in Iceland in April–May, usually before plant growth has begun; for these capital breeders, winters with earlier warmth in Ireland give them an advantage in the short Icelandic nesting season, when young must be able to fly south between October and December. In the winters when Whooper Swans were most numerous at the Kilcolman Reserve, more also departed earlier, perhaps to avoid continued competition at the two times it may have been most intense—during premigratory fattening and then in Iceland when seeking the best nesting sites (Stirnemann et al. 2012).

OVERSUMMERING

Some migratory birds do not depart their wintering site at all when they should be returning for another breeding season. Oversummering has been recorded in 15 families but is most widespread among plovers and sandpipers that nest at high northern latitudes and winter far to the south. On the Banc d'Arguin, the oversummering population is

13% of the shorebirds present in winter; in Suriname it is 11–18%, in Kenya 15–20%, and in South Africa 11.5%. Most of the oversummering birds are of larger, long-lived species—fewer than 1% of the wintering Little Stints (*Calidris minuta*) remain in summer on the Banc d'Arguin, while nearly 50% of the much larger Black-bellied Plovers (*Pluvialis squatarola*) and Whimbrels (*Numenius phaeopus*) do not leave (van Dijk et al. 1990).

Oversummering may be justified for any shorebird that is not in fit condition to depart on the long migration north at the end of winter. The Arctic nesting season, from early June until late August, requires a prompt arrival. Birds that initiate breeding late are not likely to fledge young before food diminishes with the early onset of the Arctic winter. The smaller, more short-lived shorebirds are least likely to oversummer, as found on the Banc d'Arguin, because the amount of food they must find to fuel their prealternate molt and their flight north is much less than for larger species. In addition, the time they need to complete a reproductive cycle is less than for larger birds. Finally, because the larger species live longer, they lose less by missing a breeding season.

Several factors may make a shorebird unprepared to depart at the end of winter. Injuries, illness, sterility, hormonal imbalance, reduced prey availability, foraging inexperience in immature birds, and parasite loads may all be involved. Poor condition for departure may be indicated by failure to undertake or complete a prealternate molt, which requires increased feeding. In June on the Banc d'Arguin, 75% of the oversummering Whimbrels and 97% of the Eurasian Curlews (*Numenius arquata*) are still molting. Some 70% of the remaining Dunlins (*Calidris alpina*), 58% of Black-bellied Plovers, and 44% of Sanderlings are in basic plumage, not having molted at all (van Dijk et al. 1990). In northern Venezuela, most oversummering Greater Yellowlegs (*Tringa melanoleuca*) have a delayed or no prealternate molt (McNeil et al. 1993).

Most oversummering shorebirds are immature and are probably less able foragers. Of the species that could easily be aged on the Banc d'Arguin in June, 95% of Ringed Plovers (*Charadrius hiaticula*), 83% of Ruddy Turnstones, and 76% of Eurasian Oystercatchers were first-year or subadult birds (van Dijk et al. 1990). A study in South Africa comparing feeding success of Curlew Sandpipers (*Calidris ferruginea*) in March, before most depart, found that immature birds had 50% of the intake rate of adults. In Mauritania (20° N) and Mauritius (12° S), immature Curlew Sandpipers deposit some fat by March, but in South Africa (33° S) they do not. For inefficient feeders with very long migrations broken at only a few staging areas, the extra time they would have to spend to gain as much as experienced adults would set them farther behind at every stop. This is reflected in the pattern of more oversummering shorebirds at the southern end of their winter ranges (Hockey et al. 1998).

At the Langebaan Lagoon on the Atlantic coast of Cape Province, South Africa, at 33° S, most migrant shorebirds present at any season are immatures. The total of oversummering birds is 11.5% of the wintering population, but this varies by species from 2.5% to 30.1%. Recoveries of banded birds show that many come from the Taymyr Peninsula in Siberia. For Black-bellied Plovers, Eurasian Curlews, Bar-tailed Godwits, Curlew Sandpipers, Red Knots, and Ruddy Turnstones, there is a three-year pattern to the numbers of oversummering birds; these are greatest in the year following high

At the southernmost end of the winter range of many shorebirds, immatures like these Curlew Sandpipers may remain over the following summer rather than make the long trip back to their breeding range.

productivity in Taymyr. The most successful years for these Arctic nesters are at the peak of the three-year population cycle for lemmings (*Dicrostonyx torquatus* and *Lemmus sibiricus*), when Arctic foxes (*Alopex lagopus*) prey more on these rodents than on the eggs and young of ground-nesting shorebirds and waterfowl. Oversummering numbers of Whimbrels and Greenshanks (*Tringa nebularia*), which breed in the temperate zone, do not show this fluctuation. It may be that the higher numbers of shorebirds in South Africa following summers of greatest production in the Arctic increase competition to a level that leaves more immature birds unfit to migrate north. They remain at Langebaan Lagoon through the South African winter, where they are joined by immature Ruddy Turnstones that have shifted from the more exposed and storm-prone open coastline of western Cape Province (Underhill 1987).

Finally, a study in northern Venezuela, where most oversummering shorebirds are also immature first-year birds, found that many were heavily infested with a range of debilitating internal parasites. These can penetrate the intestines, preventing any premigratory fattening, as well as the thorax, abdomen, and liver. Birds collected between February and June showed more extensive infestations with each month, indicating how capacity to migrate may diminish the longer a bird remains. The smaller number of oversummering adults suggests that surviving birds may acquire some immunity resulting from previous infections (McNeil et al. 1993).

Oversummering is also a regular feature in the life cycle of a few larger long-lived birds in which subadults have little chance of breeding successfully. Herring Gulls (*Larus argentatus*) and other large gulls, including Great Black-backed Gulls (*L. marinus*), do not attain adult plumage or breed until they are four or five years old. North American Herring Gulls winter farthest south in the year of their birth and increasingly close to their natal colony in each successive winter; they do not return in these years to the latitude where they were born (Pierotti and Good 1994). During those summers, they may work their way north in stages to the area where they will spend the following winter. Anyone enjoying an Atlantic beach in summer will see the variety of age-specific plumages in the immature

The different plumage of Herring Gulls each year until they reach adulthood reveals how many of each age class spend the summers at various latitudes of their winter range before ultimately going north to breed.

gulls, but more recoveries of banded immature birds are necessary to determine how far most of them are from their birthplace, which may be as far north as Arctic Canada. It is not clear that limited space in colonies—for actual nesters or for subadults—was ever the evolutionary reason gulls did not mature or return to breed for several summers, since in the twentieth century both Herring and Great Black-backed Gulls substantially expanded their breeding ranges southward, establishing many new colonies at sites that had always been available.

Among birds of prey, most young Ospreys (*Pandion haliaetus*) of migratory populations remain at their initial wintering site for 18 months; a few return to their natal area in their second year. Eastern North American Ospreys winter in the West Indies and South America, those from the Midwest from Mexico to South America, and western birds in

Young Ospreys of migratory populations spend at least 18 months at their initial wintering site; many remain there an additional year.

Mexico and Central America (Poole at al. 2002). On coasts, they favor estuaries, especially those with tidal flats at river mouths, where schooling fish in shallow water are easy prey, but Ospreys also winter on the rivers and lakes of the Amazon basin. Adults are known to return in successive years to the same wintering area where they spent their first two winters as immatures. The oversummering immature birds may, however, move locally to preferred habitat after adults depart the following spring (Poole 1989:59–61). Ospreys occasionally nest within their winter range in South Africa (Dean and Tarboton 1983); perhaps these are birds from comparable Northern Hemisphere latitudes that during their initial 18-month stay are prompted by living through the same photoperiod, albeit in reverse.

DEPARTURE

Finally comes the actual departure. Birds ready to leave will make their first flight in favorable weather, waiting out any period of rain, fog, contrary winds, or other factors that would impede them. The time of day any bird leaves its wintering site is usually the same as the time it will launch itself on the later laps of its migration. Thus, soaring birds like raptors that fly only over land depart by day so they can ride on thermals. Swallows, which feed on the wing as they travel, are also typically diurnal migrants, but those leaving their winter range on the south side of the Sahara cross it at night, when the air is cooler; the swallows would in any case find few flying insects over the desert during the day. Birds traversing large expanses where they cannot land, such as the Sahara or the Gulf of Mexico, usually begin at night but may need to continue into the following day to reach a place to rest and feed. The millions of passerines and other birds that depart from Mexico and Central America in spring to fly over the Gulf of Mexico leave after dusk and reach land on the north coast of the Gulf the following afternoon; arriving in daylight gives them a few hours to feed in the trees on the barrier islands or, if the winds are favorable, to continue farther inland to more extensive woodlands.

The birds that preferentially fly at night usually cease feeding and rest before dusk, launching themselves an hour after sunset. Nocturnal migration has several advantages, especially for long-distance migrants: lower temperatures prevent overheating and dehydration; less energy is required to fly in cooler, denser air; wind speeds are lower; on clear nights, the stars may be used for navigation; and aerial predators like hawks and falcons are not active. Few birds fly through the entire night, unless they are over water or desert; most land some hours before dawn, giving them time to rest and then the following day to feed.

Diurnal migrants usually fly for a few hours early in the morning, sometimes starting an hour before sunrise, and then again near the end of the day. Cooler air and less wind may be relevant to them as well, although these birds rarely fly as high as nocturnal migrants. Finally, some birds, such as waterfowl, may fly at any hour; they typically take off around sunset, while geese more than ducks are often seen overhead during the day. The extent to which these birds may have fed at night, as some waterfowl frequently do all winter, may correlate with their choice of departure hour.

Some birds travel singly, while waterfowl are usually in the pairs they established during the winter, and these may join others to form flocks. Geese and cranes leave with the young of the previous year that have survived the entire winter with them. Other birds that may feed solitarily by day gather into groups of greater or lesser cohesion just before their departure. Just as the geese and cormorants that fly in a *V* or an echelon, a line with a slight diagonal, gain some lift from the upward-directed airflow created by the wing beats of the bird ahead of them, smaller birds that form dense flocks—shorebirds, blackbirds, starlings, and finches, among others—get the same benefit when closer together and are thus not likely to undertake a long flight before they have joined a flock. On landing to feed and rest, the association may evaporate, and the next flocks to form may contain different birds, depending on each individual's readiness.

Actual departure and any social organization are hardest to observe in birds that take off after dark. In Colombia, Blackburnian Warblers (*Setophaga fusca*) in high Andean secondary forest of native trees spend the winter as members of local mixed-species flocks, at some sites with up to seven Blackburnians in a flock but more often only one or two. Where these rapidly diminishing forests are still extensive, the Blackburnians in some flocks increase in spring, numbering as many as 50 (De La Zerda Lerner and Stauffer 1998). Evidence that flocks of warblers may in fact depart as a cohesive group is scarce and cannot be inferred from the arrival of flocks at landfalls, because these assemblages could have formed when the birds were aloft, connected by flight calls. Occasionally, warblers landing together on Gulf Coast barrier islands show signs that they have left Central America together. Two groups of Prothonotary Warblers (*Protonotaria citrea*), one of four, the other of seven, caught together in the same mist net on a barrier island in Mississippi all bore residue from orange sections on their foreheads. In Central America,

Small groups of migrating Prothonotary Warblers caught together in mist nets on a Mississippi barrier island in spring all had residue from orange sections on their foreheads, indicating that they probably fed together before they took off from Central America and stayed together through the night and the following day of their long flight over the Gulf of Mexico.

Prothonotaries are known to feed on insects attracted to open oranges; it seems likely these birds were all in the same plantation and flew north together (Moore 1990).

In contrast, among diurnal migrants that also gather in flocks before departure, the first major movements may be conspicuous. Mississippi Kites (*Ictinia mississippiensis*) winter from Bolivia and southern Brazil south to northern Argentina. There, in small, thinly scattered groups, they feed on flying insects. On February 20, 2002, a densely packed flock of at least 10,000 Mississippi Kites was seen steadily flying north over Fuerte Esperanza, in Chaco Province in northwestern Argentina. They flew at 80–100 m off the ground in a ribbon 70 m wide and 400 m long, or 40 kites wide and 250 long. This flight occurred on an overcast afternoon with southerly winds that followed several days of rain. Here, at the southern edge of their winter range, the kites may have gathered over days and waited in an ever-larger group for the rain to end and favorable winds to assist them. This flock may have been 5% of the Mississippi Kite's entire population (Areta and Seipke 2006).

Changes in behavior by the day and by the hour in birds departing their wintering site are best known from shorebirds, easily visible as they feed or rest on tidal flats and conspicuous when they ascend above their usual flying height over the flats or the sea. At Farewell Spit, at the northwestern tip of South Island, New Zealand, at 40° S, Bar-tailed Godwits and Red Knots are common wintering shorebirds. They leave for Australia in March. Most take off on an evening when tides are rising and the birds cannot continue feeding; departing flocks of knots average 40 birds, and godwits 53. They prefer a slight tailwind but sometimes fly with headwinds of up to 12 km per hour. Most departures are thus after a passage of a low-pressure system or at the approach of a high-pressure system, which both produce favorable southerly winds. During one monitored period of five days with northwest winds of 28–37 km per hour, no birds departed (Battley 1997).

The first signs of migratory restlessness at Farewell Spit on a day in early March with appropriate winds are a few small groups of knots flying east over the tidal flats for about 4 km, ascending slowly but then turning west and returning to land where they began, all in about five minutes. In one flock of 990 roosting knots facing into the wind, about 26 birds in full breeding plumage were seen walking through the flock and calling. Over the next hour, several groups of as many as 44 flew away from the flock and settled 200 m away but were not seen to migrate. Finally, a group of 28 from this smaller flock took off and ascended and departed; three from this group returned within two minutes and resumed feeding with the remaining birds. Among the godwits, small restless groups took short flights away from a roosting flock and then did not disperse to feed over the tidal flats like the others. When each of the separate groups of knots and of godwits took off, they made a small series of zigzags in the first few minutes of their ascent and then formed either a *V* or an echelon (Battley 1997).

On the Banc d'Arguin, where departure behavior of 13 shorebird species was studied, all migrants took off in late afternoon or early evening, irrespective of the tidal cycle. This may give birds orientation information both from the setting sun and, soon after, from the stars at a latitude where there are never clouds. In addition, they avoid solar radiation, higher temperatures, and water loss. Finally, they may reach their first staging area farther north on the African coast by the next morning. Nearly all departing flocks flew north-northwest into the prevailing wind as they ascended; with telescopes, they could be seen up to a height of 1.5 km and were still ascending. At this altitude and above, the winds are southwesterly; here, if the birds continue orienting northwest, wind drift will take them in the right direction to parallel the African and European coastlines (Piersma et al. 1990b).

While still on the tidal flats, many flocks of each species fly up and land again before definitely taking off—some ascending high enough to attain a formation but suddenly circling and returning. All species are very vocal, using repeated series of staccato calls. The 20% of takeoffs that are aborted may be of birds trying to recruit others, since these flocks are typically smaller than those that actually depart. The average flock in its final configuration numbers 37 birds—nearly always of a single species, since the flight benefits of flocking depend on birds being the same size and flying at the same speed. Aloft, these flocks sometimes split into two or more groups or are joined by other birds. Larger species tend to form larger flocks and move at higher airspeeds. When the birds take off in a cluster they are highly vocal, but as soon as they form a *V* or an echelon they become silent (Piersma et al. 1990b).

On Laysan Island in the Pacific, where departing Bristle-thighed Curlews must continue flying 4000 km to reach western Alaska, most adults leave in early May. From the territories they have held since the prior July or August on different parts of the island, they come to a sand flat on the north shore where as many as 156 have been observed at once. Over a few days as groups leave, the number at this staging area declines rapidly; after May 13, only oversummering birds remain on Laysan. While the birds are on the sand flat, they are restless and very vocal, with flocks flying back and forth between

Flocks of Bristle-thighed Curlews leaving Laysan Island for Alaska circle several times to gain altitude and then form a *V* or an echelon as they start their 4000 km flight north. Young birds wait until the end of their second winter to depart.

there and the south side of the island and occasionally out to sea and back. Unlike the shorebirds in New Zealand and Mauritania, curlews here leave at any hour from 9:00 a.m. until after 5:00 p.m. Before taking off, they stand quietly in a loose flock facing north to the ocean, occasionally walking or vocalizing but not feeding or drinking. Some have been there as briefly as seven minutes, others more than an hour, with larger flocks remaining longer than small ones. The birds then grow restless, with a few flying up and landing again if others do not follow. When a group finally takes off, it circles the island a few times as it ascends, calling, and then heads north or north-northwest, with the loose cluster forming itself into a *V* or an echelon. Sometimes, when the surface winds come from the east or northeast, these flights are aborted, usually within 10 minutes and before the birds are out of sight. Other times, a few of the birds in a departing flock return, usually subadults that will oversummer, waiting through another winter for their own turn (Marks and Redmond 1994).

SUMMARY

All migratory birds need to depart their wintering area at the appropriate time to return to their breeding site when they can maximize their reproductive potential. Since favorable conditions in the breeding range, especially at high latitudes, may be brief, accurate departure timing at the end of winter is essential. For short-distance migrants, local conditions may be similar enough to what these will be at their destination, but the farther birds winter from their breeding range, the more they must depend on internal mechanisms to set the spring migration schedule. For most birds, changes in day length are the cue that prompts hormonal stimulants inducing migrants first to increase food consumption and then to depart. Birds wintering near the equator, where changes in day length are minimal, seem to have timetables that are most strictly under internal control.

Three key processes take place in wintering birds before they depart: molt, more in some species and populations than in others, fattening, and gonadal development. Molt is most likely controlled endogenously or by food availability. Away from the equator, fat deposition is more directly linked to changes in day length, taking place mainly in the last 10–12 days before departure. This, in turn, stimulates migratory restlessness. The extent of weight gain depends on whether migrants can easily refuel after each short flight or whether they will be traveling long distances over water, desert, or other habitats that offer no opportunities. While gonads may begin to develop during winter, this is usually last in the sequence, because migratory birds that have already increased their weight do not benefit from the expansion of any organs not assisting their flight.

Actual departure varies within species wintering in the same region, by sex, age, physical condition, and destination. Males generally depart sooner, to arrive at their nesting site ahead of females and establish a territory. Adults of both sexes, more experienced and perhaps in better condition than first-year birds, may also depart sooner. Finally, the birds traveling least far, to where spring arrives sooner, will leave before those going to higher latitudes where spring comes much later.

Some birds choose not to leave their wintering area. Oversummering has been recorded in 15 families but is most widespread in shorebirds, especially among the larger species that have particularly long migrations to High Arctic breeding sites. Individuals in poor condition because of injury, hormonal imbalance, reduced prey availability, or some other problem may not be stimulated to migrate. Because these larger species live longer, they can afford to miss a nesting season if that increases their long-term survival.

Finally comes the actual departure. Birds ready to leave will make their first flight in favorable weather, at the same time of day they will take off on further legs of their migration. Thus, soaring birds and those that can feed on the wing depart in daylight, while nocturnal migrants take off after dark. Some nocturnal migrants that are solitary or form loose groups while feeding may gather in flocks before departure; whether these have much cohesion during the actual flight or persist over subsequent days is unknown. Shorebirds, which regularly migrate in flocks, gather in nervous, noisy groups on beaches, sometimes taking off and returning until all the flock seems ready. In some parts of the world, they depart at the end of the afternoon, getting some orientation information from the setting sun and then from the stars that will soon follow.

CHAPTER 9

CONSERVATION

THE POPULATIONS AND DISTRIBUTIONS of birds change over time, slowly or rapidly, affected by many circumstances. When, thousands of years ago, people first began altering the landscape for their own benefit, they began adding their impact to natural changes. This has only accelerated and diversified with the growth of the human population and the advances in technology that enable our species to make further changes to the world around us. Today, the increasing transformation and loss of natural lands on which birds and all other forms of life depend, as well as changes in the chemistry of air, water, and soil as a result of the dispersal of pollutants, often far from their source, and, finally, the global impacts of a warming climate are all accelerating the breakdown of ecosystems and challenging the ability of their inhabitants to adapt and survive. All over the world, the populations of many bird species are declining, some to a level indicating likely extinction if measures cannot be taken to restore the features on which these birds depend.

For birds living in seasonal environments, both the ones remaining in the same place and those moving over the course of the year, one of the first priorities of conservation science, therefore, is understanding which seasons and locales are having the greatest effect on any population changes. Among resident birds, long-term declines are generally due to loss of the habitat on which they depend year round rather than to any strictly seasonal threats; extreme winter weather may have a short-term impact, but populations usually rebound within a few years. The populations of migratory birds, however, are influenced by conditions in more than one part of the world, and annual and long-term changes may be due to factors in the breeding or winter ranges or along migration routes—or a combination of these.

The distinctive aspects of migratory and sedentary life cycles must be considered. Among woodland passerines, young of migratory species survive their first year half as well as adults, and young of sedentary species usually only one-quarter as well; the challenges of a cold winter where adult birds occupy the best sites may therefore be greater for inexperienced birds than two migrations and a season in a new environment (Sherry and Holmes 1995). For some highly migratory birds of prey, mortality is most intense on migration, but these brief periods of the annual cycle are balanced by longer portions of the year in breeding and wintering areas; there, the per-day rates may be lower but the total seasonal rates of loss equal those experienced on migration. Satellite tracking of Ospreys (*Pandion haliaetus*), Marsh Harriers (*Circus aeruginosus*), and Montagu's Harriers (*C. pygargus*) from Sweden and the Netherlands to West Africa found that mortality was six times higher per day during migration, but that this was equaled by mortality during the much longer stationary periods (Klaassen et al. 2014).

The conservation biologist Edward O. Wilson (2016:57–58) has helped popularize the useful acronym HIPPO for the major factors, in order of impact, challenging the preservation of biological diversity:

- Habitat loss
- Invasive species
- Pollution
- Population growth
- Overharvest

This sequence does not necessarily match in priority the phenomena experienced by birds in winter, but it is a useful framework for categorizing what they encounter. The complex interplay of these five factors in the breeding range, on migration, and during winter requires that each species and discrete population be studied individually to determine causes of decline and appropriate conservation measures. For many species and regions, there is still too little information from which to draw conclusions. Some species are easier to study, some regions better populated with scientists. The themes and examples given here can be only a sample of what is currently known—and likely to be revised as challenges accelerate, solutions are implemented, and additional research brings a deeper understanding.

HABITAT LOSS

SONGBIRDS IN THE NEOTROPICS

In the Caribbean islands, Mexico, and Central America, where most North American landbird migrants winter, forests have been substantially reduced in the last several decades, and the population declines of these species are often attributed to this portion of the annual cycle. But for the many species restricted to nesting in mature forest that in winter exploit widely available second growth as well as forest, the loss and fragmentation of forests in North America may be an equal or greater factor. In the Guanica Commonwealth Forest in Puerto Rico, a tract of high-quality tropical dry forest that has not changed in decades, the three commonest wintering warblers—American Redstart (*Setophaga ruticilla*), Black-and-white Warbler (*S. varia*), and Ovenbird (*Seiurus auricapillus*)—had by 2012 all declined by two-thirds compared with their populations 20 or more years earlier. Their survival rate over the winter did not diminish, suggesting that the decline was not due to any change in local conditions. Four other warblers, Nothern Parula (*Setophaga americana*), Prairie (*S. discolor*), Hooded (*S. citrina*), and Worm-eating (*Helmitheros vermivorum*), regular in the forest into the 1990s, were not found at all during some winters of the following decades (Faaborg et al. 2013).

Despite the persistence of some wintering sites intact, the overall impact of forest conversion in the Neotropics to farms, ranches, and other forms of development is severe, because the region's entire landmass is so much less than the breeding range used by these migrants. And since many of the wintering species occupy only a fraction of this region, their entire winter range may be only one-fifth or one-tenth of their breeding range, and the suitable habitat within that fraction—losses to development aside—even less (Terborgh 1989:79). Thus, in a region where the wintering birds are far more densely concentrated than during the breeding season, the loss of valuable habitat has much greater impact. Much of the research on birds wintering in the Neotropics has therefore focused on impacts of converting natural habitat to agriculture and on the ability of passerines to use land in different forms of cultivation.

The hope is to avoid more cases like the probable extinction of Bachman's Warbler (*Vermivora bachmanii*), for which the last confirmed sightings were in 1958–1961. Bachman's Warblers once bred in canebrakes and bottomland forests from Arkansas and Missouri east to South Carolina. While that habitat has diminished, much survives, but in the warbler's winter range, Cuba and the Isle of Pines, 85% of the land is now open, agricultural, or urban and suburban. There, Bachman's Warblers were found in a variety of wooded habitats; these, however, may long since have become too fragmented to support a viable population. Much about this bird's decline is speculative, but the loss of habitat in its small winter range has been much greater than in what was once its extensive breeding range (Hamel 1995).

A survey of 11 natural and modified habitats in Panama, representative of the variety of land conversion types throughout the range of wintering Neotropical migrants, found that sugarcane fields and plantations of Caribbean pine (*Pinus caribea*) had virtually no conservation value for birds. Shade coffee plantations and gallery forest corridors had the highest value, but even the best of these were not adequate for true forest specialists (Petit and Petit 2003). In the Tuxtla Mountains of Veracruz, Mexico, forest fragments held 46 species, including, in order of abundance among the wintering birds, Hooded Warbler, Wood Thrush (*Hylocichla mustelina*), Kentucky Warbler (*Geothlypis formosa*), American Redstart, Wilson's Warbler (*Cardellina pusilla*), and Summer Tanager (*Piranga rubra*), which collectively were 50% of the individual birds present. At shade coffee and cacao plantations, 44 species were found, with most of the same migrants accounting for 50% of the total numbers. In contrast, in fields of jalapeño and corn, which held only 18 species, the commonest were Dickcissels (*Spiza americana*), Indigo Buntings (*Passerina cyanea*), and Scissor-tailed Flycatchers (*Tyrannus forficatus*). Pastures had even fewer (Estrada and Coates-Estrada 2005).

Loss of habitat in the Bachman's Warbler's limited winter range in Cuba and the Isle of Pines is the likely cause of its extinction.

Several other studies, from Mexico to Costa Rica and in Colombia, have confirmed that shade coffee and cacao provide habitat closest to natural forests. Foraging generalists and those that feed in the canopy are relatively abundant, but understory birds are less common, in part because coffee plants host fewer arthropods than the trees most frequently used as shade plants (Tejeda-Cruz and Sutherland 2004). In Chiapas, Mexico, shade coffee supports 75–80% of the species native to pine-oak woods (Perfecto et al. 1996). There is, however, always the caveat that shade coffee does not support resident species that are particularly sensitive to habitat disturbance and fragmentation. And, throughout the coffee-growing zone, farmers continue to convert to the higher-yielding sun coffee, which hosts very few birds (Perfecto et al. 1996). In Tabasco, Mexico, where cacao plantations are shaded by as many as 60 tree species, creating a multistoried effect, the wintering migrants are mainly forest species, albeit not as many as in actual forest. The plantations adjacent to forest hold more migrants than do isolated groves (Greenberg et al. 2000).

In the northern Andes of Colombia, where forest is replaced not only with crops and pasture but also by tree plantations, the Blackburnian Warbler (*Setophaga fusca*) is the commonest wintering species. It is found in large remaining forests as well as in small patches, hedgerows, and riparian habitat, but never in nonnative pine plantations and rarely in those of eucalyptus (De La Zerda Lerner and Stauffer 1998). Other parts of the Colombian Andes have been converted from forest to silvopastures, a mix of grazing pasture with trees. Compared with shade crops such as coffee and cardamom, silvopastures support smaller and less diverse flocks with fewer Neotropical migrants, only 20% of flock members. A further indication of the silvopasture's modest value for migrants is that, of Blackburnian Warblers, the commonest migrant in the silvopastures, 27% were male, compared with 40–51% in nearby shade crops and secondary forest; as with many other warblers, wintering male Blackburnians occupy the highest-quality habitat. Only treeless pasture supports fewer birds than silvopastures in the Colombian Andes (McDermott and Rodewald 2014).

To preserve and expand habitat used by wintering birds, conservation organizations are working with local communities to encourage cultivation of beneficial crops and restoration of natural landscapes. Certified shade-grown coffee can now be bought in the United States and many other countries. In Guatemala, the American Bird Conservancy (ABC) and FUNDAECO are helping farmers establish hardwood tree plantations and grow additional shade crops such as black pepper and cardamom in areas used by the declining Golden-winged Warbler (*Vermivora chrysoptera*) (American Bird Conservancy 2016). Similarly, in Colombia, ABC and two local partners, Fundación ProAves and Fondo para la Acción Ambiental y la Niñez, planted a conservation corridor 9.6 km long using 500,000 seedlings of native trees from nurseries of nearby reserves. This benefits more than 150 bird species, including wintering Cerulean (*Setophaga cerulea*) and Golden-winged Warblers (American Bird Conservancy 2015).

Birds that winter on natural grasslands are also losing ground. As with North American forest birds wintering in the Neotropics, the wintering area available to migratory grassland birds is much smaller—16.3 million ha—than the 89.4 million ha of breeding area. About

85% of the migratory grassland birds of the northern Great Plains of the United States, including 7 diurnal raptors, 3 owls, and 10 sparrows and buntings, winter in Mexico's Chihuahuan Desert. Among these are McCown's Longspur (*Rhynchophanes mccownii*), with a breeding population that has declined 5.2% annually between 1966 and 2011, and the Chestnut-collared Longspur (*Calcarius ornatus*), which has declined 4.2% annually in the same period (Pool et al. 2014).

Between 2006 and 2011, the Valles Centrales, a region of Chihuahuan Desert in Mexico comprising at least 2.7 million ha, had a 6% annual increase in cropland, equaling a total loss of 69,240 ha of grasslands and shrublands. If that rate continues, the expansion could eliminate the remaining low-slope valley grasslands in the Valles Centrales by 2025. Conversion of grasslands to crops is also occurring in four other Chihuahuan Desert states of northern Mexico (Pool et al. 2014). Part of the conversion to cropland is driven by the failure of local cattle ranches after intense grazing has depleted the soil. To restore the grasslands that support both cattle and birds, Pronatura Noreste, a Mexican conservation organization, and ABC have worked with ranchers to initiate rotational grazing systems that mimic the original use of the land by herds of bison. When these are implemented, grasslands recover rapidly. Nearly 175,000 ha of ranchland in the Valles Centrales have thus far regained their value for wintering birds (Sander 2015).

AFRICA

Most European and Asian migrants to Africa winter in arid savannas. Drought and desertification are believed to be the major causes of declining populations (Newton 2004). An analysis of 30 pairs of closely related European birds, with one a resident or short-distance migrant and the other a migrant to the Sahel, found that between 1970 and 2000 the populations of the long-distance migrants had all declined more than those of the less migratory species (Sanderson et al. 2006). Unlike the Neotropical migrants, which are mostly sedentary once they reach their destination, many European and Asian birds that winter in African savannas typically move several times over the course of the season to occupy a series of distant sites—thereby requiring a broader conservation approach. Montagu's Harriers wintering in the Sahel of Senegal usually move four times over six months as the region dries and grasshoppers, their major prey, diminish. They are faithful in future years to this sequence of wintering sites. During the second half of winter, they spend more time flying in search of food, indicating that conditions have worsened. Harriers in territories with the poorest conditions depart later and arrive later at their breeding site, where they are then also likely to be limited to poorer habitat (Schlaich et al. 2016).

Other species wintering in Africa may require substantially larger territories than in the breeding season. Individual Egyptian Vultures (*Neophron percnopterus*) wintering in the Sahel have home ranges 33 times larger than they do as pairs breeding in Europe. A comparison of corticosterone levels in feathers of Egyptian Vultures taken during the breeding season in Spain and in winter in West Africa showed that this indicator of stress was substantially higher in the feathers grown in Africa (Carrete et al. 2013).

The mangrove forests on the coast of West Africa host 10% of the global population of the Subalpine Warbler and many other wintering European birds. As in other parts of the tropics, mangrove forests here are being converted to various agricultural purposes that do not support birds.

While many wintering landbirds find the agricultural landscape south of the Sahara sufficiently similar to the original natural savanna, other birds concentrate in the narrow belt of mangrove forests on the West African coast between 20° N and 7.5° N. These fairly impenetrable forests have only recently been surveyed and their conservation importance revealed. The 8400 sq km of mangroves forming a thin line down the long coast support an estimated 5 million European Reed Warblers (*Acrocephalus scirpaceus*), some 30–50% of the European population, and 0.9 million Subalpine Warblers (*Sylvia cantillans*), 10% of the species' population, as well as substantial numbers of other wintering warblers. These species collectively outnumber resident insectivorous birds in the mangroves. Even here, however, birds are not immune from the effects of drought, since local conditions fluctuate, matching those in the Sahel; during dry years, rivers carry less water to the coast and the salinity in the mangrove forests is more than the trees can tolerate. In the following spring, migrant mortality crossing the Sahara is greater than in wet years. In addition, mangrove forests in West Africa, as elsewhere, are being converted to various agricultural purposes (Zwarts et al. 2014).

SHOREBIRDS

Worldwide, shorebird populations have declined substantially. Of the 237 populations with trend data, 52% are in decline and 8% are increasing (Delany and Scott 2006). The causes vary from species to species and region to region and span the birds' entire annual cycle. A horizon scan of conservation issues affecting highly migratory shorebirds considered ongoing, episodic, and long-term future threats; it concluded that for wintering populations the chief immediate concerns come from habitat conversion and various forms of pollution. The many species that in winter—and perhaps even more so on migration—concentrate at a few extremely productive estuarine sites are especially vulnerable. There, reduction in sediment flow caused by dams, as well as land reclamation to construct ports and industrial facilities and the more widespread drainage of adjacent wetlands for agriculture, may all play a role (Sutherland et al. 2012).

For the shorebird species that disperse more widely in winter, in mangrove forests and grasslands, conversion to intensified agriculture is also reducing habitat. In Asia, especially Thailand, the Philippines, and India, destruction of mangrove forests for building development, agriculture, and commercial shrimp farming is widespread (Sutherland et al. 2012). At the Kadalundi-Vallikkunnu Community Reserve, an internationally important estuarine wetland in southwestern India, censuses in mudflats, mangroves, sand beaches, and shallow water from 2005 to 2012 found that of 18 wintering shorebird species, 8, including 3 of the 4 most abundant, declined as waste disposal, illegal sand mining, and disturbances from nearby development increased. The shorebirds shifted from their preferred mangroves and mudflats to less productive sand beaches (Aarif et al. 2014).

Some long-term surveys of shorebirds at unaltered wintering sites that nevertheless show population declines suggest that for those populations the key factors are elsewhere. In Namibia, at the two largest coastal wetlands in southern Africa, 4 of the 12 long-distance migrants—Common Ringed Plover (*Charadrius hiaticula*), Ruddy Turnstone (*Arenaria interpres*), Little Stint (*Calidris minuta*), and Red Knot (*C. canutus*)—declined between 1990 and 2013, while the resident and short-distance migrants increased or were stable (Simmons et al. 2015).

In Australia, at Moreton Bay, southeast Queensland, which supports 18 migrant and 6 resident shorebird species, 7 of the migrants declined 43–79% between 1992 and 2008, but none of the resident shorebirds declined (Wilson et al. 2011). A 24-year survey (1983–2006) at several other sites in southeastern Australia found that small wintering shorebirds from Asia and Alaska declined 73%, large ones 29%, and resident Australian shorebirds 81%. Of the 10 most important wetland sites, 8 were inland, where extraction of fresh water and loss of wetlands to river management were significant factors. In some years, the inland lakes had no water at all; hence they had the greatest impact on the resident species (Nebel et al. 2008).

In the Americas, as in the Eastern Hemisphere, declines of wintering shorebird populations are associated primarily with habitat loss. In some areas, it is not clear whether birds can move from one degraded portion of an extensive wintering site to another without overcrowding. In the Upper Bay of Panama, where tidal flats support approximately 250,000 shorebirds of several species, surveys in 1998 and 2003 found similar numbers during a period when parts of the bay were lost to urbanization, factories, paving, and landfill; factors elsewhere in the annual cycle may already have reduced their populations below what the Bay of Panama could naturally support (Sanchez et al. 2006). In southern South America, the habitat of wintering grassland specialists is subject to intensified grazing practices that replace native grasses with types best suited to cattle raising (Sutherland et al. 2012). In addition, much pastureland there is being converted to cultivation of more economically profitable corn and soybeans (Terborgh 1989:151).

The wintering sites of some declining species are still not entirely known. Surveys conducted every five years by the US Fish and Wildlife Service for wintering Piping Plovers (*Charadrius melodius*) continue to reveal important locations. The 2011 survey discovered more than 1,000 previously unknown wintering plovers in the Bahamas; these birds come from the Atlantic coast breeding population from Newfoundland to North Carolina, which numbers 3,000 of the world total of 8,000 Piping Plovers. Some of the plovers that were individually banded were found as many as five years later on the beach where they had initially been banded, or within 6 km (Gratto-Trevor et al. 2016). Thanks to efforts of Bahamian conservation organizations and the National Audubon Society, the most important of the newly found sites was designated a national park. The 2016 survey found an additional 200 plovers in the Bahamas and nearly 100 on the few islands of Turks and Caicos that were visited (Wojtowicz 2016).

Just as many Neotropical migrant landbirds can adapt to certain forms of agriculture, some other cultivated and managed habitats can serve as alternative feeding habitats for shorebirds. These include rice fields, salt ponds, coastal grazing marshes, and fishponds—provided they are managed for appropriate amounts of water at different seasons. In Cuba, where rice is the most important crop after sugar, rice farms on the south coast support significant numbers of 11 shorebird species and 3 duck species during winter. These farms are adjacent to coastal wetlands; together they form a conservation unit (Mugica et al. 2007). In the Mediterranean basin, salt ponds cover over 100,000 ha and are often classified as functional wetlands because their biological richness is so great. Many shorebirds use them interchangeably with adjacent tidal flats. While 75% of the Mediterranean salt ponds have some form of legal protection, many of them are private property and are abandoned, thereby losing their value for lack of water management. Others are converted to aquaculture tanks too deep for shorebirds (Masero 2003).

Suitable roosting sites are also an essential feature that can be managed for shorebirds wintering at estuaries where they must leave feeding areas for higher ground when the tide covers mudflats. At large estuaries, the distance to safe high ground can be far, and travel back and forth energetically costly. For Red Knots in the Dutch Wadden Sea, the commute consumes 10% of their daily energy expenditure (Piersma et al. 1993). In the Tagus Estuary of Portugal, a study found that the density of Dunlins (*Calidris alpina*) on foraging areas declined significantly with distance to the nearest roosts, which are often salt ponds. Fewer than 20% of the birds foraged more than 5 km from the roosts they used. The loss of some roosts around the estuary would make half the available feeding areas too distant to be used by Dunlins and possibly by other wintering shorebirds. Studies elsewhere, in the United States and Europe, have identified the same problem of loss of roosting sites eliminating shorebird use in otherwise productive estuaries. Reclaimed wetlands, maintained salt ponds, and dredged sand flats above the high tide line at appropriate distance from feeding areas could all enhance wintering shorebird populations (Dias et al. 2006).

Elsewhere, roosting sites may need to be bought and protected by conservation organizations. Isla de Chiloé, Chile, is the wintering area for 99% of the south-central and western Alaskan population of Hudsonian Godwits (*Limosa haemastica*) and is important also for Whimbrels (*Numenius phaeopus*). The National Audubon Society and the Centro de Estudio y Conservación del Patrimonio Natural have secured a roosting site of 6.25 ha that holds thousands of shorebirds during high tides. The organizations are also working with local landowners to develop best management practices for watersheds and agriculture in order to reduce pollution of the tidal flats where the birds feed (Gonzalez 2014).

WATERFOWL

Few of the world's ecosystems have been altered as systematically and comprehensively as wetlands. In the middle and lower latitudes of the Northern Hemisphere, these are the primary habitat of wintering waterfowl. A global survey of wetland status from 2000 includes the following examples of wetland loss, due overwhelmingly to the expansion of agriculture, in areas important for wintering waterfowl (O'Connell 2000):

- North America: 54% of US wetlands lost; 35% of Mexican wetlands lost
- Europe: Great Britain, 84% of lowland raised bog lost
- Middle East: Israel, Lebanon, Syria, almost all natural freshwater wetlands drained for agriculture
- Neotropics: East Caribbean, 50% of 220 coastal wetlands damaged; Colombia, 80% of mangrove wetlands lost in Magdalena Delta
- Africa: Ghana, several major rivers polluted; Lake Chad, up to 30% choked by aquatic reeds; Tunisia, 84% loss of wetlands from Merjedah catchment
- Asia: Japan, almost all rivers affected by impoundment for reservoirs and by eutrophication; Bangladesh, 3.7 million ha of wetland threatened by diversion of water in India

Surveys of US wetlands have noted that by the mid-1970s only 40 million ha remained of the estimated original 87 million ha of wetlands in the Lower 48 States, with 87% of the loss resulting from agriculture. On the Atlantic coast, the most significant losses have been between Connecticut and Maryland and in the Southeast, especially Florida. In Louisiana, coastal marshes have been sinking underwater for lack of sediment replenishment, due to the channeling of the Mississippi, and California has lost 90% of its wetlands, mostly to agriculture (Stewart et al. 1988). In California's Central Valley alone, where 5.5 million waterfowl winter, 95% of the wetlands are gone because of flood control, urbanization, and agriculture (Hagy et al. 2014). Some of the lost natural wetlands of the Central Valley, especially in the Sacramento Valley, have been compensated for by rice farms that have shifted from burning winter stubble to flooding the fields, thereby creating new habitat for waterfowl. On the Gulf Coast of Texas and Louisiana, however, conversion from rice agriculture to soy and cotton farming has eliminated substantial waterfowl habitat (Davis et al. 2014).

Perhaps surprisingly, despite these and other losses around the country of wetlands used in winter, a 2014 review noted that numbers for most waterfowl species are at or above their long-term average, and a few species are at all-time highs—bearing in mind that these "all-time" averages are based on surveys beginning in the twentieth century, after much of the North American landscape had been transformed. Nevertheless, the current strong numbers are due to effective management of breeding habitat and to the flexibility of many waterfowl species that have recently adapted in winter to feeding on agricultural land where traditional natural landscapes have been lost (Hagy et al. 2014). In other parts of the world such as the arid parts of Africa and the Middle East, where waterfowl hunters are not a political force and where there are few alternatives for expanding agriculture other than to divert water from wetlands or to actually occupy them, the loss of these once-extensive and little-exploited habitats is likely having a strong effect on wintering waterfowl.

Changes in winter habitat have also produced significant increases in several Northern Hemisphere waterfowl. Snow Geese (*Chen caerulescens*) populations increased at an annual rate of 5% between 1973 and 1997 as they expanded their winter exploitation of agricultural lands, including fields of rice, corn, and hay. This opened a vast new range at the time milder winters were reducing snow cover on the fields. During this period, traditional coastal wetland wintering sites decreased and hunting limits were raised (with the hope of reducing the midcontinental population by half [Blohm et al. 2007]), but the gains from the newly adopted food sources have continued to outpace the loss of wetlands. The increase in goose harvest, even when close to a million, has not reduced the population (Alisauskas et al. 2011). With more birds surviving the winter, breeding colonies have expanded, increasing productivity and thereby also contributing to the growing population that has damaged habitat in both breeding and wintering areas (Mowbray et al. 2000).

Early in the twentieth century, only 20 pairs of Whooper Swans (*Cygnus cygnus*) nested in Sweden. Protective measures were established in 1927, and by 2014 there were 5,400 pairs nesting, and approximately 10,000 swans, including some from Russia and Finland, were wintering in Sweden. Year-round protection, milder winters, and a similar shift in winter feeding sites have all contributed to the increase. In 1995, 70% of the wintering swans fed in wetlands; in 2005, 65% were exploiting grasslands, grain fields, and stubble (Nilsson 2014). A similar shift to agricultural lands in Denmark, where eutrophication

Wintering populations of Whooper Swans have prospered in Sweden as they have shifted to feeding from grasslands and agricultural fields, rather than in the wetlands they historically used.

from agricultural runoff has diminished the value of traditional aquatic feeding areas, has also brought an increased wintering population (Laubek 1995).

Populations and winter ranges of other waterfowl have changed without an obvious cause. Dark-bellied Brant Geese (*Branta bernicla bernicla*), which breed primarily on the Taymyr Peninsula in Arctic Russia, winter in coastal marshlands of western Europe, where they feed on eelgrass (*Zostera*) and algae (*Ulva* and *Enteromorpha*). Unlike the Snow Geese and Whooper Swans, they have not shifted their diet, but the world population has fluctuated substantially in recent decades, from 206,000 in 1983 to 329,000 in 1992 and 211,000 in 2011, as has the winter distribution. In 1982, 30% of the population wintered in France; by 2012 this had grown to 50%, with a fivefold increase at certain sites even as the extent of eelgrass there had diminished significantly (Valéry and Schricke 2013).

INVASIVE SPECIES

Plants and animals from other parts of the world often have great success in new environments to which they have been brought, deliberately or accidentally—but often at great cost to the local species that have evolved without any resistance to these invaders and cannot adapt or compete. Birds and other animals on islands that have long been isolated are especially vulnerable; many once-remote islands now harbor rats, cats, dogs, goats, rabbits, snakes, and other predatory or competitive animals, as well as insects carrying diseases like avian malaria and plants that choke native vegetation. These have had especially great impacts on the permanent resident birds and on others like seabirds that come to the islands only to nest. The effects of invasive plants and animals on the lands used by wintering birds are also varied and pervasive but do not directly affect breeding success. Invasive plants may affect habitat quality for wintering birds by reducing food or shelter, while invasive animals may compete for food or may prey directly on the birds.

Freshwater wetlands host substantial numbers of invasive plants and animals. In the lower Great Lakes coastal marshes, invasive plants are now considered the primary cause of wetland degradation. A wide variety of invasives have displaced native species that are important foods for wintering waterfowl. The invasive plants exclude native plants, reduce diversity, and modify wetland processes. Different invasive species require different management plans; for most, a multiyear Integrated Pest Management strategy using a combination of mowing, smothering, drowning, herbicides, biological control agents, controlled burning, and reseeding with native species must be deployed. Whatever immediate successes may result, however, the measures may well need to be reapplied, because most of the invasive wetland plants are wind dispersed and can easily return (Hagy et al. 2014).

What little survives of native North American prairie grasslands is also subject to invasive plants. Fire suppression and overgrazing by cattle increase the likelihood of invasives establishing themselves. These areas need to be managed by regular burning, mowing, and controlled grazing, with different regimes to retain or replicate the habitat requirements of various grassland birds, which are today one of the most imperiled groups in North America. In winter, most migratory grassland birds of the northern prairies seek more southerly versions of the same habitat structure they used in the breeding season. The coastal prairie region, which once filled 3.8 million ha from south-central Louisiana to the lower Texas coast, has now been reduced to less than 1% of its original area still in relatively pristine condition. A study in Texas that censused wintering grassland birds in plots of seven different natural and managed conditions, from native flora to those dominated by exotics, found that the three least common species—Sprague's Pipit (*Anthus spragueii*), Sedge Wren (*Cistothorus platensis*), and LeConte's Sparrow (*Ammodramus leconteii*)—all required the most pristine habitats, each with different fire regimes, plant heights, and litter depth (Saalfeld et al. 2016).

Estuaries, critical for wintering shorebirds, have the greatest number of nonnative species of all coastal ecosystems, and these continue to increase. At least 400 introduced marine species are established in US estuaries along the Atlantic, Gulf, and Pacific coasts. In Bodega Bay, California, where European green crabs (*Carcinus maenas*) became established in 1993, they were found to consume significant numbers of polychaetes and clams, important foods for Dunlins and several other species of wintering shorebirds. Where green crabs have reduced both polychaetes and larger clams, Dunlins consume the smaller and less valuable juvenile clams. While annual surveys of Dunlins in Bodega Bay between 1984 and 2008 have shown a steady decline, which may be due to many factors in other places and other seasons, new disruptions from invasive species on a wintering site add further stress (Estelle and Grosholz 2012).

Of all the invasive animals that directly prey on birds during all seasons, the most pervasive and usually least considered is the domestic cat (*Felis catus*). Recent estimates of the continental US cat population concluded there were more than 80 million house cats and another 30–80 million entirely feral cats. Studies on how many birds are killed each year by these cats put the range between 1.3 and 4.0 billion, with about 69% killed by unowned cats. These numbers are higher than the mortality estimates from any other anthropogenic source, including collisions with windows, buildings, communication towers, or vehicles, or from pesticide poisoning. Cats take even greater numbers of small mammals each year, as well as frogs and snakes, but not all of these may be as readily available in winter as birds (Loss et al. 2013).

Even well-fed house cats kill birds when let outside, as do feral ones in Trap-Neuter-Return colonies, where they are fed by volunteers. Since there are always more house cats escaping or being released, neutering wild ones cannot alone reduce the numbers of feral cats (Longcore et al. 2009). The only effective solution is to keep cats indoors or in screened outdoor enclosures, where they are protected from disease, predators such as coyotes, and collisions with cars. The American Bird Conservancy has taken the lead in public education and advocacy on this challenging issue, which, unlike most invasive species problems confronting birds in winter, requires modification of widespread human behavior rather than habitat restoration (Lebbin et al. 2010:296–298).

POLLUTION

Pollutants are the nonliving equivalent of invasive species in that they usually reach natural ecosystems through human activity. Their impacts, however, are more varied, from directly poisoning individual birds that ingest them to preventing reproduction to damaging or destroying the habitat and food sources on which birds depend, over long periods or through sudden catastrophic events. Because chemical pollutants are more pervasive than invasive plants and animals, and more difficult to remove from the environment, especially when ingested forms move up through the food chain, they have greater impact on bird survival at all seasons. And because many pollutants can travel far from their source, through air or water or in the bodies of prey such as migratory fish and in birds themselves, they may affect wintering birds that are not exposed during their breeding season.

PESTICIDES

Pesticides, including fungicides, herbicides, insecticides, and rodenticides, as well as antibiotics, are all now known to have much greater effects on nontarget species than was anticipated as these were developed. The direct effects include endocrine disruption, immunotoxicity or neurotoxicity, and altered metabolism that can result in altered growth or development, abnormal behavior, and decreased survival. Indirect effects run from reduction of food supplies to altered ecosystem functions (Constantini 2015). While in the United States and other developed countries at least some insecticides, such as DDT, have been banned, many with known negative impacts continue to be used, and some of the pesticides restricted in some parts of the world are still legal—and legal in much greater doses—in other nations. Chilean salmon farms use 120,000 times as much antibiotic per ton of salmon produced as do Norwegian farms (McGrath 2015).

Wetlands probably receive and absorb more forms of pesticide than any other habitat. In North America, the bays and marshes on the Texas coast may be the most contaminated of habitats used by significant numbers of wintering birds. After every rainfall, the pesticides applied on adjacent croplands are washed into the wetlands, where they are joined by petroleum hydrocarbons, heavy metals, and other industrial pollutants from factories and refineries nearby—producing an even more toxic brew in which it is difficult to identify the impact of each chemical. In winter, several million waterfowl depend on these wetlands, as do endangered Whooping Cranes (*Grus americana*), shorebirds, and many others (Cain 1988).

Among the pesticides in Texas coastal wetlands are organophosphates that, when ingested in small quantities, change waterfowl behavior so that birds do not feed, court, or migrate. Cotton defoliants, herbicides, and discharges from oil and gas wells may be responsible for the die-off of American shoal grass (*Halodule wrightii*) on which Redheads (*Aythya americana*) feed; between 1956 and 1975, the distribution of wintering Redheads in the Laguna Madre dropped from 84% of the population to 24% as the shoal grass declined

Even in their protected wetlands on the Gulf Coast of Texas, wintering waterfowl and Whooping Cranes are exposed to a wide variety of pollutants washed in after every rain from croplands, factories, and refineries nearby.

over 50% and was replaced by plants Redheads do not consume (Cain 1988). More recently, the total North American population of Redheads has fluctuated substantially, but the continued decline of shoal grass in the Laguna Madre may lead to a further shift of its winter range, since this species, unlike others in its genus, has not expanded its diet to include other winter foods (Woodin and Michot 2002).

Another organophosphate, monocrotophos, is responsible for the death of thousands of Swainson's Hawks (*Buteo swainsoni*) wintering in La Pampa Province, Argentina. This pesticide is used to control grasshoppers, a primary food of the hawks. During 1995, 3,000 Swainson's Hawks were found dead in an alfalfa field of 120 ha; the pesticide used there was not identified, but monocrotophos was applied in another alfalfa field where 982 hawks were found and in a sunflower field that within three days caused the death of an additional 700 hawks (Goldstein et al. 1996). These and other extensive killings prompted Argentinean conservation organizations and the American Bird Conservancy to advocate banning the use of monocrotophos. After they met with the manufacturer, it agreed to stop distributing the pesticide in the hawk's winter range.

OIL

The extraction, refining, and transportation of petrochemical products affect birds in several ways. On the Texas coast, wetlands have been contaminated and altered by discharges of petroleum compounds in effluents from wells, refineries, and storage pits. Vegetation, larval shrimp, crabs, and other foods important to wintering waterfowl and shorebirds are killed. Where discharges are not permitted into surface waters, refineries create evaporation pits that may cover several hectares and have a film of oil on their surface. These pits attract waterfowl and other migrants that mistake the pits for pools of water; the birds are then killed by the chemical mix. Other toxic components of oil and grease accumulate in sediments. Residues as high as 9,000 mg/kg (ppm) have been found in the Gulf Intracoastal Waterway along the Aransas National Wildlife Refuge, winter home of the majority of the world's Whooping Cranes. At the J. D. Murphree Wildlife Management Area near Port Arthur, oil and grease residues of 104,000 ppm were found (Cain 1988).

Oil spills from wells and ships at sea cause immediate and long-term harm to birds. Their immediate impacts include fouling of feathers that can cause smothering, drowning, and hypothermia, death from ingestion, and the loss of key food species. The 1999 *Erika* oil spill off the Atlantic coast of France killed 80,000–150,000 seabirds, of which 80% were Common Murres (*Uria aalge*). A study at the Common Murre colony on Skomer Island, Wales, between 1985 and 2004 found that the mortality of adults during the winters of four major oil spills in the English Channel and the Bay of Biscay was double that of other years. Immature murres, which may number 50% of the total population and winter farther from their colony of origin, spending several years at sea, were likely to have been more affected, but that is more difficult to demonstrate because they do not always return as adults to their natal colony (Votier et al. 2005). But in the years following the four oil spills, nearly twice as many immatures aged four to six years old settled for their first, still nonbreeding, summer at Skomer as in other years, perhaps because of reduced competition at the breeding colony. Thus, while in the short term there may be immature birds available to replace significant numbers of adults lost to occasional oil spills or other causes, repeated oil contamination of the wintering area could reduce the number of young birds that will later become the breeding population (Votier et al. 2008).

Impacts of a single oil spill can continue for years after the oil may dissipate on the water surface. In March 1989, the *Exxon Valdez* oil spill released 42 million liters of crude oil into Prince William Sound, Alaska. Research 10 years later found that Harlequin Ducks (*Histrionicus histrionicus*) wintering on shores of islands that had been heavily oiled had higher mortality than those on islands that had not been oiled. Also in 1999, adult Pigeon Guillemots (*Cepphus columba*) had higher levels of the detoxification enzyme CYP1A in their livers, the result of feeding on benthic invertebrates that were still contaminated (Peterson et al. 2003).

Floating oil at sea from offshore drilling sites and ship discharges can affect a significant number of seabirds even in the absence of major spills. Magellanic Penguins (*Spheniscus magellanicus*) breeding in southern Argentina migrate north sometimes as much as 3000 km to winter in the coastal waters of northern Argentina, Uruguay, and southern Brazil, usually swimming within 250 km of the shoreline. Since they travel only through water, they are more exposed to contamination from oil, even in small floating patches, than are flying birds. They are also vulnerable to the heavy shipping traffic and to entanglement in fishing nets. A survey of penguins recovered on beaches and offshore in the winter

Magellanic Penguins wintering in coastal waters of northern Argentina, Uruguay, and southern Brazil are at risk from heavy shipping traffic, entanglement in fishing nets, and contamination from floating oil residues.

range found that 7% had oil on their plumage. Almost none of the coastal waters in the penguin's winter range are protected, and offshore oil drilling is increasing on the Atlantic coast of South America (Stokes et al. 2014).

AQUATIC AND MARINE DEBRIS

Small doses of lead can cause neurological dysfunction and can sicken or kill birds. Today, most of the lead encountered by birds is in the form of shotgun pellets and ammunition fragments. If birds that have received lead shot or have ingested lead are consumed by other birds, the lead can cause further damage. In the United States, lead shot was banned for use in waterfowl hunting in 1991 but is still permitted for shooting upland game birds. Exposure to lead is probably greatest in winter, because shotgun pellets and ammunition fragments that may have been used years ago in popular hunting sites persist in lake bottoms where they may still be ingested by foraging waterfowl. And now that more waterfowl are feeding on farmland where pheasant, quail, and doves are shot, they are likely to be picking up bits of lead there as well. Lead sinkers used in fishing are still unrestricted; they can be consumed by fish-eating birds and by waterfowl foraging in wetlands. In coastal waters, lead poisoning is a major cause of death for wintering loons (Lebbin et al. 2010:336–337).

The oceans have become the dumping ground for a wide range of debris, mostly composed of plastic, Styrofoam, metal, glass, rubber, and other materials light enough to float on or near the surface but unlikely ever to degrade or decompose. Birds mistake small pieces for food or may consume fish that have already made that mistake. Illegal but widespread dumping by ships puts more of these materials at sea, adding to what reaches the open ocean when washed from land and carried by currents. In addition to choking birds that pick up the debris or filling their stomach, the chemicals in plastics disrupt hormones, causing further damage or death. Nearly half of all seabird species are known to have ingested plastic debris (Lebbin et al. 2010:340–341). Because more of these birds are at sea during the winter, often in more heavily trafficked waters, their exposure to these hazards is probably greatest then.

POPULATION

The global growth of the human population is ultimately responsible for all the conservation issues that affect birds and natural ecosystems. Human needs for food, shelter, energy, and, with a rise in material standards, more goods with their associated debris, as well as new forms of leisure and recreation that compete with birds for space, all have their impact. The increasing concentration of people in urban and adjacent suburban areas that were until recently relatively natural landscapes is often mirrored—in reverse—by bird populations, even when the development does not reduce the habitat actually used by birds. A study of wintering waterfowl-habitat associations at 32 sites in Narragansett Bay, Rhode Island, ranging from urban to rural found that the fewest birds were found where ducks were hunted and where residential development was greatest. Shoreline areas with extensive grasslands, shrubs, or forest supported many more ducks of more species (McKinney et al. 2006).

Increased food production can have impacts beyond the many already described from conversion of natural landscapes to farms, plantations, and ranches. In some parts of Spain, wintering Red Kites (*Milvus milvus*), already in decline from illegal poisoning, congregate to feed at refuse dumps where livestock carcasses, especially those of pigs, have been discarded. Some 25% of kites feeding primarily on carcasses at two sites were found to have *Salmonella* bacteria, which can cause death, persistent disease, and infertility in survivors. These kites also had bacteria associated with sewage and sludge derived from municipal solid waste as well as residues of veterinary drugs, while kites sampled in another region where they preyed primarily on wild rabbits had little or no pathogenic bacteria (Blanco et al. 2006).

DISTURBANCE AT WINTER RECREATION DESTINATIONS

The increase of leisure and affluence enjoyed by more people today than ever before has led to development and expansion of resorts and recreational sports at all latitudes. Where once people took their major, perhaps only, vacation during the summer, now many are able to take winter holidays as well. Thus, in addition to the proliferation of beach destinations used by people seeking winter warmth, ski resorts and other facilities for snow sports have expanded to meet this demand. Some birds wintering in coastal or mountainous landscapes that until recent decades were inaccessible or little used by people during winter are now exposed to disturbance as well as habitat loss.

Most shorebirds winter at latitudes where day length is less than during their breeding season and where tides may further limit the hours their feeding sites are accessible. Species wintering at beaches, mudflats, and estuaries where they may occupy space sought by people for recreation or for harvesting shellfish are often subject to more disturbance than the birds can tolerate and still feed at the rates required to maintain daily energy levels. The human disturbance is often in addition to what shorebirds receive from birds of prey, to which they must also respond by flying to escape and then returning or finding another suitable site for feeding or roosting. Tolerance of disturbance varies by species, by productivity of the site where they are feeding, and with weather.

A three-year study of Eurasian Oystercatchers (*Haematopus ostralegus*) in the Réserve Naturelle of the Baie de Somme on the Atlantic coast of France found that in winters with good feeding conditions, the birds could be put to flight by disturbance 1.0–1.5 times per daylight hour without any reduction in fitness, based on previous British studies of weight loss and starvation in wintering oystercatchers. But when feeding conditions are poor and weather severe, they can be disturbed only 0.2–0.5 times per hour. However, the oystercatchers in the reserve were routinely put to flight up to 1.73 times per daylight hour, both by people who stayed on the flats and by raptors. In protected portions of the Baie de Somme and in similar areas, managers can control at least the frequency of human disturbance, but this is problematic at sites where visitation is not regulated (Goss-Custard et al. 2006).

In California, where winters are mild and the human population on the coast is large, the disturbance to shorebirds on beaches may be constant. At two beaches near Monterey, Sanderlings (*Calidris alba*) were disturbed by people or their dogs once every 15 minutes on average. Approach by walkers or runners to within 30 m caused Sanderlings to stop feeding and run; closer approaches often prompted them to fly (Thomas et al. 2003). Similarly, wintering Snowy Plovers (*Charadrius alexandrinus*) on beaches in Santa Barbara moved or flew when approached within 40 m; they were disturbed by people or off-leash dogs an average of 2.2 times per hour on weekends and 1.4 times per hour on weekdays, causing them to decrease their feeding rate. Plovers fed most in the morning, when people and dogs were fewer, and then roosted until after dark, when they resumed feeding. Several beaches in Santa Barbara that were once used for nesting and then, as human visitation increased, only for wintering have now been abandoned entirely by Snowy Plovers (Lafferty 2001). Both these studies recommended the creation of restricted-access areas, beachgoer education, and enforcement of regulations on dogs.

Roosting sites used at high tide are as important to shorebirds as feeding areas. At Roebuck Bay in northwestern Australia where tourism is increasing, wintering Great Knots (*Calidris tenuirostris*) and Red Knots use a variety of sites, depending on tidal conditions and time of day. The roost most used during daytime high tides of intermediate height was subject to frequent disturbance by both birds of prey and humans. The energy costs of repeated short alarm flights during the course of a day there (an average of 3.36 per hour) were more than the cost of flying 25 km to the nearest alternative site, but the knots did not leave, perhaps because many short flights were less significant than the risk of heat stress in a prolonged flight of that distance; the knots used more-distant roost sites at night, when the flight costs were lower. For both species, daily costs of alarm flights while at daytime roosts and the longer flights to night roosts consumed 17.3–28.7% of the birds' total energy budget. Whether with more human disturbance knots can increase their energy intake to compensate for any greater costs depends on digestive capacity and prey availability (Rogers et al. 2006).

Construction of ski lifts and ski runs has substantially altered the habitat of birds living in regions where winter sports are popular, while resort facilities accommodating visitors may increase the presence of generalized predators not previously able to survive cold winters at high elevations. In the Alps, where Black Grouse (*Tetrao tetrix*) are declining, they live almost year round in the narrow timberline belt where most ski facilities are concentrated. A comparison of 15 ski resorts and 15 natural sites in the Valais region of the Swiss Alps found that Black Grouse were 36% less common near ski resort sites than in natural areas. In the Valais, 25–50% of potentially suitable Black Grouse habitat is affected around the main ski resorts alone (Patthey et al. 2008).

Black Grouse are also exposed to disturbance far beyond the resorts by free-riding skiers and snowboarders, who leave the ski runs to explore slopes on their own. They inadvertently flush resting birds from their snow burrows, where grouse normally spend more than 80% of their time in winter, leaving only at dawn and dusk to feed. The escape flights consume energy and expose grouse to cold and predators. An analysis of fecal matter from the burrows of grouse in the Swiss Alps that were flushed at midday in areas used by free-riding skiers and snowboarders found higher levels of corticosterone, the hormone produced in stress conditions, in the grouse that were most often disturbed (Arlettaz et al. 2007).

This technique was also used in the Black Forest of Germany to assess impacts of winter disturbance over three winters on Capercaillie (*T. urogallus*), another grouse declining rapidly because of habitat loss. Its winter energy resources are limited by a diet of low-quality conifer needles. Corticosterone levels in Capercaillie near winter recreation areas were higher than in those at undisturbed sites. Some birds in resort areas shifted their home range away from the places most heavily trafficked by people, and sites with tourist infrastructure were occupied only in the off-season. Since the topography and vegetation that Capercaillie require are also ideal ski resort terrain, however, the birds have little room to move (Thiel et al. 2008).

VERTICAL ELEMENTS IN HORIZONTAL LANDSCAPES

Birds that inhabit two-dimensional landscapes such as oceans or grasslands and other open countryside often have difficulty avoiding or adjusting to the presence of large vertical structures. These come in forms as varied as wind turbines, communication towers, and oil and gas rigs, all of which are increasing their occupation of flat ecosystems where birds have had no experience with obstacles in the third dimension. Loons, sea ducks, and alcids all have relatively short, pointed wings and high wing loadings that enable them to fly rapidly through unobstructed space such as over the sea, but they have less maneuverability than long-winged birds. They are vulnerable to collision with wind turbines if they are flying at the same height as the moving blades. In terrestrial environments of expansive grasslands, some species, from sparrows to grouse, are poor fliers and depend on concealment to avoid predators; they avoid areas with trees and artificial structures where birds of prey may perch.

In the Atlantic, the first wind farms were created in the early 1990s off Denmark and Sweden. Comparisons of seabird populations before and after two wind farms were established in Danish waters found that Northern Gannets (*Morus bassanus*), sea ducks, and alcids increased their avoidance of the area and as far as 4 km beyond, while gulls and terns increased their use of the waters around the turbines (Fox and Petersen 2006). Common Eiders (*Somateria mollissima*) experimentally attracted by decoys to the waters within and around one Danish wind farm generally avoided flying near the turbines but did not respond differently whether the blades were moving and generating noise or were turned off; their avoidance was only of the structures themselves. Most eiders reduced the frequency of both flights and landings on the sea at 200 m from the wind park. Only 2% of the eiders flew at the height swept by the blades and none were above it; 91% flew less than 10 m over the water surface. Despite avoiding the turbines when in flight, eiders approached them by swimming in order to feed in the shoals beneath them (Larsen and Guillemette 2007).

Today, there are plans for more wind farms in Danish waters as well as off the coasts of Germany and the United Kingdom. Some wind farms may have as many as 1,000 turbines. A study in the German portion of the North Sea measured factors affecting impacts on birds; these included flight maneuverability, flight altitude, percentage of time flying, nocturnal flight activity, sensitivity to disturbance by ships and helicopters, flexibility of habitat use, and conservation status for the bird species found there. Of the wintering birds, loons were the most sensitive, followed by White-winged Scoter (*Melanitta fusca*) and Great Cormorant (*Phalacrocorax carbo*), with gulls and Northern Fulmar (*Fulmarus glacialis*) the least sensitive. Based on the winter distribution of these birds, coastal waters were found to be more significant than offshore ones (Garthe and Huppop 2004).

On the North American side of the Atlantic, wind farm development is less advanced. The first wind turbines began operation in 2017 off the coast of Rhode Island. A survey assessing the risk to seven groups of marine birds off Rhode Island and Long Island found that shallow inshore waters were the greatest conservation priority but that certain offshore areas were also significant, especially for gannets. Currently proposed development projects are in areas of low conservation priority and would not substantially affect marine birds in any season (Winiarski et al. 2014).

In the interior of North America, expansive grasslands are considered one of the best places to site wind farms. Grassland birds, however, are the most rapidly declining group of birds on the continent. Construction of wind turbines, as with oil rigs, natural gas extraction facilities, and communication towers and lines, results in habitat fragmentation or loss. Even when any of these are completed, sensitive species cannot tolerate the resulting tall landscape features where predators might perch; they tend to avoid or abandon areas having any vertical structures. The traffic related to operation and maintenance is an additional disturbance.

Sensitivity among grassland birds varies by species, by season, and by habitat particulars; bird communities in pristine grasslands are less tolerant than are those in modified agricultural landscapes. In addition, the grassland birds that forage in open areas tend to be strong fliers and are more often in groups that can more easily spot predators; they are more tolerant of natural or artificial vertical elements nearby. In contrast, the species that are solitary, poor fliers, and dependent on concealment are most often in denser vegetation farther from any elevated perches predators could use (Stevens et al. 2013). These characteristics may be useful predictors for assessing the impacts of vertical structures on birds in many grassland ecosystems.

A study in north-central Texas, an important region for wintering grassland birds, found these behavioral patterns confirmed in a comparison of responses of several species to wind turbines in pastures, hay fields, and winter wheat fields. Eastern and Western Meadowlarks (*Sturnella magna* and *S. neglecta*), Sprague's Pipits, and Savannah Sparrows (*Passerculus sandwichensis*), which all form loose flocks in winter, were not displaced in any of the habitats. LeConte's Sparrows, solitary and flushing only reluctantly, were limited to hay fields and lightly grazed fields dominated by native vegetation; they generally stayed at least 400 m from any turbine (Stevens et al. 2013).

The same ecological and predator-avoidance strategies may explain the intolerance of Greater Sage-Grouse (*Centrocercus urophasianus*) to tall structures of any kind (Lammers and Collopy 2007)—even when the disturbance from road building, construction, and maintenance are not factors. Sage-grouse inhabit some of the lands in the West most sought after for energy development. In 2005, a previously unknown wintering area to which 1,500–2,000 sage-grouse fly from as far as 160 km away was discovered in Wyoming's upper Green River Basin. This unprotected site has been proposed for gas development with 3,500 wells and 363 km of roads (Thuermer 2015).

In Great Britain, where farmland is in some areas the only open terrestrial landscape suitable for wind turbines, four functional groups of wintering birds have been identified as potentially affected—seed eaters, corvids, game birds, and Eurasian Skylarks (*Alauda arvensis*). When two wind farms were monitored over the course of a winter, the only

Greater Sage-Grouse winter in many unprotected areas that have been proposed for extensive gas development. They cannot tolerate the disturbance or the presence of vertical structures on which predators may perch.

species found to avoid the turbines was the Ring-necked Pheasant (*Phasianus colchicus*), which of all the 33 species observed is the closest ecologically and in maneuverability to the sage-grouse. The other species were as likely to be in farm fields with or without turbines (Devereux et al. 2008). Several other European studies found that the wintering birds most consistently avoiding the area from 200 to 400 m around wind farms were Eurasian Wigeon (*Mareca penelope*), Northern Lapwing (*Vanellus vanellus*), Eurasian Golden-Plover (*Pluvialis apricaria*), and geese in the genus *Anser* (Hotker 2006).

Raptors are generally considered the birds most vulnerable to collision with rotating turbine blades. In some locations, this is especially true during migration, when large numbers of raptors may pass through landscape corridors with topography that produces the ideal winds for themselves as well as for turbines. A 10-year study of raptor mortality at two wind farms in Cádiz, Spain, however, found that all nine species present suffered more deaths from collisions in winter. Mortality rates did not correlate with species abundance and were not significant except for Griffon Vultures (*Gyps fulvus*). These were the most vulnerable, even though they were less common in winter than during other seasons. Compared to the other raptors present, the vultures are weak fliers, dependent on wind and thermal updrafts and therefore less maneuverable than birds using powered flight. Because thermal updrafts are less common in winter, a result of lower soil temperatures and less sun, this is the season vultures have greatest difficulty gaining altitude and avoiding the turbines, especially where hills have only gentle slopes, which produce weaker updrafts (de Lucas et al. 2008).

Communication towers are another vertical structure proliferating in natural landscapes. In the United States and Canada, they are estimated to kill 6.8 million birds per year, mostly nocturnal migrants during spring and autumn. Passerines constitute 97.4% of the estimated mortality, with warblers being 58.4% of all mortality, followed by vireos (13.4%), thrushes (7.7%), and sparrows (5.8%). For 29 of the species, the annual tower mortality is greater than 1% of the total population, and 0.5% for an additional 15 species. Thirteen of the 20 North American species killed most frequently are categorized as either Endangered or Birds of Conservation Concern by the US Fish and Wildlife Service. As with the raptors in Spain, mortality is not proportional to population abundance; some species are killed at greater rates than others—and some of these are also high on the list of species most frequently killed by other human additions to the landscape, such as feral cats and glass windows (Longcore et al. 2013). While the winter mortality from communication towers may be low in comparison, facultative migrants that travel at night in response to unseasonable cold are vulnerable. Much of the information available on facultative migration comes from dead birds collected beneath communication towers (Terrill and Ohmart 1984).

OVERHARVEST

Throughout the world, wherever there are wetlands, waterfowl are hunted. Where hunting is regulated, it is usually restricted to autumn and winter. Most nations monitor their waterfowl populations, and agencies with scientific expertise determine the daily and seasonal limits per hunter for each species. Among waterfowl specialists, debate continues as to whether the impact of the annual harvest is compensatory—removing a number of birds that would otherwise die from natural causes over the autumn and winter—or additive, by reducing the population below what it would otherwise be at the end of the season. The extent to which hunting is either compensatory or additive varies by species and locale as well as by age and sex and also by year, since the productivity of waterfowl populations fluctuates and local hunting conditions are affected by weather.

A study of hunting impacts on dabbling ducks in California's Central Valley focused on five species that constitute 80% of the annual duck harvest in the state. Harvest was high in October, when the hunting season began, and significantly lower in November and December, with a marked increase in January. The proportion of adults taken rose substantially between fall and winter. This may be because adults are later fall migrants, while immature birds are also less experienced. Hunting during the winter is likely to have a greater impact on future populations than the autumn take, since in California by January 90% of female Mallards (*Anas platyrhynchos*), Northern Pintails (*A. acuta*), and Northern Shovelers (*Spatula clypeata*) have formed pairs, as have 80% of female Green-winged Teal (*S. carolinensis*). The breakup of pairs can reduce productivity, especially of adults; in Mallards, adult females are twice as effective at hatching their clutch of eggs as are first-time breeders. Thus, here as in many regions of the United States where most ducks have paired by January, a shorter hunting season may protect more adults and increase their future productivity. Some hunters, however, are calling for a longer season (Miller et al. 1988).

In parts of Central America and Europe, the hunting season is longer, but there are no data on how the likely disruption of pairs may affect productivity. In Europe, this may be especially important for geese, which winter in families with young still dependent on their parents. Late shooting of adults may not only prevent re-pairing before the following breeding season but also reduce survival of young birds left with one or no parents (Anderson et al. 1988).

While the harvest of waterfowl may in most cases be sustainable, another group of birds is suffering from accidental harvest—bycatch—that is insufficiently regulated or enforced and is on a scale that can only be additional to natural winter mortality. In all the world's oceans, large numbers of seabirds are caught and drowned in trawls, nets, and longline hook fisheries as they pursue the discards from factory ships, the bait on hooks, and fish caught beneath the surface in nets. In some fisheries, longlines may be 100 km long and have thousands of baited hooks. Mortality occurs throughout the year, but for some species it is most intense during winter, when many seabirds are far from their breeding sites and concentrate in nutrient-rich portions of the ocean where fishing fleets of many nations deploy larger ships and ever more extensive gear to maximize their catch.

Set and drift gillnets have the greatest number of documented species accidentally caught, 92 and 88, respectively. Longlines and handlines using baited hooks have together caught 127 species. Collectively, these four gear types are known to have caught 193 species. Of these, 63 are classified as Critically Endangered, Endangered, or Vulnerable by the US Fish and Wildlife Service. The global species total is likely higher, since the waters of the Arctic, West African currents, the Indian Ocean, and Asia have been little studied. The synthetic fibers now used in nets and lines increase mortality because these are less visible to birds and do not decompose; when the gear has been lost or discarded at sea, it continues to catch and drown birds and other animals. This is known as "ghost-fishing" (Pott and Wiedenfeld 2017).

Seabird bycatch is not a new phenomenon. Between 1952 and 2001, as many as 34 million wintering Sooty Shearwaters (*Ardenna grisea*) and Short-tailed Shearwaters (*A. tenuirostris*) were killed in the North Pacific by drift nets set by Japanese, Korean, and Taiwanese vessels (Uhlmann et al. 2005). A United Nations ban in 1992 on most forms of drift netting has reduced the toll substantially, but even the current mortality may not be sustainable. Elsewhere, however, the toll on seabirds is increasing in the many portions of the global ocean where fishing is unregulated, a substantial number of boats are unmonitored and illegal, and fleets deploy larger nets and longer lines.

Accurate numbers are difficult to secure, because few vessels have observers, tracking methods vary in different fisheries and regions, and data compilations are from different years. Annual mortality from global longline fisheries alone is estimated at 160,000 to 320,000 birds, mostly albatrosses, shearwaters, and petrels (Gianuca et al. 2017). Some 400,000 birds die each year in gillnets, mainly in the northwest Pacific, Iceland, and Baltic Sea; there, loons, alcids, and diving ducks are most vulnerable, while in the Southern Hemisphere several small penguin species are most frequently caught (Zydelis et al. 2013).

In subpolar regions of the Southern Hemisphere, the majority of seabirds killed by longlines are males and adults of both sexes, while more females and immatures are killed in subtropical waters; this may reflect different sex-based distributions, especially in the nonbreeding season. Female Wandering Albatrosses (*Diomedea exulans*) from South Georgia are found more often than males in subtropical waters where they are killed by longline tuna fisheries, while males are commoner in subpolar waters. For species in which bycatch mortality creates an unequal sex ratio, the impact on population is compounded because most seabirds return to the same mate each spring and require biparental care to raise young; a shortage of either sex reduces the number of birds that can mate and breed, especially in species like albatrosses where pair formation is a lengthy process and normally lasts for life. Similarly, because seabirds reproduce slowly and begin breeding only when several years old, the loss of adults prevents a declining population from regenerating (Gianuca et al. 2017).

The 1992 United Nations moratorium on the use of drift nets more than 2.5 km long on the high seas has been effective, but set nets and smaller drift nets are still used in nearshore waters of all continents except Antarctica (Zydelis et al. 2013). The European Union banned the use of drift nets by European Union fleets in 1998. Practical measures in net fisheries include making nets more visible, restricting their use at certain seasons, places, and depths, and replacing them with longlines.

Similarly, mortality from longlines can be reduced by putting weights near the hooks so they sink below the depth where most birds would pursue them, adding bird-scaring streamers, or restricting settings to night hours, when seabirds are less active.

Implementation of these measures by some fleets has substantially reduced bycatch—without diminishing fish catch—but the total extent of implementation is not known.

Wandering Albatrosses and many other seabirds are attracted to fishing vessels, where they are drowned after being caught on hooks or entangled in nets.

In the New Zealand fishery, using weighted lines reduced mortality of White-chinned Petrels (*Procellaria aequinoctialis*), one of the most difficult species to deter from baited hooks, by up to 98.7% (Robertson et al. 2006). The Convention on the Conservation of Antarctic Marine Living Resources now has complete observer coverage in that region on the longline fishing fleets of the 24 participating nations and the European Union; it has found that these measures can reduce bycatch to practically zero. The Agreement on the Conservation of Albatrosses and Petrels (ACAP), a multilateral agreement covering 31 species, entered into force in 2004 and is observed by 13 nations, but not some major fishing states such as Canada, Japan, and the United States. It continues to develop and promote means to reduce bycatch.

SUMMARY

For birds living in a seasonal environment, conservation requires an understanding of when during the annual cycle—whether it is spent in one place or ranges across continents—birds are vulnerable to population losses. Many birds are exposed to distinctive threats during winter.

Around the world, habitat loss is overwhelmingly the leading cause of bird declines. The highly compressed ranges of birds wintering in the Neotropics, for some species as little as one-tenth of their breeding range, means that loss of habitat there, particularly from the conversion of forests to agricultural land, has a greater impact than comparable losses in North America. Coffee and cacao, when grown under shade trees, can support many migrants through the winter, although few understory and true forest specialists. Mexican grasslands, winter home to many birds of the Great Plains, become useless to birds when converted to row crops. There, restoring grasslands and rotating cattle on them so the land is not overgrazed provides necessary habitat. In Africa, the desertification of the Sahel, where many Eurasian birds winter, is due to a combination of climate change and direct human impact through overgrazing and tree cutting. Here, many birds shift their location as often as four times during the winter, while most Neotropical winterers are sedentary. For shorebirds, the loss of wetlands, especially the extremely productive estuaries where large numbers concentrate on migration and through the winter, is the most severe threat. Rice fields, salt ponds, fishponds, and some other managed areas can provide alternative habitat for some species. Waterfowl have also declined all over the world from loss of wetlands; in prosperous parts of the world, some of the remaining wetlands where birds winter may be managed for their benefit, and some species have also adapted to feed on agricultural land.

Invasive species compound habitat loss when nonnative plants choke out natural vegetation, and animals from elsewhere compete with or directly consume wintering birds. Of all invasive species, house cats, including feral ones, are the most destructive, killing an estimated 1.3–4.0 billion birds annually in the United States alone. Birds may be especially vulnerable in winter, when the frogs, snakes, and small mammals on which cats also prey are less available.

Pollutants such as pesticides, oil, and aquatic and marine debris can travel far from their source and, if consumed, move up the food chain, thereby affecting a sequence of birds and other animals. Wetlands probably receive and absorb more pollutants than any other habitat, exposing wintering birds from Whooping Cranes to waterfowl to endocrine disrupters, neurotoxins, and other chemicals that result in abnormal behavior and reduced survival. Many pesticides now banned in the United States and other developed countries are still widely used in tropical regions where many birds winter; the wintering species that concentrate at the edges of agricultural areas are especially vulnerable. At sea, residue of spills from oil wells and ships remains for years; it can coat feathers or be ingested either directly or through food that wintering seabirds, widely dispersed over the oceans, consume. Similarly, seabirds mistake pieces of plastic and other debris for food and may choke or starve as their stomach fills with it. On agricultural land and in wetlands, waterfowl consume lead shot, which can sicken and kill them.

The global growth of the human population, with its increasing needs for food, other resources, space, and production of more forms of debris, is ultimately responsible for all the conservation issues that affect birds and natural ecosystems. The increase of leisure and affluence is sending more people to recreation sites, from beaches to mountain slopes, that until recently were undisturbed habitats of wintering birds. Demand for more energy has led to the construction of wind turbines and oil and gas rigs at sea and on land in flat habitats where birds have no experience with tall vertical structures and avoid them.

Finally, some birds are overharvested, either through deliberate hunting, as particularly affects wintering waterfowl, or incidentally as bycatch in various types of oceanic fishing gear from nets to longlines of baited hooks that especially attract albatrosses, shearwaters, petrels, and alcids. Much of this mortality could be reduced by use of equipment that prevents birds from being caught on hooks and lines.

CHAPTER 10

CLIMATE CHANGE

IN ADDITION TO ALL the traditional conservation issues, some well known for a century or more but mostly local in origin and impact, the world is now undergoing a global modification that affects every ecosystem and all its denizens, even those remote from the sources of the problem. Edward O. Wilson (2016:57) categorizes the increase of greenhouse gas emissions, primarily from the burning of forests and fossil fuels, as a form of pollution; this is accurate, bearing in mind that its effects exacerbate all the other challenges covered by the acronym HIPPO.

Today, atmospheric levels of carbon dioxide and methane, the two leading greenhouse gases, exceed natural levels of the last 650,000 years. In the twentieth century, the Earth's climate warmed by 0.3–0.6° C (IPCC 1996), surface sea temperatures are now, on average, 0.3° C higher, and seas have risen an average of 10–20 cm. Averages, however, mask more substantial local changes: in western North America, mean air temperatures have risen 1–2° C since the late 1940s; in the northeastern United States, mean annual temperature has increased 0.25° C per decade since 1970; the North Sea is 0.6° C warmer than in the early 1960s. Changes in the twenty-first century have already been and are predicted to be both more rapid and more intense, including a global average temperature increase of 2–4.5° C, with the deciduous forests of eastern North America having

January temperatures 2.5–6.6° C higher than today. Some of the natural absorption of greenhouse gases will cease: since 2000, less carbon dioxide is being absorbed by land and water systems; forests in Costa Rica and Panama are now growing at slower rates than in 1985 (Cox 2010:10–43).

There are already visible changes in the distribution and seasonal timing of the major events in birds' annual cycles. In the last few decades, many species have shifted their range at one season or in all seasons to stay within the region where their food sources thrive. Seabirds that depend on cold-water krill and fish have moved poleward, birds on mountains have climbed closer to the summits, and widespread continental species have shifted or expanded to new regions, while some species with very specific ecological requirements are now found in a smaller part of their recent range. Likely also are changes in population densities of both predators and prey, because of gradual changes such as temperature rise as well as extreme events including droughts, fires, and intense storms. Because the many complex and interacting components of every ecosystem will never all respond at the same rate to local climate changes, the impacts on birds will vary in ways that today can hardly be predicted, and new communities will be formed that may be more advantageous or challenging for their members.

Migratory birds may be at greater risk, because they depend on more sites. But because of their mobility, some may be more flexible than sedentary species with limited distribution and restricted habitats. Partial and short-distance migrants will likely have shifts in the relative proportion of individuals that migrate, as well as in the distance they travel. Among long-distance migrant species, those with high connectivity may have more difficulty shifting their winter range, especially if adjacent areas also no longer meet their requirements; populations that disperse more widely in winter are more apt to find suitable sites. For all migrants, there may be changes in their route and direction to newly favorable destinations, as well as in timing and speed, both in arrival at and departure from wintering sites.

WINTER RANGE SHIFTS

Any birdwatcher with more than a few years of experience is likely to have noticed the regular presence of species that were once rare or unknown in winter. For many birds, nighttime low temperatures may define the upper limit of their winter range. This has also been calculated as the isotherm line where the energy needed to compensate for a colder environment is not more than 2.5 times a bird's basal metabolic rate (Root 1988). As minimum nighttime temperatures have risen in recent decades, this line has moved poleward for many birds.

Precipitation and other related environmental factors are also important, reflecting the ecology of each species. For ground foragers, extent of snow cover is critical; in the eastern and central United States, the lower frequency and shorter duration of snow cover have enabled Turkey Vultures (*Cathartes aura*), American Robins (*Turdus migratorius*), Eastern Bluebirds (*Sialia sialis*), Common Grackles (*Quiscalus quiscula*), and White-throated

Sparrows (*Zonotrichia albicollis*), among others, to remain common north of their mid-twentieth-century winter range. The rise in coastal water temperatures, with their associated impacts on fish and shellfish, has similarly led to the northward shift of many wintering waterfowl.

A study based on Christmas Bird Count results from 1975 through 2004 found that for 184 migratory and 70 sedentary North American species that winter principally at latitudes 30–50° N, all had shifted northward either the boundaries or centers of abundance of their winter range, in some cases both. The more southerly species expanded their northern limit most, while among the northerly species the northward shift of their center of abundance was more significant. The greatest advances were in the upper Midwest and northeastern United States. Over these 40 years, the latitudinal boundaries moved northward an average of 1.48 km per year, while the average northward shift in center of abundance was 1.03 km per year (La Sorte and Thompson 2007).

Further use of the same Christmas Bird Count data concentrating on 59 species and extended through 2009 found that northward expansion was linked to the rise of minimum winter temperatures, which gained most at higher latitudes. Expansion, however, proceeded at different rates for different species. Progress depended not only on their own tolerance but on the presence or movement of their food sources. The birds that expanded northward most rapidly were species with already northerly ranges and broad niches; those least likely to expand had poor dispersal abilities, strong ties to very specific habitats, or already-established ecological competitors to the north. Thus, the components of winter communities shift at different rates, depending on the ability of each species to respond and adapt (La Sorte and Jetz 2012).

In northern Europe, the same phenomena have been observed, and thanks to more extensive data, a longer timeline can be drawn. A 1949 review using data going back to the mid-nineteenth century found northward expansions of both residents and short-distance migrants. During this period, the northern and central European climate grew more maritime in both temperature and humidity. Of 25 species that had their northern breeding limit in southwestern Finland, 44% increased in numbers and expanded their range (Kalela 1949). Another Finnish study, using data from 1960 to 2014 for 29 breeding species, found that the winter range of several species also moved northward—as much as 21.2 km per year for the Northern Lapwing (*Vanellus vanellus*) and 15 km per year for the Chaffinch (*Fringilla coelebs*)—thereby diminishing the distance these birds travel on migration, because their breeding range was not advancing northward as rapidly (Potvin et al. 2016).

A review of resident and migratory birds breeding in Britain compared distributions in 1968–1972 and 1988–1991 and found that, on average, birds had extended their range 18.9 km northward during that time (Thomas and Lennon 1999). Some European species have made much greater strides: White Storks (*Ciconia ciconia*), which previously migrated to sub-Saharan Africa, began wintering in southern Europe in the 1980s, and Common Cranes (*Grus grus*), formerly wintering no closer than southern Spain, have established a wintering population in northern France (Cox 2010:233).

A British study focused on the winter distribution of nine shorebird species in 1984–1985 and 1997–1998 found that eight of the nine shifted their range eastward and northward along the nonestuarine coasts as rain, low temperatures, and high winds all decreased; in that period, winters became 1.5° C milder. These shifts reflect the tendency of shorebirds to winter as close to their breeding area as conditions permit; thus, looking to the future, predictions based on likely rates of warming on the coasts of western Europe and Scandinavia suggest that by 2080 the winter ranges of these species could be substantially north of where they are today. One limiting factor of potentially more-northerly winter shorebird concentrations, however, is the shorter day length; shorebirds feed less efficiently at night and may, beyond certain latitudes, be unable to meet their energy requirements (Rehfisch et al. 2004).

For both coastal and oceanic birds, changes in sea-surface temperature may have the greatest influence on winter range, because these can cause changes in the entire length of the marine food chain from phytoplankton to the prey consumed by birds. Between 1983 and 1991, as temperatures warmed in the Peruvian and California Currents, Sooty Shearwaters (*Ardenna grisea*) shifted their winter range to the west and central North Pacific, where water remained cooler. Prior to the 1990s, an estimated three to five million Sooty Shearwaters wintered in the California Current; between 1985 and 1994, primarily after 1989, that number dropped by 90% as shearwaters moved farther north (Spear and Ainley 1999). During the same period, between 1987 and 1998, Pink-footed Shearwaters (*A. creatopus*), which breed on islands off the Chilean coast and in winter concentrate in warmer waters than Sooty Shearwaters, more than doubled in number off the coast of southern California (Hyrenbach and Veit 2003).

SHIFTS IN TIME SPENT IN WINTER RANGE

Experienced birdwatchers will also have seen changes in spring arrival and autumn departure dates, which for many birds mean they have altered the time they spend in their winter range. Globally, spring events at middle and higher latitudes have advanced an average of 2.3 days per decade since approximately 1970 (Parmesan and Yohe 2003), effectively reducing the length of winter. The onsets of various autumn phenomena have begun later, although in less easily measured ways. Some birds have shifted their own cycle at the same rate, others less.

In the northeastern United States, at Concord, Massachusetts, the first spring arrival dates for 22 species recorded as recently as 2007 are close to those observed by Henry

Sooty Shearwaters and many other seabirds have shifted their winter ranges as ocean temperatures have risen and their traditional areas no longer support the food chain on which the birds depend.

David Thoreau in 1851–1854, with significant advances only in migrants that typically winter in the southern United States rather than in the tropics (Ellwood et al. 2010). At the Cayuga Lake basin of central New York between 1903 and 1993, 39 migratory landbirds advanced their spring arrival dates an average of 5.5 days/50 years; 35 other species showed no change (Oglesby and Smith 1995).

A comparison of first arrival dates for 83 migrant species in Fargo, North Dakota, using data from 1910–1950 and 2004–2014 found that more than half arrived earlier in this century, some as much as 31 days sooner. But their alignment with advances in the growing season, which was about 4–5% earlier in 2004–2014, varied primarily by the distance the birds had migrated. Short-distance migrants (Yellow-headed Blackbird [*Xanthocephalus xanthocephalus*], Cedar Waxwing [*Bombycilla cedrorum*], and Eastern Bluebird) that wintered closest to North Dakota tended to come ahead of the traditional growth stage, while the mid- and long-distance migrants (Sora [*Porzana carolina*], Wood Thrush [*Hylocichla mustelina*], and Gray-cheeked Thrush [*Catharus minimus*]) arrived later relative to the growth stage (Travers et al. 2015).

On the central Arctic coast of Alaska, at Anachlik Island in the Colville River Delta on the Beaufort Sea, between 1964 and 2013, 16 migrants, from loons to Snow Buntings (*Plectrophenax nivalis*), arrived between 3 and 10 days earlier, and an average of 6 days earlier. Species that traveled the greatest distance and over the sea did not advance their schedule as much as short-distance migrants traveling over land. The first arrival of these species was 1.03 days earlier for every 1° C annual rise in temperature, but except for the White-fronted (*Anser albifrons*) and Snow (*Chen caerulescens*) Geese, none of the advances in arrival date kept up with the local advance of spring (Ward et al. 2016).

Among diurnal raptors, the time spent in the winter range has diminished. For 14 species, spring migration data (1972–2012) from the Whitefish Point Bird Observatory in Michigan and autumn data (1972–2011) from the Hawk Ridge Bird Observatory in Minnesota, both on the south side of Lake Superior, showed that the median migration date advanced by 1.3 days per decade in spring and was delayed in autumn by 2.3 days per decade, nearly twice the spring rate. As seen in other migrants, species wintering at middle and high latitudes had the greatest shift. Golden Eagles (*Aquila chrysaetos*) and Rough-legged Hawks (*Buteo lagopus*) spent more than 30 fewer days south of the observatories in recent years. This may be because they breed at high latitudes, where warming has been greatest, and their mammalian prey is now available over more of the year. In addition, these species may both face greater competition in their winter range, where raptors are at higher density and more species are present. In contrast, Northern Goshawks (*Accipiter gentilis*), which feed primarily on midsized forest birds, had relatively little change in arrival and departure dates, perhaps because their avian prey, much of it migratory, has not extended its seasonal presence significantly (Van Buskirk 2012). Similarly, in eastern North America, data from seven raptor-migration watch sites from 1985 to 2012 found that for 16 species, the peak autumn migration dates of short-distance migrants were one day later for every decade, which coincided with an increase in average temperatures. Only the transequatorial migrants—Osprey (*Pandion haliaetus*), Peregrine Falcon (*Falco peregrinus*), and Broad-winged Hawk (*Buteo platypterus*)—remained on their traditional schedule (Therrien et al. 2017).

In Europe, many resident and short-distance migrants now respond to warmer local conditions by beginning their breeding season sooner than they did a few decades ago. In England, resident Great Tits (*Parus major*) advanced egg laying 15 days between 1985 and 2005, keeping pace with emergence of the caterpillars they feed their young (Both 2010). In Finland, Eurasian Sparrowhawks (*Accipiter nisus*) advanced their spring arrival by 11 days between 1979 and 2007. Juvenile sparrowhawks, born a few days earlier than decades ago, have advanced their autumn departure by 0.38 day per year during this period, while adults, which molt their flight feathers before migrating, are not leaving much earlier than in the past (Lehikoinen et al. 2010).

Birds wintering in sub-Saharan Africa, however, have advanced their schedule relatively less. As in other hemispheres, these long-distance migrants can receive no cues to conditions at or near their breeding area; they continue to follow a schedule that is under endogenous control or set by changes in day length and by the local environment. Variations from year to year in the quality of winter habitat may influence departure date. The Sahel, where many European birds winter, began recovering from a severe drought in 1980, and the last few decades have been much more favorable to birds there. A study correlating rainfall and extent of winter vegetation in West Africa between 1982 and 2000 with the spring arrival times in Spain of White Storks, Common Swifts (*Apus apus*), Eurasian Cuckoos (*Cuculus canorus*), Barn Swallows (*Hirundo rustica*), and Nightingales (*Luscinia megarhynchos*) found that birds returned sooner—and therefore probably departed their wintering sites earlier— following winters with more rainfall. Wetter winters generate more vegetation and more food for these species, enabling more to survive and putting them in better body condition by winter's end than in drier years (Gordo and Sanz 2008). Similarly, of 20 species wintering in sub-Saharan Africa and breeding in Oxfordshire, England, 17 advanced their arrival dates an average of eight days between 1971 and 2000 (Cotton 2003).

Earlier arrival in spring does not, however, necessarily mean that birds remain longer at their breeding site. Many of the birds that are able to begin nesting earlier also now return to their winter home sooner. The decisive factor seems to be whether birds raise more than one brood in each breeding season. Many resident species do, but fewer migrants do, especially long-distance ones. Of 20 species nesting in Denmark, including terns, hawks, and passerines, the species that routinely attempt more than one brood extended their breeding season by 0.43 day per year between 1970 and 2007, while the single-brooded species shortened their season by 0.44 day per year, leaving earlier (Møller et al. 2010). All the species that arrived in Oxfordshire earlier also departed an average of eight days sooner at the end of summer, thereby maintaining the same length of time on their wintering grounds (Cotton 2003). Shorebirds from Siberia and the Pacific coasts have also now advanced both their wintering arrival and departure dates in Australia (Beaumont et al. 2006). The earlier that shorebirds can now begin and complete nesting in the accelerating Arctic spring, the sooner they may depart for their winter range. Throughout the Arctic, most adult shorebirds typically leave their nesting area ahead of their young, as soon as these can fend for themselves, perhaps to reduce competition for dwindling food supplies at the end of summer as well as to get a head start to wintering sites where they may claim a territory.

A survey of 65 landbird species moving in autumn through a mountain pass in the Swiss Alps during 1958–1999 found a distinction between most of the long-distance migrants ultimately going south of the Sahara and the species wintering in southern Europe and North Africa. Of the 25 long-distance migrants, 20 advanced their European departure, while 5 delayed it. Among the short-distance migrants, 28 delayed and 12 advanced their departure date. Species wintering in sub-Saharan Africa that reach and pass through the Sahel sooner avoid some of the dry season, which begins in September. Short-distance migrants that delay a return to wintering areas, or begin wintering closer to their breeding range, gain three potential advantages: they may be able to extend their breeding season to produce another brood, their shorter travel distance involves less risk, and they can return earlier the following season to begin nesting sooner. In fact, the only long-distance migrants passing through the Swiss Alps that delayed their autumn departure were species that at least sometimes attempt more than a single brood. Similarly, the short-distance migrants that usually raise two broods departed later than the species that produce only one brood (Jenni and Kéry 2003).

For both short- and long-distance migrants, sexual selection pressures may lead some birds to leave their wintering area sooner in order to arrive at their breeding site earlier. Using degree of sexual dimorphism as an indicator of the intensity of sexual selection by females among nine migrant species common at two bird observatories in Germany and Denmark, a study found that between 1976 and 1997 the more dimorphic species advanced their spring arrival times most, irrespective of wintering latitude (Spottiswoode et al. 2006).

The local winter environmental changes and sexual selection pressures that have led to some birds departing their wintering sites earlier in recent decades have not influenced all species equally. Of 100 European migrant species for which there are good data on mean spring arrival dates between 1960 and 2006, the species with stable or increasing populations all advanced their migration dates, while the ones that seem to have least advanced their departure from wintering areas are the species that have most declined, irrespective of whether they wintered in sub-Saharan Africa or around the Mediterranean basin. As found in other studies, the multibrooded species advanced their schedule more than did single-brooded species. The more northerly breeding birds that did not advance their arrival dates declined more than those of lower latitudes, perhaps because their migration timetable puts them farther out of synchrony with the advance of spring and food-peak dates at their destination, and because this gives any competing resident species a greater head start (Møller et al. 2008).

A comparison of 193 separate populations of migratory birds breeding in the Nearctic and Palearctic found that the Nearctic species most in decline between 1980 and 2006 were those with the greatest differences between the increase in the average temperatures in their breeding and winter ranges, leading to migrants arriving later relative to the advance of spring. Among Palearctic migrants, between 1990 and 2000 species with greater distance between wintering and breeding areas declined more than did short-distance migrants; they too are likely to reach their breeding area too late to take advantage of food peaks (Jones and Cresswell 2010). The two factors may reflect the same phenomena, since birds most distant in winter from their breeding range are inevitably in regions with climates that are responding at different rates than the warming at higher latitudes. In addition, the bulk of European birds wintering in sub-Saharan Africa are farther from their breeding range than are most Nearctic migrants in the Neotropics.

The best-studied species in which the inability to shift winter departure times has led to a decline in breeding success is the Pied Flycatcher (*Ficedula hypoleuca*). It winters in sub-Saharan Africa and nests in much of Europe to northern Scandinavia and Finland, requiring at least three weeks to migrate north. Between 1987 and 2003 in the Netherlands, a warming climate has led to the increasingly early emergence of the caterpillars that are the chief food flycatchers give to nestlings. But the flycatchers have not arrived in time to initiate nesting soon enough to harvest the caterpillars. While in recent years the birds have begun laying eggs five days after they arrive—the minimum time required for egg production—that alone has not put them on the same schedule as their prey. Pied Flycatchers have advanced their breeding by 0.5 day per year while caterpillars have advanced their emergence 0.75 day per year (Both 2010). Monitored populations have declined by 90% in areas with the earliest food peak and only 10% in areas with the latest food peak. Declines of Pied Flycatcher populations elsewhere in Europe have been attributed to the same mismatch of arrival time and caterpillar abundance (Both et al. 2006). In Finland, data from 1943 to 2003 show that Pied Flycatchers, even though now arriving slightly earlier, delay their egg-laying dates relative to the advancing schedule of their food sources, indicating that the birds may not be able to find enough food to fuel egg production; as a result, their clutch size has decreased (Laaksonen et al. 2006).

Could Pied Flycatchers evolve to leave their winter range earlier? Experiments exposing this species and two other European winterers in central and southern Africa, the Garden Warbler (*Sylvia borin*) and the Common Redstart (*Phoenicurus phoenicurus*), to the winter light regime of Africa north of the Sahara found that spring migratory activity and testicular growth advanced significantly for all three. Thus, they can respond physiologically to changes in the usual stimulus for winter departure (Coppack et al. 2003). But, while there are a few recent records of Pied Flycatchers wintering around the Mediterranean, the most its current sub-Saharan winter range could shift north without leapfrogging over the desert is 10 degrees (Lehikoinen and Sparks 2010:93). For a species like the Pied Flycatcher in which individuals maintain a winter territory and return to it in subsequent years, such a shift would happen extremely slowly.

The mismatch for long-distance migrants between spring arrival time and availability of prey has not yet been studied or quantified for many species. The focus of the research thus far has concentrated on food for nestlings rather than adults. Even in the Pied Flycatcher, during the breeding season adults consume a much wider variety of arthropods than they feed their young and are thus themselves little affected by the earlier emergence of caterpillars. Far less known is the potential extent of mismatches for birds, especially those with specialized diets, arriving in their wintering areas if the season there has also advanced. Species traveling far across the equator to encounter spring conditions may find fruit, arthropods, and larger animals with seasonal cycles on an increasingly different schedule. The birds that have shifted but not extended the time they spend on their breeding sites may be more successful at both ends of their migratory route than the species that linger before their autumn migration.

CHANGING WINTER COMMUNITIES

Because the plant and animal species within every ecosystem will respond individually to changing climatic conditions, new communities will evolve as each species expands or contracts its range at its own rate. Both aerial insects and plants with seeds dispersed by wind or by birds will likely shift their range more rapidly than creatures that walk or crawl, and more rapidly than long-lived trees with seeds slow to germinate or to grow far from their source. Different aspects of local climate shifts will affect plants and animals differently and therefore shape the components of future winter communities. For some, average or extreme temperatures may be the critical factor, for others the amount, intensity, or timetable of precipitation. At any given location, food sources may increase or decrease, competition may intensify or be relieved, and predators may be attracted or decline.

Among birds, carnivorous, omnivorous, and insectivorous species are most sensitive to cold weather and are likely to prosper in warmer winters, while coastal and marine birds are relatively insensitive to temperature but suffer when warmer seas diminish their prey. Specialists in winter fruits or seeds will be vulnerable to changes in their timing as well as to increased competition. For raptors feeding on small mammals, less snow cover increases access to prey but may over time reduce populations of small animals that thrive under snow. Long-distance migrants dependent on wetlands and foliage that supports insects benefit from increases in winter precipitation (Pearce-Higgins and Green 2014:107–118). Arctic-nesting shorebirds wintering in the Southern Hemisphere may need to travel farther, not less, to reach newly suitable wintering areas; their success will depend on the availability of stopover sites.

New communities may create or intensify competition among species that previously had little interaction in winter. Populations of ecological generalists are expected to increase, while specialists may find fewer suitable locales. Experienced resident birds are likely to outcompete new winter neighbors, such as an influx of short-distance migrants and former migrants. But nonmigratory birds have relatively low dispersal abilities—a few kilometers for small birds, more for larger species—so these may not be able to shift their winter range as easily as migrants, especially where suitable habitat is separated by ecological barriers that nonmigratory birds cross with difficulty (Brotons and Jiguet 2010).

In both the American and Old World tropics, wintering passerines compete relatively little with resident birds. The migrants often dominate and exploit different feeding or habitat niches, but these birds will not be in place like residents to adjust their schedules to changes in the timing of food availability. Reductions in precipitation, predicted for many tropical regions, may disproportionately reduce food resources for the birds living in forest edges and the more open habitats favored by many migrants, because these are likely to dry out faster than forest interiors. Because migratory birds have more flexible basal metabolic rates than sedentary tropical birds that live in habitats with little temperature variation, however, the migrants are likely to have more resilience to extreme heat (Şekercioğlu et al. 2012).

These long-distance migrants may be at a greater competitive disadvantage when they return to their breeding range, where milder winters will have enabled more residents to survive and both residents and short-distance migrants to begin nesting earlier. A comparison of the landbird communities of the Lake Constance region in central Europe between 1980–1981 and 1990–1992 found that the increasingly warm winters of more recent years led to a slight increase in the number and proportion of breeding short-distance and resident species and a strong decrease in the number and proportion of long-distance migrants. Conditions in the Sahel, where the long-distance migrants from Lake Constance winter, did not change significantly during those years, indicating that winter survival rates were not a factor in the decline of these birds (Lemoine and Bohning-Gaese 2003).

RAPID EVOLUTION IN RESPONSE TO CLIMATE CHANGE

New environmental conditions could quickly render maladaptive some traits that have been selected for during thousands of years of a stable climate. Gradual changes in local environments will likely be less disruptive than any increase in extreme weather events such as droughts or floods that eliminate large proportions of any species' population. Some birds have already shown physical changes caused by global warming. Between 1961 and 2006, a period of increasing average temperatures at all seasons, body weight decreased in more than 100 species of breeding, migratory, and wintering birds regularly measured and banded at the Powdermill Nature Reserve in western Pennsylvania. Smaller body size within a population of any species is associated with warmer temperatures (Bergmann's rule). The greatest declines in size among the birds measured at Powdermill were in long-distance migrants and in species with northerly breeding distributions, two overlapping categories. Whether selection for smaller size was operating most intensely during summer, winter, or migration cannot be determined; over the 45 years of the study, many of these species lived through increasingly warm seasons at all times of year (Van Buskirk et al. 2010).

Similarly, of 14 British birds measured at two locations in England from 1968 to 2003, 4 resident species showed decreases in body weight (Yom-Tov et al. 2006). Specimens of 4 resident passerines in Israel collected between 1950 and 1999 showed a significant decline in body mass during this period (Yom-Tov 2001). As with the North American birds, the relative impact on body size of increasing warmth in each season cannot be determined, but winter temperatures will set the lower limit to which any bird's body mass may reduce. Smaller size may enable birds to withstand greater heat, while it may diminish their capacity to maintain adequate body temperature through the coldest winter nights.

Some physical changes are clearly due to factors during the nonbreeding season that have changed the benefits or costs of various anatomical features. The effects of natural selection during winter may then be accelerated by sexual selection. In other cases, however, traits that are traditionally favored in mate choice may run counter to features beneficial to birds as climate changes (Møller et al. 2010:170–172). Thus far, only one study has demonstrated the link between natural selection and sexual selection of a feature affected by climate. Among male Barn Swallows wintering in South Africa and nesting in Denmark, changes in tail length demonstrate the potential for rapid selection of certain traits. These birds migrate 12,500 km each spring, typically over 36–50 days. Between 1984 and 2003, length of the outermost tail feathers of yearling males increased by an average of 11.4 mm, while female average tail length increased by 3.3 mm. These feathers are grown in South Africa between December and March. Given that tail length has been increasing, the males with longer tail feathers are presumed to survive their lengthy migration in years with adverse environmental conditions, while shorter-tailed males may not; females migrate more slowly, so their tail size may be a less critical factor. Once back in Denmark, females selectively mate with males having longer tails, a feature female Barn Swallows have always favored, and pass this attribute to the next generation. In this study, the only area with known adverse conditions was Algeria, where swallows are well on their way north, but the results demonstrate the potential for rapid selection effects from climate changes at any stage of the nonbreeding season (Møller and Szép 2005).

THE FUTURE

The extent and duration of future climate change depends on the amount of greenhouse gases released into the atmosphere and how long they remain; different gases have different rates of dissolution. Predictions of the level by which global average temperatures will rise continue to shift as the capacity of scientists to model future scenarios advances—and as the slowness of governments and international organizations to limit greenhouse gas emissions makes it increasingly difficult to bring levels to within the range of minimal damage. The trend in analyses has been to find an accelerating rate of temperature rise. Today, with the global average temperature the highest it has been in 650,000 years, a rise of another 2° C will make the world climate warmer than it has been in 2.5 million years (Pearce-Higgins and Green 2014:9). Since temperatures will rise more in some regions than others, the impacts on ecosystems will vary.

Predicted changes include the following (Hurrell and Trenberth 2010):

- Greatest temperature increases will occur at high northern latitudes and over land, with less warming over the southern oceans and North Atlantic.
- Snow cover will contract and more Arctic permafrost will thaw.
- Sea ice coverage will shrink, with large parts of the Arctic Ocean no longer having year-round ice cover by the mid-twenty-first century.
- Average sea levels may rise 30–40 cm, excluding possible collapse of ice sheets.
- Hot extremes, heat waves, and heavy precipitation will become more frequent.
- Precipitation will increase in high latitudes and decrease in most subtropical land regions.

- Droughts will be longer and more intense.
- Sea level pressure will increase over the subtropics and middle latitudes and decrease over high latitudes, moving storm tracks poleward.

A global model that combined data on the elevational ranges of landbirds, four habitat-loss scenarios, and an intermediate estimate of surface warming of 2.8° C projected 400–550 likely extinctions and an additional 2,150 species at risk of extinction by 2100. Worldwide, every degree of additional warming projected an additional extinction of 100–500 species. Tropical birds are at greatest risk, both the species living on mountains as habitats move upward and those in extensive lowlands like the Amazon and Congo basins where there is no higher place to go. Most of the species this study found at future risk are not currently considered vulnerable to other conservation threats (Şekercioğlu et al. 2007). Another projection, focused on 314 species in the United States and Canada that are already classified as endangered or threatened by climate change, used several climate scenarios to 2080 to assess impacts on their current habitat. It found that 165 of these would lose habitat in their breeding range, 81 only in their winter range, and 66 in both (Schuetz et al. 2015).

In addition to the incremental losses of habitat from gradual changes in climate, extreme events such as hurricanes, winter storms, floods, and droughts are predicted to increase in number or intensity or both, and these may have devastating local effects. The most frequent course of hurricanes, over the Caribbean islands, Mexico, and Central America as well as the United States, aligns with many of the places where North American wintering birds are most concentrated. Some migrants have already arrived during hurricane season, and others will reach these sites to find them transformed. Hurricane Maria, which struck Puerto Rico in September 2017, completely destroyed the El Yunque National Forest, blowing down most of the trees, stripping leaves and mosses from the remaining ones, and causing landslides. Much else on the island was left in similar condition. Staff of the International Institute of Tropical Forestry on Puerto Rico estimated it would take 100 years for the forest to recover—presuming no similar hurricanes occur over that period (Ferré-Sadurni 2017). The projected increase in hurricane intensity, with more storms moving beyond the Caribbean and Atlantic coast of North America, will also affect birds at sea. Cory's Shearwaters (*Calonectris diomedea*) travel over 25,000 km every year between their Mediterranean breeding sites and winter range in the South Atlantic; fewer birds return in spring following years with more hurricanes along their route (Genovart et al. 2013).

At higher latitudes, winter storms are projected to become more frequent. These may prevent birds from feeding or may reduce their ability to withstand exposure. The European Shag (*Phalacrocorax aristotelis*) is especially vulnerable to winter gales because of its incomplete waterproofing and very small fat reserves, both of which reduce buoyancy and help it swim swiftly underwater but provide little insulation in cold and wet weather. Population drops at the shag colony on the Isle of May in southeastern Scotland between 1961 and 2006 were often strongly correlated with heavy rainfall and strong easterly winds in February, when many birds died of exposure or starvation. In addition to mortality resulting from poor insulation, shags cease feeding when winds are strongest. In contrast, shag colonies on Scotland's west coast have less winter mortality because the more varied contours of the coastline provide shelter from winds of every direction (Frederiksen et al. 2008).

Flooding of restricted habitats by heavy rainfall can prevent birds from feeding. During October–December 2000 more rain fell in England than in any year since records began in 1766. That winter an isolated population of Bearded Tits (*Panurus biarmicus*) in northwest England declined by 94%, when a prolonged flood in November and December covered the *Phragmites* litter layer in the reed bed where these birds feed; this was followed by 23 days of cold weather, including 4 days of continuous snow cover. After the flood, the surviving tits were 20% lighter than during previous winters. The population did not double until 2004 and was then still only a small fraction of what it had been in 2000 before the rains began (Wilson and Peach 2006).

The Sahel, where wintering birds' survival and fitness are closely linked with the extent of precipitation during the prior rainy season, May to September or October, is projected to have more dry periods. Recent droughts are a preview of the potential future impact of even a single unusual winter. The extremely dry winter of 1968–1969 in the western Sahel is thought responsible for the 77% drop in Whitethroats (*Sylvia communis*) returning to the United Kingdom the following spring. There were similar declines that year in western Europe. Unlike some other European winterers, including Willow Warbler (*Phylloscopus trochilus*) and Pied Flycatcher, that begin the season in this region but move farther south after several weeks, Whitethroats remain in West Africa all winter (Winstanley et al. 1974). Despite much better conditions in the Sahel the last few decades, Whitethroats in the United Kingdom have increased only slightly since the mid-1980s (Wormworth and Şekercioğlu 2011:140).

These gradual and occasional brief dramatic effects of climate change potentially affect population trends and possible declines toward extinction in two ways: reducing

Short-term extreme winter events such as flooding and freezing can decimate vulnerable populations like these Bearded Tits, which in one area declined by 94% during the severe British winter of 2000.

fecundity during the breeding season and lowering survival the rest of the year. Warming winter conditions may link the two factors with unexpected consequences. Eurasian Oystercatchers (*Haematopus ostralegus*) have higher survival rates during warm winters, when they can more easily meet their daily energy requirements. But their main prey species in spring and summer, ragworms (*Nereis diversicolor*), are less abundant after warm winters, and this will reduce fecundity. For a long-lived species like the oystercatcher, population stability will therefore depend on a mix of colder and warmer winters (van de Pol et al. 2010).

Breeding success of Gray Jays (*Perisoreus canadensis*) declined at the southern edge of their range in Algonquin Provincial Park in Ontario, Canada, between 1980 and 2006 as mean air temperatures rose 0.4° C per decade, because the start of freezing weather was increasingly delayed and much of the food the jays hoarded during autumn rotted before they retrieved it for themselves or their young later in winter (Waite and Strickland 2006).

The projected decrease in rainfall in subtropical regions will have widespread impacts on birds wintering at these latitudes in every hemisphere, as already demonstrated by population fluctuations during recent dry periods. These often lead to poorer reproduction the following year. Bank Swallows (*Riparia riparia*) from Hungary winter in the southern Sahel of West Africa, where they feed over wetlands; at Hungarian colonies censused from 1986 to 1992, their breeding population dropped by half following the dry West African winter of 1990–1991, when they returned two weeks later than in other years and very few swallows attempted a second brood (Szép 1995). In Jamaica, American Redstarts (*Setophaga ruticilla*) wintering in dry scrub between 2002 and 2005 usually had less food available and were in poorer condition than redstarts elsewhere on the island in mangrove forests. In the exceptionally rainy spring of 2004, the scrub redstarts' body condition was equal to that of mangrove birds and they departed on the same schedule. But between 1995 and 2005 rainfall in Jamaica declined 17%, conforming to climatological predictions; in the future more of the redstarts are likely to be in the poorer condition typical of scrub birds in most winters. This will delay their return to North America, increase the distance they travel, and reduce their breeding success (Studds and Marra 2007).

Breeding success of Gray Jays has declined in Ontario where rising winter temperatures have caused some of the food the jays store to rot before they retrieve it.

Sea-level rise may substantially reduce the habitat available to birds of coastal wetlands, estuaries, and beaches. On the east coast of the United States, the sea will penetrate farther and at greater depth into coastal wetlands from the mid-Atlantic states all the way around the Gulf of Mexico. In Chesapeake Bay, sea level is projected to rise 43 to 74 cm in this century, eliminating many tidal marshes and eelgrass beds that support major wintering populations of several duck species. On the Pacific coast, the combination of sea-level rise and decreased water flow in rivers will affect estuaries used by both waterfowl and shorebirds as stopover and wintering sites (Cox 2010:173). A model projecting the extent of future intertidal wetlands at four US sites critical to shorebirds (Willapa Bay, Humboldt Bay, San Francisco Bay, and Delaware Bay) found that with a conservative warming scenario of 2° C, 20–70% of these wetlands would be lost within this century. Losses would be exacerbated at sites with steep topography or seawalls preventing wetlands from moving inland (Galbraith et al. 2002).

A model used to assess the quality of Poole Harbour, Dorset, on the south coast of England, a site of national and international importance for several wintering shorebird species ranging in size from Dunlin (*Calidris alpina*) to Eurasian Oystercatcher and Eurasian Curlew (*Numenius arquata*), found that they would all be seriously affected by sea-level rise. By 2080, sea level is projected to rise on this coast an estimated 19–79 cm. Since Poole Harbour has only a narrow outlet to the sea, tides recede slowly and the increase in water will reduce the time intertidal areas would be available for feeding. An advance of mean tidal height by 10 cm would reduce survival rates in all shorebird species, and a 40 cm advance would prevent any birds from surviving the winter (Durell et al. 2006).

On tropical and some temperate coasts where shorebirds winter, sea-level rise is enabling mangroves to colonize salt marshes and tidal flats. Even where mangroves have not eliminated shorebird habitat, many birds will not feed or roost near mangroves because these can conceal predators. From New Zealand through Australia and north to Japan, this is an increasing problem, with mangroves likely to expand their range on coasts in both the Northern and Southern Hemispheres as temperatures rise (Sutherland et al. 2012). On the Pacific coast of South America and in parts of Argentina the beaches used by wintering shorebirds are backed by high bluffs rather than wetlands; as sea level rises, the exposed portion of the beach will become narrower and narrower. Eventually, the bluffs may be eaten away by the sea and will collapse, forming a new layer of beach, but the small animals on which shorebirds feed may not be able to colonize this new level of sediment rapidly.

In the tropics, resident landbirds, many with restricted ranges and poor mobility, are projected to suffer high rates of extinction as shifts in elevational range push montane species into smaller and smaller areas, often competing with new colonizers from lower elevations, while lowland birds in flat regions have no place to move. Migrants to the tropics will be affected by the same shifts in habitat suitability, which will be exacerbated by changes in seasonal rainfall and the increase of extreme weather events such as heat waves and hurricanes. The Caribbean islands, Central America, and equatorial South America are projected to become significantly drier, likely reducing plant productivity and insect numbers, as well as increasing the frequency—once never present, now regular in wet tropical forests—of dry-season fires. Arid zones in the tropics and subtropics, the wintering sites for other species, will lose some of their water sources. Small desert birds will require 150–200% more water during the hottest parts of the day to survive projected temperature increases. Finally, once-restricted avian diseases may proliferate, with migrants potentially carrying them from one region to another (Şekercioğlu et al. 2012).

From pole to pole, seabird survival and fecundity are most affected by sea-surface temperature and its impact on the marine food chain. At high latitudes, even modest warming of the nutrient-rich cold-water currents reduces productivity below sustainable levels. A study from 1989 to 2002 of four alcids nesting at 70° N on Hornøya, an island off northern Norway in the Barents Sea, during a period of variable sea-surface temperatures found that survival between breeding seasons was lower when water temperatures were higher. Warm sea temperatures reduce numbers of the fish alcids consume. The greatest mortality occurs during autumn, when the birds are molting at sea and flightless, and therefore least able to search for richer feeding areas. In winter, they are more mobile. Since sea-surface temperature during the breeding season had no effect on survival, its likely continued rise in autumn and winter will be a critical factor for these far-northern alcids (Sandvik et al. 2005).

At 56° N, the Black-legged Kittiwake (*Rissa tridactyla*) colony on the Isle of May in southeastern Scotland declined by more than 50% between 1990 and 2002. The primary food given to nestlings is sand eels (*Ammodytes marinus*) spawned during the previous winter. Years with high winter sea-surface temperatures result in a scarcity of sand eels, putting both adult and young kittiwakes in poor condition by the end of the breeding season. Overwinter survival is then low, and returning birds, which first feed on the remaining sand eels from the previous year before shifting to the new generation, continue in poor condition. Here, too, the expected continued rise in sea-surface temperature could produce a downward spiral of both survival and fecundity (Frederiksen et al. 2004).

Mayes Island, off Kerguelen Island at 48° S in the Indian Ocean, is one of six sub-Antarctic islands where Blue Petrels (*Halobaena caerulea*) nest. They feed on krill and fish at sea between Kerguelen and the northern limit of Antarctic pack ice. Mortality in summer is very low; 70% occurs in winter. A censusing project between 1986 and 2001 found lower annual survival following winters with high sea-surface temperatures caused by El Niño events in the Pacific; these typically result in warmer waters around Kerguelen three or four years later and reduce krill production. In one unusual cycle, waters were warm from 1994 to 1997, and the Blue Petrel population decreased by 40%; far fewer birds returned to nest, and they produced fewer young than in years when waters were cold. Population fluctuated over the 15 years of the study, and mortality was greatest during winters of warm water when Blue Petrels were numerous and therefore competing more intensely for scarcer krill and fish, not only among themselves but also with other seabirds. Winters in which Blue Petrel populations were lower did not have such high rates of mortality if waters were warm, but these years left fewer adults to reproduce (Barbraud and Weimerskirch 2003).

At 66° S, three Antarctic seabirds nesting on the Île des Pétrels, Pointe Géologie Archipelago, Terre Adélie, between 1963 and 2002 showed a strong correlation in both population and breeding success with extent of sea ice in prior winters, the season critical for krill reproduction. Southern Fulmars (*Fulmar glacialoides*), Snow Petrels (*Pagodroma nivea*), and Emperor Penguins (*Aptenodytes forsteri*) all feed on Antarctic krill (*Euphausia superba*),

Emperor Penguin survival is closely correlated with Antarctic winter sea-surface temperature: as temperatures rise and sea ice is less extensive, penguins find less food.

but each species feeds when krill are at different ages. Thus the correlations between extent of winter sea ice, which varies from year to year, and seabird survival and fecundity show up as many as six and seven years later, when the krill reach the age consumed by each of these species. Populations of all three, however, declined over the study period, the fulmar and petrel gradually after 1980, the penguin by nearly half between 1975 and 1980. In years with low krill availability, the fulmar and the petrel do not attempt to nest; in years with abundant krill, all three species fledge more young. As the Southern Ocean continues to warm, diminishing the extent of winter sea ice, seabird fecundity will be reduced (Jenouvrier et al. 2005).

Elsewhere in Terre Adélie, at the Emperor Penguin colony near the Dumont d'Urville Station, winter temperatures between 1950 and 2000 were stable during the 1950s and 1960s but began to vary extensively and were high through the 1970s and early 1980s and then variable until 2000, while summer temperatures remained stable. As at the Île des Pétrels, the penguin population dropped by 50% in the late 1970s and remained stable at the lower level at least to 2000. There was a close negative correlation between winter sea-surface temperature and survival: survival was lower when temperatures were higher and sea ice less extensive. At the same time, since Emperor Penguins incubate and then feed chicks during the winter, hatching success is higher in winters when sea ice is less extensive and adults need not travel so far between their nest and open water to feed. As Antarctica continues to warm, it is not clear whether winter temperatures can remain within a range that maintains the right balance for the Emperor Penguin's complex needs (Barbraud and Weimerskirch 2001).

SUMMARY

Impacts of climate change have already affected the distributions and populations of birds, and these are likely to increase as changes in temperature, rainfall, ocean currents, and other factors accelerate in this century. Since each element of every ecosystem will respond differently, new communities will be formed that may be more advantageous or challenging for their members. Migratory birds, which depend on more sites over the course of the year, will be exposed to more changes, but their mobility may enable them to adapt while species of limited distribution may be squeezed into a shrinking range of suitable habitat.

The most visible impact of climate change on wintering birds has been the shifts in their range. Many species of middle latitudes are migrating less far, if at all, and their range in both winter and summer has moved poleward. Ground foragers can now survive the winter in places where snow cover used to make feeding difficult. Waterbirds are not limited by areas that once froze for months of winter. As spring arrives earlier and autumn commences later, many migrants have also altered their migration schedule and the time they spend in wintering areas. Short-distance migrants especially are leaving their wintering site sooner. Some now return there later, while others, mainly species that never attempt a second brood, leave their breeding areas after their young have fledged even though that is now sooner on the calendar than before.

Long-distance migrants, with schedules that are set by changes in day length or under endogenous control, have shifted their schedule far less. On arrival in spring, however, they find the season more advanced, often to the point where the food they give their young is no longer widely available. Resident birds in the same habitat have begun nesting weeks earlier, remaining in sync with the emergence of the prey they feed their young. The migrants least able to accelerate their schedule are the species most declining. Still unknown is whether migrants now reach their wintering area at a time that no longer matches the production of the food they depend on there; this may be especially critical for birds that cross the equator to latitudes where the austral spring is also advancing more rapidly.

In addition to evolving new migratory timetables and winter ranges, some birds have also shown physical responses to climate change. In eastern North America, more than 100 species now weigh less on average than a few decades ago; smaller body size is associated with warmer temperatures. Sexual selection as well as survival pressure may also influence the rapid evolution of new traits fostered by changing winter climate, since this is the season when many birds molt into the plumage they will display during the following breeding season.

Current prediction of future rates of climate change impacts, including greatest temperature increases at high latitudes, shrinkage of sea ice, sea-level rise, increase in hot extremes, heat waves, heavy precipitation, and droughts, with precipitation generally decreasing in most subtropical land regions, can be linked to potential extinction rates of birds. Tropical birds are most vulnerable, but some analyses also pinpoint the risks birds face in each season, identifying loss of winter habitat. The projected decrease in rainfall in subtropical regions will have widespread effects on birds wintering at these latitudes. Sea-level rise will reduce coastal estuaries and wetlands on which waterfowl and shorebirds depend. Warmer sea-surface temperatures change the distribution and reduce the productivity of the food chain on which seabirds depend. Studies from high latitudes in all oceans have shown that changes in winter production have more impact on seabird survival than summer changes. In addition to these incremental shifts, the anticipated increase in extreme events such as hurricanes, winter storms, floods, and droughts will affect the winter range and survival of many species.

BIBLIOGRAPHY

Aarif, K.M., S.B. Muzaffar, S. Babu, and P.K. Prasadan. 2014. Shorebird assemblages respond to anthropogenic stress by altering habitat use in a wetland in India. *Biodiversity and Conservation* 23(3):727–740.

Able, K.P., and J.R. Belthoff. 1998. Rapid "evolution" of migratory behaviour in the introduced house finch of eastern North America. *Proc. Roy. Soc. Lond. B* 265:2063–2071.

Adkisson, C.S. 1996. Red Crossbill (*Loxia curvirostra*). In *The Birds of North America*, no. 256 (A. Poole and F. Gill, eds.). Academy of Natural Sciences, Philadelphia, and American Ornithologists' Union, Washington, DC.

Aharon-Rotman, Y., K.L. Buchanan, N.J. Clark, M. Klaassen, and W.A. Buttemer. 2016. Why fly the extra mile? Using stress biomarkers to assess wintering habitat quality in migratory shorebirds. *Oecologia* 182(2):385–395.

Alisauskas, R.T. 1998. Winter range expansion and relationships between landscape and morphometrics of midcontinent Lesser Snow Geese. *Auk* 115(4):851–862.

Alisauskas, R.T., R.F. Rockwell, K.W. Dufour, E.G. Cooch, G. Zimmerman, K.L. Drake, J.A. Leafloor, T.J. Moser, and E.T. Reed. 2011. Harvest, survival, and abundance of midcontinent Lesser Snow Geese relative to population reduction efforts. *Wildlife Monographs* 179:1–42.

Alonso, J.C., L.M. Bautista, and J.A. Alonso. 2004. Family-based territoriality vs flocking in wintering common cranes *Grus grus*. *J. Avian Biol.* 35:434–444.

American Bird Conservancy. 2015. *Annual Report*, p. 9.

———. 2016. *Annual Report*, p. 6.

Amlaner, C.J., Jr., and N.J. Ball. 1983. A synthesis of sleep in birds. *Behaviour* 87:85–119.

Anderson, M.G., G.R. Hepp, F. McKinney, and M. Owen. 1988. Workshop summary: Courtship and pairing in winter. Pp. 123–131 in M.W. Weller, ed., *Waterfowl in Winter*. Minneapolis: University of Minnesota Press.

Andreev, A.V. 1991. Winter habitat segregation in the sexually dimorphic Black-billed Capercaillie *Tetrao urogalloides*. *Ornis Scandinav.* 22:287–291.

Archaux, F., P-Y. Henry, and G. Balanca. 2008. High turnover and moderate fidelity of White Storks *Ciconia ciconia* at a European wintering site. *Ibis* 150:421–424.

Areta, J.I., G.G. Mangini, F.A. Gandoy, F. Gorleri, D. Gomez, E.A. Depino, and E.A. Jordan. 2016. Ecology and behavior of Alder Flycatchers (*Empidonax alnorum*) on their wintering grounds in Argentina. *Wilson J. Ornithol.* 128(4):830–845.

Areta, J.I., and S.H. Seipke. 2006. A 10,000 Mississippi Kite flock observed in Fuerte Esperanza, Argentina. *Ornitologia Neotropical* 17:433–437.

Arlettaz, R., P. Patthey, M. Baltic, T. Leu, M. Schaub, R. Palme, and S. Jenni-Eiermann. 2007. Spreading free-riding snow sports represent a novel serious threat for wildlife. *Proc. Roy. Soc. Lond. B* 274(1614):1219–1224.

Askins, R.A., D.N. Ewert, and R.L. Norton. 1992. Abundance of wintering migrants in fragmented and continuous forest in the U.S. Virgin Islands. Pp. 197–206 in J.M. Hagan III and D.W. Johnston, eds., *Ecology and Conservation of Neotropical Migrant Landbirds*. Washington, DC: Smithsonian Institution Press.

Askins, R.A., J.F. Lynch, and R. Greenberg. 1983. Population declines in migratory birds in eastern North America. Pp. 1–57 in D.M. Power, ed., *Current Ornithology*, vol. 7. New York: Plenum Press.

Atkinson, E.C. 1993. Winter territories and night roosts of Northern Shrikes in Idaho. *Condor* 95:515–527.

Audubon, J.J. 1967. *The Birds of America*, vol. 6. New York: Dover Publications.

Avery, M.L. 1995. Rusty Blackbird (*Euphagus carolinus*). In *The Birds of North America*, no. 200 (A. Poole and F. Gill, eds.). Academy of Natural Sciences, Philadelphia, and American Ornithologists' Union, Washington, DC.

Back, G.N., M.R. Barrington, and J.K. McAdoo. 1987. Sage Grouse use of snow burrows in northeastern Nevada. *Wilson Bull.* 99(3):488–490.

Balda, R.P., and A.C. Kamil. 1992. Long-term spatial memory in Clark's Nutcracker *Nucifraga columbiana*. *Animal Behaviour* 44:761–769.

Balph, M.H. 1979. Flock stability in relation to social dominance and agonistic behavior in wintering Dark-eyed Juncos. *Auk* 96:714–722.

Baltosser, W.H., and S.M. Russell. 2000. Black-chinned Hummingbird (*Archilocus alexandri*). In *The Birds of North America*, no. 495 (A. Poole and F. Gill, eds.). The Birds of North America, Inc., Philadelphia.

Barbraud, C., and H. Weimerskirch. 2001. Emperor Penguins and climate change. *Nature* 411(6834):183–186.

———. 2003. Climate and density shape population dynamics of a marine top predator. *Proc. Roy. Soc. Lond. B* 270(1529):2111–2116.

Bateson, P.P.G., and R.C. Plowright. 1959. The breeding biology of the Ivory Gull in Spitsbergen. *Brit. Birds* 52(4):105–114.

Battley, P.F. 1997. The northward migration of Arctic waders in New Zealand: Departure behaviour, timing and possible migration routes of Red Knots and Bar-tailed Godwits from Farewell Spit, north-west Nelson. *Emu* 97:108–120.

Bautista, L.M., and J.C. Alonso. 2013. Factors influencing daily food-intake patterns in birds: A case study with wintering Common Cranes. *Condor* 115(2):330–339.

Beaumont, L.J., I.A.W. McAllan, and L. Hughes. 2006. A matter of timing: Changes in the first date of arrival and last date of departure of Australian migratory birds. *Global Change Biology* 12:1339–1354.

Bednekoff, P.A., H. Biebach, and J. Krebs. 1994. Great Tit fat reserves under unpredictable temperatures. *J. Avian Biol.* 25(2):156–160.

Benkman, C.W. 1992. White-winged Crossbill. In *The Birds of North America*, no. 27 (A. Poole, P. Stettenheim, and F. Gill, eds.). Academy of Natural Sciences, Philadelphia, and American Ornithologists' Union, Washington, DC.

Bennett, S.E. 1980. Interspecific competition and the niche of the American Redstart (*Setophaga ruticilla*) in wintering and breeding communities. Pp. 319–335 in A. Keast and E.S. Morton, eds., *Migrant Birds in the Neotropics: Ecology, Behavior, Distribution, and Conservation*. Washington, DC: Smithsonian Institution Press.

Bensch, S., D. Hasselquist, A. Hedenstrom, and U. Ottosson. 1991. Rapid moult among Palaearctic passerines in West Africa—an adaptation to the oncoming dry season? *Ibis* 133:47–52.

Benson, A-M., and K. Winker. 2001. Timing of breeding range occupancy among high-latitude passerine migrants. *Auk* 118(2):513–519.

Berger, A.J. 1966. Mid-winter nesting of the American Robin in western Pennsylvania. *Auk* 83:668.

Berner, T.O., and T.C. Grubb Jr. 1985. An experimental analysis of mixed-species flocking in birds of deciduous woodland. *Ecology* 66(4):1229–1236.

Berthold, P. 2001. *Bird Migration: A General Survey*. 2nd ed. Oxford: Oxford University Press.

Berthold, P., A.J. Heilbig, G. Mohr, and U. Querner. 1992. Rapid microevolution of migratory behavior in a wild bird species. *Nature* 360:668–670.

Berthold, P., W. van den Bossche, Z. Jakubiec, C. Kaatz, M. Kaatz, and U. Querner. 2002. Long-term satellite tracking sheds light upon variable migration strategies of White Stork (*Ciconia ciconia*). *J. Ornithol.* 143:489–495.

Biebach, H. 1977. Das Winterfett der Amsel (*Turdus merula*). *J. Ornithol.* 118:117–133.

Billerman, S.M., G.H. Huber, D.W. Winkler, R.J. Safran, and I.J. Lovette. 2011. Population genetics of a recent transcontinental colonization by breeding Barn Swallows (*Hirundo rustica*). *Auk* 128(3):506–513.

Black, J.M., and M. Owen. 1989. Parent-offspring relationships in wintering barnacle geese. *Animal Behaviour* 37:187–198.

Blake, J.G., and B.A. Loiselle. 1992. Habitat use by Neotropical migrants at La Selva Biological Station and Braulio Carrillo National Park, Costa Rica. Pp. 257–272 in J.M. Hagan III and D.W. Johnston, eds., *Ecology and Conservation of Neotropical Migrant Landbirds*. Washington, DC: Smithsonian Institution Press.

Blanco, G., J.A. Lemus, and J. Grande. 2006. Faecal bacteria associated with different diets of wintering Red Kites: Influence of livestock carcass dumps in microflora alteration and pathogen acquisition. *J. Applied Ecology* 43(5):990–998.

Blem, C.R. 1990. Avian energy storage. Pp. 59–113 in D.M. Power, ed., *Current Ornithology*, vol. 7. New York: Plenum Press.

Blem, C.R., and J.F. Pagels. 1984. Mid-winter lipid reserves of the Golden-crowned Kinglet. *Condor* 86:491–492.

Blohm, R.J., D.W. Sharp, P.I. Padding, R.W. Kokel, and K.D. Richkus. 2007. Integrated waterfowl management in North America. Pp. 199–203 in G.C. Boere, C.A. Galbraith, and D.A. Stroud, eds., *Waterbirds around the World*. Edinburgh: Stationary Office.

Bluhm, C.K. 1988. Temporal patterns of pair formation and reproduction in annual cycles and associated endocrinology in waterfowl. Pp. 123–185 in R.F. Johnston, ed., *Current Ornithology*, vol. 5. New York: Plenum Press.

Bock, C.E., and L.W. Lepthien. 1976. Synchronous eruptions of boreal seed-eating birds. *Am. Nat.* 110:559–571.

Bohning-Gaese, K. 2005. Influence of migrants on temperate bird communities: A macroecological approach. Pp. 143–153 in R. Greenberg and P.P. Marra, eds., *Birds of Two Worlds: The Ecology and Evolution of Migration*. Baltimore: Johns Hopkins University Press.

Both, C. 2010. Food availability, mistiming, and climatic change. Pp. 129–147 in A.P. Møller, W. Fiedler, and P. Berthold, eds., *Effects of Climate Change on Birds*. Oxford: Oxford University Press.

Both, C., S. Bouwhuis, C.M. Lessells, and M.E. Visser. 2006. Climate change and population declines in a long-distance migratory bird. *Nature* 441:81–83.

Boulet, M., H.L. Gibbs, and K.A. Hobson. 2006. Integrated analysis of genetic, stable isotope, and banding data reveal migratory connectivity and flyways in the northern Yellow Warbler (*Dendroica petechia*; *aestiva* group). *Ornithological Monographs*, no. 61:29–78.

Boulet, M., and D.R. Norris. 2006. The past and present of migratory connectivity. *Ornithological Monographs*, no. 61:1–13.

Boxall, P.C., and M.R. Lein. 1982. Possible courtship behavior by Snowy Owls in winter. *Wilson Bull.* 94(1):79–81.

———. 1989. Time budgets and activity of wintering Snowy Owls. *J. Field Ornithol.* 60(1):20–29.

Brittingham, M.C., and S.A. Temple. 1988. Impacts of supplemental feeding on survival rates of Black-capped Chickadees. *Ecology* 69:581–589.

———. 1992. Does winter bird feeding promote dependency? *J. Field Ornithol.* 63(2):190–194.

Brodeur, S., J.-P.L. Savard, M. Robert, P. Laporte, P. Lamothe, R.D. Titman, S. Marchand, S. Gilliland, and G. Fitzgerald. 2002. Harlequin duck *Histrionicus histrionicus* population structure in eastern Nearctic. *J. Avian Biol.* 33:127–137.

Brodsky, L.M., and P.J. Weatherhead. 1985. Time and energy constraints on courtship in wintering American Black Ducks. *Condor* 87:33–36.

Brooke, M. 2004. *Albatrosses and Petrels across the World*. Oxford: Oxford University Press.

Brooke, R.K., and P. Herroelen. 1988. The nonbreeding range of southern African bred European Bee-eaters *Merops apiaster*. *Ostrich* 59:63–66.

Brooks, D.J., ed. 1992. *Handbook of the Birds of Europe, the Middle East, and North Africa*, vol. 6. Oxford: Oxford University Press.

Brotons, L., and F. Jiguet. 2010. Bird communities and climate change. Pp. 275–294 in A.P. Møller, W. Fiedler, and P. Berthold, eds., *Effects of Climate Change on Birds*. Oxford: Oxford University Press.

Brown, C.R. 1997. Purple Martin (*Progne subis*). In *The Birds of North America*, no. 287 (A. Poole and F. Gill, eds.). Academy of Natural Sciences, Philadelphia, and American Ornithologists' Union, Washington, DC.

Brown, C.R., and M.B. Brown. 1995. Cliff Swallow (*Hirundo pyrrhonota*). In *The Birds of North America*, no. 149 (A. Poole and F. Gill, eds.). Academy of Natural Sciences, Philadelphia, and American Ornithologists' Union, Washington, DC.

Brown, D.R., and J.A. Long. 2007. What is a winter floater? Causes, consequences, and implications for habitat selection. *Condor* 109:548–565.

Brown, D.R., and T.W. Sherry. 2008. Solitary winter roosting of Ovenbirds in core foraging area. *Wilson J. Ornithol.* 120(3):455–459.

Brown, D.R., P.C. Stouffer, and C.M. Strong. 2000. Movement and territoriality of wintering Hermit Thrushes in southeastern Louisiana. *Wilson Bull.* 112(3):347–353.

Brown, L.H., E.K. Urban, and K. Newman. 1982. *The Birds of Africa*, vol. 1. London: Academic Press.

Brown, M.B., and C.R. Brown. 1988. Access to winter food resources by bright- versus dull-colored House Finches. *Condor* 90:729–731.

Burger, A.E. 1984. Winter territoriality in Lesser Sheathbills on wintering grounds at Marion Island. *Wilson Bull.* 96(1):20–33.

Burnham, K.K., and I. Newton. 2011. Seasonal movements of Gyrfalcons *Falco rusticolus* include extensive periods at sea. *Ibis* 153:468–484.

Burns, K.C., and J. van Horik. 2007. Sexual differences in food re-caching by New Zealand robins *Petroica australis*. *J. Avian Biol.* 38:394–398.

Cain, B.W. 1988. Wintering waterfowl habitat in Texas: Shrinking and contaminated. Pp. 583–596 in M.W. Weller, ed., *Waterfowl in Winter*. Minneapolis: University of Minnesota Press.

Callopy, M.W. 1977. Food caching by female American Kestrels in winter. *Condor* 79:63–68.

Cama, A., P. Josa, J. Ferrer-Obiol, and J.M. Arcos. 2011. Mediterranean Gulls *Larus melanocephalus* wintering along the Mediterranean Iberian coast: Numbers and activity rhythms in the species' main winter quarters. *J. Ornithol.* 152(4):897–907.

Carey, M.J., R.A. Phillips, J.R.D. Silk, and S.A. Shaffer. 2014. Trans-equatorial migration of Short-tailed Shearwaters revealed by geolocators. *Emu* 114:352–359.

Carlisle, J.D., G.S. Kaltenecker, and D.L. Swanson. 2005. Molt strategies and age differences in migration timing among autumn landbird migrants in southwestern Idaho. *Auk* 122(4):1070–1085.

Carr, J.M., and S.L. Lima. 2014. Wintering birds avoid warm sunshine: Predation and the costs of foraging in sunlight. *Oecologia* 174(3):713–721.

Carrete, M., G.R. Bortolotti, J.A. Sanchez-Zapata, A. Delgado, A. Cortés-Avizanda, J.M. Grande, and J.A. Donazar. 2013. Stressful conditions experienced by endangered Egyptian vultures on African wntering areas. *Animal Conservation* 16(3):353–358.

Catry, P., A. Campos, V. Almada, and W. Cresswell. 2004. Winter segregation of migrant European robins *Erithacus rubecula* in relation to sex, age, and size. *J. Avian Biol.* 35(3):204–209.

Cawthorne, R.A., and J.H. Marchant. 1980. The effects of the 1978/79 winter on British bird populations. *Bird Study* 27:163–172.

Chamberlin, M.L. 1980. Winter hunting behavior of a Snowy Owl in Michigan. *Wilson Bull.* 92(1):116–120.

Chan, K. 2001. Partial migration in Australian landbirds: A review. *Emu* 101:281–292.

Chesser, R.T. 1994. Migration in South America: An overview of the austral system. *Bird Conservation International* 4:91–107.

———. 1997. Patterns of seasonal and geographical distribution of austral migrant flycatchers (Tyrannidae) in Bolivia. *Ornithological Monographs*, no. 48:171–204.

———. 2005. Season distribution and ecology of South American austral migrant flycatchers. Pp. 168–181 in R. Greenberg and P.P. Marra, eds., *Birds of Two Worlds: The Ecology and Evolution of Migration*. Baltimore: Johns Hopkins University Press.

Chipley, R.M. 1976. The impact of wintering migrant wood warblers on resident insectivorous passerines on a subtropical Colombian oak woods. *Living Bird* 15:119–141.

Christie, K.S., and R.W. Ruess. 2015. Experimental evidence that ptarmigan regulate willow bud production to their own advantage. *Oecologia* 178(3):773–781.

Churchill, J.B., P.B. Wood, and D.F. Brinker. 2002. Winter home range and habitat use of female Northern Saw-whet Owls on Assateague Island, Maryland. *Wilson Bull.* 114(3):309–313.

Cimprich, D.A., F.R. Moore, and M.P. Guilfoyle. 2000. Red-eyed Vireo (*Vireo olivaceus*). In *The Birds of North America*, no. 527 (A. Poole and F. Gill, eds.). The Birds of North America, Inc., Philadelphia.

Clergeau, P. 1981. "Non-roosting" Starlings *Sturnus vulgaris* in Brittany. *Ibis* 123:527–528.

Cody, M.L. 1968. Interspecific territoriality among hummingbird species. *Condor* 70:270–271.

Constantini, D. 2015. Land-use changes and agriculture in the tropics: Pesticides as an overlooked threat to wildlife. *Biodiversity and Conservation* 24(7):1837–1839.

Conway, C.J., G.V.N. Powell, and J.D. Nichols. 1995. Over winter survival of Neotropical migratory birds in early-successional and mature tropical forests. *Conservation Biology* 9:855–864.

Cooper, S.J. 1999. The thermal and energetic significance of cavity roosting in Mountain Chickadees and Juniper Titmice. *Condor* 101:863–866.

Coppack, T., F. Pulido, M. Czisch, D.P. Auer, and P. Berthold. 2003. Photoperiodic response may facilitate adaptation to climatic change in long-distance migratory birds. *Proc. Roy. Soc. Lond. B* (Suppl.) 270:S43–S46.

Cormier, R.L., D.L. Humple, T. Gardali, and N.E. Seavy. 2013. Light-level geolocators reveal strong migratory connectivity and within-winter movements for a coastal California Swainson's Thrush (*Catharus ustulatus*) population. *Auk* 130(2):283–290.

Cotton, P.A. 2003. Avian migration phenology and global climate change. *Proc. Nat. Acad. Sci.* 100(21):12219–12222.

Cox, G.W. 2010. *Bird Migration and Global Change*. Washington, DC: Island Press.

Cox, R.R., Jr., and A.D. Afton. 1996. Evening flights of female Northern Pintails from a major roost site. *Condor* 98:810–819.

Cramp, S., ed. 1983. *Handbook of the Birds of Europe, the Middle East, and North Africa*, vol. 3. Oxford: Oxford University Press.

———, ed. 1988. *Handbook of the Birds of Europe, the Middle East, and North Africa*, vol. 5. Oxford: Oxford University Press.

Cramp, S., and D.J. Brooks, eds. 1992. *Handbook of the Birds of Europe, the Middle East, and North Africa*, vol. 6. Oxford: Oxford University Press.

Cresswell, W., and D.P. Whitfield. 1994. The effects of raptor predation on wintering wader populations at the Tyninghame estuary, southeast Scotland. *Ibis* 136:223–232.

Cristol, D.A., M.B. Baker, and C. Carbone. 1999. Differential migration revisited: Latitudinal segregation by age and sex class. Pp. 33–88 in V. Nolan Jr., E.D. Ketterson, and C.F. Thompson, eds., *Current Ornithology*, vol. 15. New York: Kluwer Academic/Plenum Publishers.

Csada, R.D., and R.M. Brigham. 1992. Common Poorwill. In *The Birds of North America*, no. 32 (A. Poole, P. Stettenheim, and F. Gill, eds.). Academy of Natural Sciences, Philadelphia, and American Ornithologists' Union, Washington, DC.

Cubie, D. 2014. Site fidelity, residency, and sex ratios of wintering Ruby-throated Hummingbirds (*Archilochus colubris*) on the southeastern U.S. Atlantic coast. *Wilson J. Ornithol.* 126(4):775–778.

Curry-Lindahl, K. 1963. Molt, body weights, gonadal development, and migration in *Motacilla alba*. Pp. 960–973 in C.G. Sibley, ed., *Proceedings of the XIII International Ornithological Congress*, vol. 2. Baton Rouge, LA: American Ornithologists' Union.

Custer, C.M., T.W. Custer, and D.W. Sparks. 1996. Radio telemetry documents 24-hour feeding activity of wintering Lesser Scaup. *Wilson Bull.* 106(3):556–566.

Dann, P. 2014. Prey availability, and not energy content, explains prey choice of Eastern Curlews *Numenius madagascariensis* in southern Australia. *Ardea* 102(2):213–224.

Davis, J.B., M. Guillemain, R.M. Kaminski, C. Arzel, J.M. Eadie, and E.C. Rees. 2014. Habitat and resource use by waterfowl in the Northern Hemisphere in autumn and winter. *Wildfowl* Special Issue 4:17–69.

Dawson, A., T.J. Nicholls, A.R. Goldsmith, and B.K. Follett. 1988. Comparative endocrinology of photorefractoriness. Pp. 634–640 in H. Ouellet, ed., *Acta XIX Congressus Internationalis Ornithologici*, vol. 1. Ottawa: University of Ottawa Press.

Dawson, W.R., and R.L. Marsh. 1989. Metabolic acclimatization to cold and season in birds. Pp. 83–94 in C. Bech and R.E. Reinertsen, eds., *Physiology of Cold Adaptation in Birds*. New York: Plenum Press.

Dean, W.R.J., and W.R. Tarboton. 1983. Osprey breeding records in South Africa. *Ostrich* 54:238–239.

Dekker, D., M. Out, M. Tabak, and R. Ydenberg. 2012. The effect of kleptoparasitic Bald Eagles and Gyrfalcons on the kill rate of Peregrine Falcons hunting Dunlins wintering in British Columbia. *Condor* 114(2):290–294.

Dekker, D., and R. Ydenberg. 2004. Raptor predation on wintering Dunlins in relation to the tidal cycle. *Condor* 106:415–419.

Delany, S., and D. Scott, eds. 2006. *Waterbird Population Estimates*. 4th ed. Wageningen, Netherlands: Wetlands International.

De La Zerda Lerner, S., and D.F. Stauffer. 1998. Habitat selection by Blackburnian Warblers wintering in Colombia. *J. Field Ornithol.* 69(3):457–465.

de Lucas, M., G.F.E. Janss, D.P. Whitfield, and M. Ferrer. 2008. Collision fatality of raptors in wind farms does not depend on raptor abundance. *J. Applied Ecology* 45(6):1695–1703.

Derrickson, K.C., and R. Breitwisch. 1992. Northern Mockingbird. In *The Birds of North America*, no. 7 (A. Poole, P. Stettenheim, and F. Gill, eds.) Academy of Natural Sciences, Philadelphia, and American Ornithologists' Union, Washington, DC.

Devereux, C.L., M.J.H. Denny, and M.J. Whittingham. 2008. Minimal effects of wind turbines on the distribution of wintering farmland birds. *J. Applied Ecology* 45(6):1689–1694.

Dhondt, A.A., M.J.L. Driscoll, and E.C.H. Swarthout. 2007. House Finch *Carpodacus mexicanus* roosting behaviour during the non-breeding season and possible effects of mycoplasmal conjunctivitis. *Ibis* 149:1–9.

Dias, M.P., J.P. Granadeiro, M. Lecoq, C.D. Santos, and J.M. Palmeirim. 2006. Distance to high-tide roosts constrains the use of foraging areas by Dunlins: Implications for the management of estuarine wetlands. *Biological Conservation* 131(3):446–452.

Dias, M.P., J. P. Granadeiro, R.A. Phillips, H. Alonso, and P. Catry. 2011. Breaking the rule: Individual Cory's Shearwaters shift winter destinations between hemispheres and across ocean basins. *Proc. Roy. Soc. Lond. B* 278(1713):1786–1793.

Dias, R.A., D.E. Blanco, A.P. Goijman, and M.E. Zaccagnini. 2014. Density, habitat use, and opportunities for conservation of shorebirds in rice fields in southeastern South America. *Condor* 116:384–393.

Diggs, N.E., P.P. Marra, and R.J. Cooper. 2011. Resource limitation drives patterns of habitat occupancy during the nonbreeding season for an omnivorous songbird. *Condor* 113(3):646–654.

Dinets, V., and M. Sanchez. 2017. Brown Dippers (*Cinclus pallasi*) overwintering at −65°C in northeastern Siberia. *Wilson J. of Ornithol.* 129(2):397–400.

Dingle, H. 2008. Bird migration in the Southern Hemisphere: A review comparing continents. *Emu* 108:341–359.

Divoky, G.J. 1976. The pelagic feeding habits of Ivory and Ross's Gulls. *Condor* 78:85–90.

Dobinson, H.M., and A.J. Richards. 1964. The effects of the severe winter of 1962/63 on birds in Britain. *Brit. Birds* 57(10):373–434.

Dodd, S.L., and M.A. Colwell. 1996. Seasonal variation in diurnal and nocturnal distributions of nonbreeding shorebirds at North Humboldt Bay, California. *Condor* 98:196–207.

———. 1998. Environmental correlates of diurnal and nocturnal foraging patterns of nonbreeding shorebirds. *Wilson Bull.* 110(2):182–189.

Dolbeer, R.A. 1982. Migration patterns for age and sex classes of blackbirds and starlings. *J. Field Ornithol.* 53(1):28–46.

Dolby, A.S., and T.C. Grubb Jr. 1999. Effects of winter weather on horizontal and vertical use of isolated forest fragments by bark-foraging birds. *Condor* 101:408–412.

Dowsett, R.J. 1988a. Migration among southern African land birds. Pp. 765–777 in H. Ouellet, ed., *Acta XIX Congressus Internationalis Ornithologici*, vol. 1. Ottawa: University of Ottawa Press.

———. 1988b. Intra-African migrant birds in south-central Africa. Pp. 778–790 in H. Ouellet, ed., *Acta XIX Congressus Internationalis Ornithologici*, vol. 1. Ottawa: University of Ottawa Press.

Dowsett-Lemaire, F. 1979. The imitative range of the song of the Marsh Warbler *Acrocephalus palustris*, with special reference to imitations of African birds. *Ibis* 121:453–468.

Drent, R., and T. Piersma. 1990. An exploration of the energetics of leap-frog migration in Arctic breeding waders. Pp. 399–412 in E. Gwinner, ed., *Bird Migration: Physiology and Ecophysiology*. Berlin: Springer-Verlag.

Dugger, B.D. 1997. Factors influencing the onset of spring migration in Mallards. *J. Field Ornithol.* 68(3):331–337.

Dugger, B.D., and K.M. Dugger. 2002. Long-billed Curlew (*Numenius americanus*). In *The Birds of North America*, no. 628 (A. Poole and F. Gill, eds.). The Birds of North America, Inc., Philadelphia.

Durell, S.E.A. Le V. dit. 2007. Differential survival in adult Eurasian Oystercatchers *Haematopus ostralegus*. *J. Avian Biol.* 38:530–535.

Durrell, S.E.A. Le V. dit., R.A. Stillman, W.W.G. Caldow, S. McGrotry, A.D. West, and J. Humphreys. 2006. Modelling the effect of environmental change on shorebirds: A case study on Poole Harbour, UK. *Biological Conservation* 131(3):459–473.

Eaton, R.F. 1934. The migratory movements of certain colonies of Herring Gulls in eastern North America. Part 3. *Bird-Banding* 5:70–84.

Ebbinge, B., A. St. Joseph, P. Prokosch, and B. Spaans. 1982. The importance of spring staging areas for Arctic-breeding geese, wintering in western Europe. *Aquila* 89:249–258.

Egan, E.S., and M.C. Brittingham. 1994. Winter survival rates of a southern population of Black-capped Chickadees. *Wilson Bull.* 106(3):514–521.

Egbert, J.R., and J.R. Belthoff. 2003. Wing shape in House Finches differs relative to migratory habit in eastern and western North America. *Condor* 105(4):825–829.

Egevang, C., I.J. Stenhouse, R.A. Phillips, A. Petersen, J.W. Fox, and J.R.D. Silk. 2010. Tracking of Arctic terns *Sterna paradisaea* reveals longest animal migration. *Proc. Nat. Acad. Sci.* 107(5):2078–2081.

Eisenmann, E. 1959. South American migrant swallows of the genus *Progne* in Panama and northern South America; with comments on their identification and molt. *Auk* 76:528–532.

Eisenmann, E., and F. Haverschmidt. 1970. Northward migration to Surinam of South American martins (*Progne*). *Condor* 72:368–369.

Elkins, N. 2004. *Weather and Bird Behaviour*. London: T. & A.D. Poyser.

Ellwood, E.R., R.B. Primack, and M.L. Talmadge. 2010. Effects of climate change on spring arrival times of birds in Thoreau's Concord from 1851 to 2007. *Condor* 112(4):754–762.

England, A.S, M.J. Bechard, and C.S. Houston. 1997. Swainson's Hawk (*Buteo swainsoni*). In *The Birds of North America*, no. 265 (A. Poole and F. Gill, eds.). Academy of Natural Sciences, Philadelphia, and American Ornithologists' Union, Washington, DC.

Estelle, V., and E.D. Grosholz. 2012. Experimental test of the effects of a non-native invasive species on a wintering shorebird. *Conservation Biology* 26(3):472–481.

Estrada, A., and R. Coates-Estrada. 2005. Diversity of Neotropical migratory landbird species assemblages in forest fragments and man-made vegetation in Los Tuxtlas, Mexico. *Biodiversity and Conservation* 14(7):1719–1734.

Evans, P.R. 1976. Energy balance and optimal foraging strategies in shorebirds: Some implications for their distributions and movements in the non-breeding season. *Ardea* 64:117–139.

Evans Mack, D., and W. Yong. 2000. Swainson's Thrush (*Catharus ustulatus*). In *The Birds of North America*, no. 540 (A. Poole and F. Gill, eds.). The Birds of North America, Inc., Philadelphia.

Faaborg, J., and W.J. Arendt. 1984. Population sizes and philopatry of winter resident warblers in Puerto Rico. *J. Field Ornithol.* 55(3):376–378.

Faaborg, J., W.J. Arendt, J.D. Toms, K.M. Dugger, W.A. Cox, and M. Canals Mora. 2013. Long-term decline of a winter-resident bird community in Puerto Rico. *Biodiversity and Conservation* 22(1):63–75.

Fayet, A.L., R. Freeman, A. Shoji, H.L. Kirk, O. Padget, C.M. Perrins, and T. Guilford. 2016. Carry-over effects on the annual cycle of a migratory seabird: An experimental study. *J. Anim. Ecol.* 85(6):1516–1527.

Fedynich, A.M., and R.D. Godfrey Jr. 1989. Gadwall pair recaptured in successive winters on the Southern High Plains of Texas. *J. Field Ornithol.* 60(2):168–170.

Fellowes, E.C. 1971. House Martins apparently roosting in nests of Striped Swallows. *Brit. Birds* 64(10):460.

Fernández, G., H. de la Cueva, and N. Warnock. 2001. Phenology and length of stay of transient and wintering Western Sandpipers at Estero Punta Banda, Mexico. *J. Field Ornithol.* 74(4):509–520.

Fernández, G., and D.B. Lank. 2012. Territorial behavior of Western Sandpipers on their nonbreeding grounds: Effect of sex and foraging interference. *J. Field Ornithol.* 83(3):272–281.

Ferré-Sadurni, L. 2017. Lush refuge in Puerto Rico left shredded by hurricane. *New York Times*, October 12.

Fischer, D.H. 1981. Winter time budgets of Brown Thrashers. *J. Field Ornithol.* 52(4):304–308.

Fitzpatrick, J.W. 1980. Wintering of North American Tyrant Flycatchers in the Neotropics. Pp. 67–78 in A. Keast and E.S. Morton, eds., *Migrant Birds in the Neotropics: Ecology, Behavior, Distribution, and Conservation*. Washington, DC: Smithsonian Institution Press.

Ford, T.B., and J.A. Gieg. 1995. Winter behavior of the Common Loon. *J. Field Ornithol.* 66(1):22–29.

Fort, J., H. Steen, H. Strøm, Y. Tremblay, E. Grønningsæter, E. Pettex, W. Porter, and D. Grémillet. 2013. Energetic consequences of contrasting winter migratory strategies in a sympatric Arctic seabird duet. *J. Avian Biol.* 44(3):255–262.

Foster, M. 2007. Winter behavior and ecology of the Alder Flycatcher (*Empidonax alnorum*) in Peru. *Ornitologia Neotropical* 18:171–186.

Fox, A.D., and I.K. Petersen. 2006. Assessing the degree of habitat loss to marine birds from the development of offshore wind farms. Pp. 801–804 in G.C. Boere, C.A. Galbraith, and D.A. Stroud, eds., *Waterbirds around the World*. Edinburgh: Stationary Office.

Fraser, K.C., A. Shave, A. Savage, A. Ritchie, K. Bell, J. Siegrist, J.D. Ray, K. Applegate, and M. Pearman. 2017. Determining fine-scale migratory connectivity and habitat selection for a migratory songbird by using GPS technology. *J. Avian Biol.* 48:339–345.

Frazier, A., and V. Nolan Jr. 1959. Communal roosting by the Eastern Bluebird in winter. *Bird-Banding* 30:219–226.

Frederiksen, M., F. Daunt, M.P. Harris, and S. Wanless. 2008. The demographic impact of extreme events: Stochastic weather drives survival and population dynamics in a long-lived seabird. *J. Anim. Ecol.* 77(5):1020–1029.

Frederiksen, M., S. Wanless, M.P. Harris, P. Rothery, and L.J. Wilson. 2004. The role of industrial fisheries and oceanographic change in the decline of North Sea black-legged kittiwakes. *J. Applied Ecology* 41(6):1129–1139.

Fretwell, S. 1969. Dominance behavior and winter habitat distribution in juncos (*Junco hyemalis*). *Bird-Banding* 40(1):1–26.

Froehlich, D.R., S. Rohwer, and B.J. Stutchbury. 2005. Spring molt constraints versus winter territoriality. Pp. 321–335 in R. Greenberg and P.P. Marra, eds., *Birds of Two Worlds: The Ecology and Evolution of Migration*. Baltimore: Johns Hopkins University Press.

Fry, C.H. 1992. The Moreau ecological overview. *Ibis* 134 suppl. 1:3–6.

Galbraith, H., R. Jones, R. Park, J. Clough, S. Herrod-Julius, B. Harrington, and G. Page. 2002. Global climate change and sea level rise: Potential losses of intertidal habitats for shorebirds. *Waterbirds* 25(2):173–183.

Garthe, S., and O. Huppop. 2004. Scaling possible adverse effects of marine wind farms on seabirds: Developing and applying a vulnerability index. *J. Applied Ecology* 41(4):724–734.

Gates, J.E., and L.W. Gysel. 1978. Avian nest dispersion and fledging success in field-forest ecotones. *Ecology* 59:871–883.

Gauthreaux, S.A., Jr. 1982. The ecology and evolution of avian migration systems. Pp. 93–168 in D.S. Farner, J.R. King, and K.C. Parkes, eds., *Avian Biology*, vol. 6. New York: Academic Press.

Gautier-Leclerc, M., A. Tamisier, and F. Cézilly. 2000. Sleep-vigilance trade-off in Gadwall during the winter period. *Condor* 102:307–313.

Genovart, M., A. Sanz-Aguilar, A. Fernandez-Chacon, J.M. Igual, R. Pradel, M.G. Forero, and D. Oro. 2013. Contrasting effects of climatic variability on the demography of a trans-equatorial migratory seabird. *J. Anim. Ecol.* 82(1):121–130.

Germain, R.R., P.P. Marra, T.K. Kyser, and L.M. Ratcliffe. 2010. Adult-like plumage coloration predicts winter territory quality and timing of arrival on the breeding grounds of yearling male American Redstarts. *Condor* 112(4):676–682.

Gianuca, D., R.A. Phillips, S. Townley, and S.C. Votier. 2017. Global patterns of sex- and age-specific variation in seabird bycatch. *Biological Conservation* 205:60–76.

Gill, B.J., and M.E. Hauber. 2012. Piecing together the epic transoceanic migration of the Long-tailed Cuckoo (*Eudynamys taitensis*): An analysis of museum and sighting records. *Emu* 112:326–332.

Gill, J.A., K. Norris, P.M. Potts, T.G. Gunnarsson, P.W. Atkinson, and W.J. Sutherland. 2001. The buffer effect and large-scale population regulations in migratory birds. *Nature* 412:436–438.

Gill, R.B., Jr., P. Canevari, and E.H. Iversen. 1998. Eskimo Curlew (*Numenius borealis*). In *The Birds of North America*, no. 347 (A. Poole and F. Gill, eds.). Academy of Natural Sciences, Philadelphia, and American Ornithologists' Union, Washington, DC.

Gillings, S., R.J. Fuller, and W.J. Sutherland. 2005. Diurnal studies do not predict nocturnal habitat choice and site selection of European Golden-Plovers (*Pluvialis apricaria*) and Northern Lapwings (*Vanellus vanellus*). *Auk* 122(4):1249–1260.

Gochfeld, M. 1985. Numerical relationships between migrant and resident bird species in Jamaican woodlands. Pp. 654–662 in P.A. Buckley, M.S. Foster, E.S. Morton, R.S. Ridgely, and F.A. Buckley, eds., *Neotropical Ornithology*. Ornithological Monograph No. 36. Washington, DC: American Ornithologists' Union.

Goldstein, M.I., B. Woodbridge, M.E. Zaccagnini, S.B. Canavelli, and A. Lanusse. 1996. An assessment of mortality of Swainson's Hawks on wintering grounds in Argentina. *J. Raptor Res.* 30(2):106–107.

Gonzalez, N. 2014. Audubon and CEPAN protect migratory bird roosting site off Chilean coast. Audubon Press Room, June 6.

Goodale, E., and G. Beauchamp. 2010. The relationship between leadership and gregariousness in mixed-species bird flocks. *J. Avian Biol.* 41(1):99–103.

Gordo, O., and J.J. Sanz. 2008. The relative importance of conditions in wintering and passage areas on spring arrival dates: The case of long-distance Iberian migrants. *J. Ornithol.* 149:199–210.

Goss-Custard, J.D., and S.E.A. Le V. dit Durell. 1984. Feeding ecology, winter mortality and the population dynamics of Oystercatchers on the Exe estuary. Pp. 190–208 in P.R. Evans, J.D. Goss-Custard, and W.G. Hale, eds., *Coastal Waders and Wildfowl in Winter.* Cambridge: Cambridge University Press.

Goss-Custard, J.D., P. Triplet, F. Sueur, and A.D. West. 2006. Critical thresholds of disturbance by people and raptors in foraging wading birds. *Biological Conservation* 127(1):88–97.

Gram, W.K. 1998. Winter participation by Neotropical migrant and resident birds in mixed-species flocks in northeastern Mexico. *Condor* 100:44–53.

Grande, J.M., M.A. Santillan, P.M. Orozco, M.S. Liébana, M.M. Reyes, M.A. Galmes, and J. Ceregheti. 2015. Barn Swallows keep expanding their breeding range in South America. *Emu* 115:256–260.

Gratto-Trevor, C., S.M. Haig, M.P. Miller, T.D. Mullins, S. Maddock, E. Roche, and P. Moore. 2016. Breeding sites and winter site fidelity of Piping Plovers wintering in the Bahamas, a previously unknown major wintering area. *J. Field Ornithol.* 87(1):29–41.

Greenberg, R. 1981. Dissimilar bill shapes in New World tropical versus temperate forest foliage-gleaning birds. *Oecologia* 49:143–147.

———. 1984. *The Winter Exploitation Systems of Bay-breasted and Chestnut-sided Warblers in Panama.* University of California Publications in Zoology 116. Berkeley: University of California Press.

———. 1987. Seasonal foraging specialization in the Worm-eating Warbler. *Condor* 89:158–168.

———. 1992. Forest migrants in non-forest habitats on the Yucatan Peninsula. Pp. 273–286 in J.M. Hagan III and D.W. Johnston, eds., *Ecology and Conservation of Neotropical Migrant Landbirds.* Washington, DC: Smithsonian Institution Press.

———. 1995. Insectivorous migratory birds in tropical ecosystems: The breeding currency hypothesis. *J. Avian Biol.* 26(3):260–264.

Greenberg, R., P. Bichier, and A. Cruz Angon. 2000. The conservation value for birds of cacao plantations with diverse planted shade in Tabasco, Mexico. *Animal Conservation* 3(2):105–112.

Greenberg, R., C.E. Gonzales, P. Bichier, and R. Reitsma. 2001. Nonbreeding habitat selection and foraging behavior of the Black-throated Green Warbler complex in southeastern Mexico. *Condor* 101:31–37.

Greenberg, R., and V. Salewski. 2005. Ecological correlates of wintering social systems in New World and Old World migratory passerines. Pp. 336–358 in R. Greenberg and P.P. Marra, eds., *Birds of Two Worlds: The Ecology and Evolution of Migration.* Baltimore: Johns Hopkins University Press.

Greenberg, R., and J. Salgado Ortiz. 1994. Interspecific defense of pasture trees by wintering Yellow Warblers. *Auk* 111(3):672–682.

Greenberg, R., J. Salgado Ortiz, and C. Macias Caballero. 1994. Aggressive competition for critical resources among migratory birds in the Neotropics. *Bird Conservation International* 4:115–127.

Greenberg, R.S., and J.A. Gradwohl. 1980. Observations of paired Canada Warblers *Wilsonia canadensis* during migration in Panama. *Ibis* 122:509–512.

Greene, E., D. Wilcove, and M. McFarland. 1984. Observations of birds at an army ant swarm in Guerrero, Mexico. *Condor* 86:92–93.

Greenwood, J.J.D., and S.R. Baillie. 1991. Effects of density-dependence and weather on population changes of English passerines using a non-experimental paradigm. *Ibis* 133 suppl. 1:121–133.

Gregoire, P.E., and C.D. Ankney. 1990. Agonistic behavior and dominance relationships among Lesser Snow Geese during winter and spring migration. *Auk* 107:550–560.

Grossman, A.F., and G.C. West. 1977. Metabolic rate and temperature regulation of winter acclimatized Black-capped Chickadees *Parus atricapillus* of interior Alaska. *Ornis Scandinav.* 8:127–138.

Grubb, T.C., Jr. 1977. Weather-dependent foraging behavior of some birds wintering in a deciduous woodland: Horizontal adjustments. *Condor* 79:271–274.

———. 1978. Weather-dependent foraging rates of wintering woodland birds. *Auk* 95:370–376.

Grubb, T.G., and R.G. Lopez. 1997. Ice fishing by wintering Bald Eagles in Arizona. *Wilson Bull.* 109(3):546–548.

Grummt, W. 1962. Erfolgreiche Winterbrut der Amsel, *Turdus merula*, in Berlin. *Vogelwarte* 21(4):295–296.

Gunnarsson, T.G., J.A. Gill, J. Newton, P.M. Potts, and W.J. Sutherland. 2005. Seasonal matching of habitat quality and fitness in a migratory bird. *Proc. Roy. Soc. Lond. B* 272:2319–2323.

Gwinner, E. 1988. Photorefractoriness in equatorial migrants. Pp. 626–633 in H. Ouellet, ed., *Acta XIX Congressus Internationalis Ornithologici*, vol. 1. Ottawa: University of Ottawa Press.

Gwinner, E.G. 1972. Adaptive functions of circannual rhythms in warblers. Pp. 218–236 in K.H. Voous, ed., *Proceedings of the XV International Ornithological Congress*. Leiden, Netherlands: E.J. Brill.

———. 1996. Circannual clocks in avian reproduction and migration. *Ibis* 118:47–63.

Hagy, H.M., S.C. Yaich, J.W. Simpson, E. Carrera, D.A. Haukos, W.C. Johnson, C.R. Loesch, F.A. Reid, S.E. Stephens, R.W. Tiner, B.A. Werner, and G.S. Yarris. 2014. Wetland issues affecting waterfowl conservation in North America. *Wildfowl* Special Issue 4:343–367.

Halkin, S.L., and S.U. Linville. 1999. Northern Cardinal (*Cardinalis cardinalis*). In *The Birds of North America*, no. 440 (A. Poole and F. Gill, eds.). The Birds of North America, Inc., Philadelphia.

Hamel, P.B. 1995. Bachman's Warbler (*Vermivora bachmanii*). In *The Birds of North America*, no. 150 (A. Poole and F. Gill, eds.). Academy of Natural Sciences, Philadelphia, and American Ornithologists' Union, Washington, DC.

Hamel, P.B., and A. Kirkconnell. 2005. Composition of mixed-species flocks of migrant and resident birds in Cuba. *Cotinga* 24:28–34.

Harrington, B. 1996. *The Flight of the Red Knot*. New York: W.W. Norton.

Harrington, B.A. 1999. The globetrotting White-rumped Sandpiper. Pp. 119–133 in K.P. Able, ed., *Gathering of Angels: Migrating Birds and Their Ecology*. Ithaca, NY: Comstock Books.

———. 2001. Red Knot (*Calidris canutus*). In *The Birds of North America*, no. 563 (A. Poole and F. Gill, eds.). The Birds of North America, Inc., Philadelphia.

Harris, M.P., S. Wanless, and D.A. Elston. 1998. Age-related effects of a nonbreeding event and a winter wreck on the survival of Shags *Phalacrocorax aristotelis*. *Ibis* 140:310–314.

Harrison, J.A., D.G. Allan, L.G. Underhill, M. Herremans, A.J. Tree, V. Parker, and C.J. Brown, eds. 1997. *The Atlas of Southern African Birds*. Vol. 1, *Non-Passerines*. Johannesburg: BirdLife South Africa.

———. 1997. *The Atlas of Southern African Birds*. Vol. 2, *Passerines*. Johannesburg: BirdLife South Africa.

Hatch, J.J. 2002. Arctic Tern (*Sterna paradisaea*). In *The Birds of North America*, no. 707 (A. Poole and F. Gill, eds.). The Birds of North America, Inc., Philadelphia.

Hays, H., P. Lima, L. Monteiro, J. DiConstanzo, G. Cormons, I.C.T. Nisbet, J. E. Saliva, J.A. Spendelow, J. Burger, J. Pierce, and M. Gochfeld. 1999. A nonbreeding concentration of Roseate and Common Terns in Bahia, Brazil. *J. Field Ornithol.* 70(4):455–464.

Heckscher, C.M., S.M. Taylor, J.W. Fox, and V. Afanasyev. 2011. Veery (*Catharus fuscescens*) wintering locations, migratory connectivity, and a revision of its winter range using geolocator technology. *Auk* 128(3):531–542.

Hedenstrom, A., S. Bensch, D. Hasselquist, M. Lockwood, and U. Ottosson. 1993. Migration, stopover and moult of the Great Reed Warbler *Acrocephalus arundinaceus* in Ghana, West Africa. *Ibis* 135:177–180.

Heinrich, B. 1989. *Ravens in Winter*. New York: Vintage Books.

———. 2003. Overnighting of Golden-crowned Kinglets during winter. *Wilson Bull.* 115(2):113–114.

Heitmeyer, M.D. 1988. Body composition of female Mallards in winter in relation to annual cycle events. *Condor* 90:669–680.

Hendricks, P. 1981. Observations on a winter roost of Rosy Finches in Montana. *J. Field Ornithol.* 52(3):235–236.

Hepp, G.R., and J.D. Hair. 1983. Reproductive behavior and pairing chronology in wintering dabbling ducks. *Wilson Bull.* 95(4):675–682.

Heredia, B., J.C. Alonso, and F. Hiraldo. 1991. Space and habitat use by Red Kites *Milvus milvus* during winter in the Guadalquivir marshes: A comparison between resident and wintering populations. *Ibis* 133:374–381.

Hewitt, D.G., and R.L. Kirkpatrick. 1997. Daily activity times of Ruffed Grouse in southwestern Virginia. *J. Field Ornithol.* 68(3):413–420.

Higgins, P.J., ed. 1999. *Handbook of Australian, New Zealand and Antarctic Birds.* Vol. 4, *Parrots to Dollarbird.* Melbourne: Oxford University Press.

Higgins, P.J., and S.J.J.F. Davies, eds. 1996. *Handbook of Australian, New Zealand and Antarctic Birds.* Vol. 3, *Snipes to Pigeons.* Melbourne: Oxford University Press.

Hill, D.P., and L.K. Gould. 1997. Chestnut-collared Longspur (*Calcarius ornatus*). In *The Birds of North America*, no. 288 (A. Poole and F. Gill, eds.). Academy of Natural Sciences, Philadelphia, and American Ornithologists' Union, Washington, DC.

Hill, G.E., R.R. Sargent, and M.B. Sargent. 1998. Recent change in the winter distribution of Rufous Hummingbirds. *Auk* 115(1):240–245.

Hitchcock, C.L., and D.F. Sherry. 1995. Cache pilfering and its prevention in pairs of Black-capped Chickadees. *J. Avian Biol.* 26:187–192.

Hockey, P.A.R., J.K. Turpie, and C.R. Velasquez. 1998. What selective pressures have driven the evolution of deferred northward migration by juvenile waders? *J. Avian Biol.* 29:325–330.

Hogstad, O. 1987a. Social rank in winter flocks of Willow Tits *Parus montanus. Ibis* 129:1–9.

———. 1987b. It is expensive to be dominant. *Auk* 104:333–336.

———. 1989. Subordination in mixed-age bird flocks—a removal study. *Ibis* 131:128–134.

Holberton, R.L., and A.M. Dufty Jr. 2005. Hormones and variation in life history strategies of migratory and nonmigratory birds. Pp. 290–302 in R. Greenberg and P.P. Marra, eds., *Birds of Two Worlds: The Ecology and Evolution of Migration.* Baltimore: Johns Hopkins University Press.

Holmes, R.T., and T.W. Sherry. 1992. Site fidelity of migratory warblers in temperate breeding and Neotropical wintering areas: Implications for population dynamics, habitat selection, and conservation. Pp. 563–575 in J.M. Hagan III and D.W. Johnston, eds., *Ecology and Conservation of Neotropical Migrant Landbirds.* Washington, DC: Smithsonian Institution Press.

Holmes, R.T., T.W. Sherry, and L. Reitsma. 1989. Population structure, territoriality and overwinter survival of two migrant warbler species in Jamaica. *Condor* 91:545–561.

Horvath, E.G., and K.A. Sullivan. 1988. Facultative migration in Yellow-eyed Juncos. *Condor* 90:482–484.

Hotker, H. 2006. *The Impact of Repowering Wind Farms on Birds and Bats.* Bergenhusen, Germany: Michael-Otto-Institut im NABU.

Hurrell, J.W., and K.E. Trenberth. 2010. Climate change. Pp. 9–29 in A.P. Møller, W. Fiedler, and P. Berthold, eds., *Effects of Climate Change on Birds.* Oxford: Oxford University Press.

Hutto, R.L. 1980. Winter habitat distribution of migratory land birds in western Mexico, with special reference to small foliage-gleaning insectivores. Pp. 181–203 in A. Keast and E.S. Morton, eds., *Migrant Birds in the Neotropics: Ecology, Behavior, Distribution, and Conservation.* Washington, DC: Smithsonian Institution Press.

———. 1985. Habitat selection by nonbreeding, migratory land birds. Pp. 455–476 in M.L. Cody, ed., *Habitat Selection in Birds.* Orlando, FL: Academic Press.

———. 1994. The composition and social organization of mixed-forest flocks in a tropical deciduous forest in western Mexico. *Condor* 96:105–118.

Hyrenbach, K.D., and R.R. Veit. 2003. Ocean warming and seabird communities of the southern California Current System (1987–98): Response at multiple temporal scales. *Deep-Sea Research II* 50:2537–2565.

Imber, M.J., and T.G. Lovegrove. 1982. Leach's Storm Petrels (*Oceanodroma l. leucorhoa*) prospecting for nest sites on the Chatham Islands. *Notornis* 29:101–108.

IPCC (Intergovernmental Panel on Climate Change). 1996. *Climate Change 1995: The Science of Climate Change.* Contribution of Working Group I to the Second Assessment Report of the IPCC. New York: Cambridge University Press.

Irving, L. 1960. *Birds of Anaktuvak Pass, Kobuk, and Old Crow: A Study in Arctic Adaptation.* US National Museum Bulletin 217. Washington, DC.

Isacch, J.P., and M.M. Martinez. 2003. Temporal variation in abundance and the population status of non-breeding Nearctic and Patagonian shorebirds in the flooding pampa grasslands of Argentina. *J. Field Ornithol.* 74(3):233–242.

Ismar, S.M.H., R.A. Phillips, M.J. Rayner, and M.E. Hauber. 2011. Geolocation tracking of the annual migration of adult Australasian Gannets (*Morus serrator*) breeding in New Zealand. *Wilson J. Ornithol.* 123(1):121–125.

Iverson, S.A., B.D. Smith, and F. Cooke. 2004. Age and sex distributions of wintering Surf Scoters: Implications for the use of age ratios as an index of recruitment. *Condor* 106:252–262.

Jackson, J.A., and H.R. Ouellet. 2002. Downy Woodpecker (*Picoides pubescens*). In *The Birds of North America*, no. 613 (A. Poole and F. Gill, eds.). The Birds of North America, Inc., Philadelphia.

Jaeger, E.C. 1949. Further observations on the hibernation of the Poor-will. *Condor* 51(3):105–109.

Jahn, A.E., V.R. Cueto, J.C. Sagario, A.M. Mamani, J.Q. Vidoz, J.L. de Casenave, and A.G. Di Giacomo. 2009. Breeding and winter site fidelity among eleven Neotropical austral migrant bird species. *Ornitologia Neotropical* 20:275–283.

Jahn, A.E., D.J. Levey, V.R. Cueto, J. Pinto Ledezma, D.T. Tuero, J.W. Fox, and D. Masson. 2013. Long-distance bird migration within South America revealed by light-level geolocators. *Auk* 130(2):223–229.

Jahn, A.E., D.J. Levey, and K.G. Smith. 2004. Reflections across hemispheres: A system-wide approach to New World bird migration. *Auk* 121(4):1005–1013.

James, R.A., Jr. 1989. Mate feeding in wintering Western Grebes. *J. Field Ornithol.* 60(3):358–360.

Jansson, C., J. Ekman, and A. von Brömssen. 1981. Winter mortality and food supply in tits *Parus* spp. *Oikos* 37(3):313–322.

Jaramillo, A.P. 1993. Wintering Swainson's Hawks in Argentina: Food and age segregation. *Condor* 95:475–479.

Jedlicka, J.A., R. Greenberg, I. Perfecto, S.M. zsPhilpott, and T.V. Dietsch. 2006. Seasonal shift in the foraging niche of a tropical avian resident: Resource competition at work? *J. Trop. Ecol.* 22:385–395.

Jenni, L., and M. Kéry. 2003. Timing of autumn bird migration under climate change: Advances in long-distance migrants, delays in short-distance migrants. *Proc. Roy. Soc. Lond. B* 270:1467–1471.

Jenouvrier, S., H. Weimerskirch, C. Barbraud, Y.-H. Park, and B. Cazelles. 2005. Evidence of a shift in the cyclicity of Antarctic seabird dynamics linked to climate. *Proc. Roy. Soc. Lond. B* 272(1566):887–895.

Jenssen, B.M., and M. Ekker. 1989. Thermoregulatory adaptations to cold in winter-acclimated Long-tailed Ducks (*Clangula hyemalis*). Pp. 147–152 in C. Bech and R.E. Reinertsen, eds., *Physiology of Cold Adaptation in Birds*. New York: Plenum Press.

Jiménez, J.E., A.E. Jahn, R. Rozzi, and N.E. Seavy. 2016. First documented migration of individual White-crested Elaenias (*Elaenia albiceps chilensis*) in South America. *Wilson J. Ornithol.* 128(2):419–425.

Jirinec, V., B.R. Campos, and M.D. Johnson. 2011. Roosting behavior of a migratory songbird on Jamaican coffee farms: Landscape composition may affect delivery of an ecosystem service. *Bird Conservation International* 21:353–361.

Job, J., and P.A. Bednekoff. 2011. Wrens on the edge: Feeders predict Carolina Wren *Thryothorus ludovicianus* abundance at the northern edge of their range. *J. Avian Biol.* 42(1):16–21.

Johnsgard, P.A. 1981. *The Plovers, Sandpipers, and Snipes of the World.* Lincoln: University of Nebraska Press.

Johnson, M.D., T.W. Sherry, A.M. Strong, and A. Medori. 2005. Migrants in Neotropical bird communities: An assessment of the breeding bird hypothesis. *J. Anim. Ecol.* 74:333–341.

Johnson, O.W., P.L. Bruner, J.J. Rotella, P.M. Johnson, and A.E. Bruner. 2001. Long-term study of apparent survival in Pacific Golden-Plovers at a wintering ground on Oahu, Hawaiian Islands. *Auk* 118(2):342–351.

Johnson, T.B. 1980. Resident and North American migrant bird interactions in the Santa Marta highlands, northern Colombia. Pp. 239–247 in A. Keast and E.S. Morton, eds., *Migrant Birds in the Neotropics: Ecology, Behavior, Distribution, and Conservation.* Washington, DC: Smithsonian Institution Press.

Jones, J., P.R. Perazzi, E.H. Carruthers, and R.J. Robertson. 2000. Sociality and foraging behavior of the Cerulean Warbler in Venezuelan shade-coffee plantations. *Condor* 102:958–962.

Jones, T., and W. Cresswell. 2010. The phenology mismatch hypothesis: Are declines of migrant birds linked to uneven global climate change? *J. Anim. Ecol.* 79(1):98–108.

Kalela, O. 1949. Changes in geographic ranges in the avifauna of northern and central Europe in relation to recent changes in climate. *Bird-Banding* 20(2):77–103.

Kallander, H., and H.G. Smith. 1990. Food storing in birds: An evolutionary perspective. Pp. 147–207 in D.M. Power, ed., *Current Ornithology*, vol. 7. New York: Plenum Press.

Ke, D., and X. Lu. 2009. Burrow use by Tibetan Ground Tits *Pseudopodoces humilis*: Coping with life at high altitudes. *Ibis* 151:321–331.

Keast, A. 1980. Spatial relationships between migratory parulid warblers and their ecological counterparts in the Neotropics. Pp. 109–130 in A. Keast and E.S. Morton, eds., *Migrant Birds in the Neotropics: Ecology, Behavior, Distribution, and Conservation.* Washington, DC: Smithsonian Institution Press.

Keith, G.S., E.K. Urban, and C.H. Fry. 1992. *The Birds of Africa*, vol. 4. London: Academic Press.

Kéry, M. 2007. Wintering Peregrine Falcons (*Falco peregrinus*) in the Peruvian Amazon. *Ornitologia Neotropical* 18:613–616.

Kessel, B. 1976. Winter activity patterns of Black-capped Chickadees in interior Alaska. *Wilson Bull.* 88(1):36–61.

Ketterson, E.D., and V. Nolan Jr. 1983. The evolution of differential bird migration. Pp. 357–402 in R.F. Johnston, ed., *Current Ornithology*, vol. 1. New York: Plenum Press.

King, J.R. 1972. Adaptive periodic fat storage by birds. Pp. 200–217 in K.H. Voous, ed., *Proceedings of the XV International Ornithological Congress.* Leiden, Netherlands: E.J. Brill.

King, J.R., and D.S. Farner. 1959. Premigratory change in body weight and fat in wild and captive male White-crowned Sparrows. *Condor* 61:315–324.

Kirk, D.A., and J.E.P. Currall. 1994. Habitat associations of migrant and resident vultures in central Venezuela. *J. Avian. Biol.* 25:327–337.

Klaassen, R.H.G, M. Hake, R. Strandberg, B.J. Koks, C. Trierweiler, K-M. Exo, F. Bairlein, and T. Alerstam. 2014. When and where does mortality occur in migratory birds? Direct evidence from long-term satellite tracking of raptors. *J. Anim. Ecol.* 83(1):176–184.

Klein, B.C. 1988. Weather-dependent mixed-species flocking during the winter. *Auk* 105:583–584.

Kluijver, H.N. 1950. Daily routines of the Great Tit, *Parus m. major* L. *Ardea* 38(3/4):99–135.

Knapton, R.W., and J.R. Krebs. 1976. Dominance hierarchies in winter Song Sparrows. *Condor* 78:567–569.

Knorr, O.A. 1957. Communal roosting of the Pygmy Nuthatch. *Condor* 59:398.

Knox, A.G., and P.E. Lowther. 2000. Common Redpoll (*Carduelis flammea*). In *The Birds of North America*, no. 543 (A. Poole and F. Gill, eds.). The Birds of North America, Inc., Philadelphia.

Koenig, W.D. 2001. Synchrony and periodicity of eruptions by boreal birds. *Condor* 103:725–735.

Koenig, W.D., and J.M.H. Knops. 2001. Seed-crop size and eruptions of North American boreal seed-eating birds. *J. Anim. Ecol.* 70(4):609–620.

Koenig, W.D., P.B. Stacey, M.T. Stanback, and R.L. Mumme. 1995. Acorn Woodpecker (*Melanerpes formicivorus*). In *The Birds of North America*, no. 194 (A. Poole and F. Gill, eds.). Academy of Natural Sciences, Philadelphia, and American Ornithologists' Union, Washington, DC.

Komar, O., B.J. O'Shea, A.T. Peterson, and A.G. Navarro-Siguenza. 2005. Evidence of latitudinal sexual segregation among migratory birds wintering in Mexico. *Auk* 122(3):938–948.

Koronkiewicz, T.J., M.K. Sogge, C. Van Riper III, and E. H. Paxton. 2006. Territoriality, site fidelity, and survivorship of Willow Flycatchers wintering in Costa Rica. *Condor* 108:558–570.

Kramer, G.R., H.M. Streby, S.M. Peterson, J.A. Lehman, D.A. Buehler, P.B. Wood, D.J. McNeil, J.L. Larkin, and D.E. Andersen. 2017. Nonbreeding isolation and population-specific migration patterns among three populations of Golden-winged Warblers. *Condor* 119:108–121.

Krams, I.I. 2001. Seeing without being seen: A removal experiment with mixed flocks of Willow and Crested Tits *Parus montanus* and *cristatus*. *Ibis* 143:476–481.

Krebs, J.R., N.S. Clayton, S.D. Healy, D.A. Cristol, S.N. Patel, and A.R. Jolliffe. 1996. The ecology of the avian brain: Food-storing memory and the hippocampus. *Ibis* 138:34–46.

Kress, S.W. 1967. A robin nests in winter. *Auk* 79(2):245–246.

Kricher, J.C., and W.E. Davis Jr. 1992. Patterns of avian species richness in disturbed and undisturbed habitats in Belize. Pp. 240–246 in J.M. Hagan III and D.W. Johnston, eds., *Ecology and Conservation of Neotropical Migrant Landbirds*. Washington, DC: Smithsonian Insitution Press.

Kruger, K., R. Prinzinger, and K.-L. Schuchmann. 1982. Torpor and metabolism in hummingbirds. *Comp. Biochem. Physiol.* 73A(4):679–689.

Kus, B.E., P. Ashman, G.W. Page, and L.E. Stenzel. 1984. Age-related mortality in a wintering population of Dunlin. *Auk* 101:69–73.

Laaksonen, T., M. Ahola, T. Eeva, R.A. Vaisanen, and E. Lehikoinen. 2006. Climate change, migratory connectivity and changes in laying date and clutch size in the Pied Flycatcher. *Oikos* 114:277–290.

Lack, D. 1968. Bird migration and natural selection. *Oikos* 19:1–9.

Lack, D., and P. Lack. 1972. Wintering warblers in Jamaica. *Living Bird* 11:129–153.

Lafferty, K.D. 2001. Disturbance to wintering Western Snowy Plovers. *Biological Conservation* 101(3):315–325.

Lammers, W.M., and M.W. Collopy. 2007. Effectiveness of avian predator perch deterrents on electrical transmission lines. *J. Wildlife Manag.* 71:2752–2758.

Lane, S.J., and M. Hassall. 1996. Nocturnal feeding by Dark-bellied Brent Geese *Branta bernicla bernicla*. *Ibis* 138:291–297.

Langrand, O. 1990. *Guide to the Birds of Madagascar*. New Haven, CT: Yale University Press.

Larkin, J.L., D. Raybuck, A. Roth, L. Chavarría-Duriaux, G. Duriaux, M. Siles, and C. Smalling. 2017. Geolocators reveal migratory connectivity between wintering and breeding areas of Golden-winged Warblers. *J. Field Ornithol.* 88(3):288–298.

Larsen, J.K., and M. Guillemette. 2007. Effects of wind turbines on flight behaviour of wintering common eiders: Implications for habitat use and collision risk. *J. Applied Ecology* 44(3):516–522.

La Sorte, F.A., and W. Jetz. 2012. Tracking of climatic niche boundaries under recent climate change. *J. Anim. Ecol.* 81(4):914–925.

La Sorte, F.A., and F.R. Thompson III. 2007. Poleward shifts in winter ranges of North American birds. *Ecology* 88(7):1803–1812.

Latta, S.C., and J. Faaborg. 2001. Winter site fidelity of Prairie Warblers in the Dominican Republic. *Condor* 103:455–468.

———. 2002. Demographic and population responses of Cape May Warblers wintering in multiple habitats. *Ecology* 83(9):2502–2515.

Latta, S.C., C.C. Rimmer, and K.P. McFarland. 2003. Winter bird communities in four habitats along an elevational gradient on Hispaniola. *Condor* 105:179–197.

Laubek, B. 1995. Habitat use by Whooper Swans *Cygnus cygnus* and Bewick's Swans *Cygnus columbianus bewicki* wintering in Denmark: Increasing agricultural conflicts. *Wildfowl* 46:8–15.

Lebbin, D.J., M.J. Parr, and G.H. Fenwick. 2010. *The American Bird Conservancy Guide to Bird Conservation.* Chicago: University of Chicago Press.

Leck, C.F. 1972. The impact of some North American migrants at fruiting trees in Panama. *Auk* 89:842–850.

———. 1980. Establishment of new population centers with changes in migration patterns. *J. Field Ornithol.* 51(2):168–173.

Lefebvre, G., and B. Poulin. 1996. Seasonal abundance of migrant birds and food resources in Panamanian mangrove forests. *Wilson Bull.* 108(4):748–759.

Lehikoinen, A., P. Saurola, P. Byholm, A. Lindén, and J. Valkama. 2010. Life history events of the Eurasian sparrowhawk *Accipiter nisus* in a changing climate. *J. Avian Biol.* 41(6):627–636.

Lehikoinen, E., and T.H. Sparks. 2010. Changes in migration. Pp. 89–112 in A.P. Møller, W. Fielder, and P. Berthold, eds., *Effects of Climate Change on Birds.* Oxford: Oxford University Press.

Leisler, B. 1990. Selection and use of habitat of wintering migrants. Pp. 156–174 in E. Gwinner, ed., *Bird Migration: Physiology and Ecophysiology.* Berlin: Springer-Verlag.

———. 1992. Habitat selection and coexistence of migrants and Afrotropical residents. *Ibis* 134 supp. 1:77–82.

Lemmon, D., M.L. Withiam, and C.P.L. Barkan. 1997. Male protection and winter pair-bonds in Black-capped Chickadees. *Condor* 99:424–433.

Lemoine, N., and K. Bohning-Gaese. 2003. Potential impact of global climate change on species richness of long-distance migrants. *Conservation Biology* 17(2):577–586.

Lewden, A., M. Petit, M. Milbergue, S. Orio, and F. Vézina. 2014. Evidence of facultative daytime hypothermia in a small passerine wintering at northern latitudes. *Ibis* 156:321–329.

Lewis, T.L., D. Esler, W.S. Boyd, and R. Zydelis. 2005. Nocturnal foraging behavior of wintering Surf Scoters and White-winged Scoters. *Condor* 107:637–647.

Ligon, J.D. 1970. Still more responses of the Poor-will to low temperatures. *Condor* 72(4):496–498.

———. 1978. Reproductive interdependence of Pinon Jays and pinon pines. *Ecological Monographs* 48:111–126.

Lofts, B., A.J. Marshall, and A. Wolfson. 1963. The experimental demonstration of pre-migration activity in the absence of fat deposition in birds. *Ibis* 105:99–105.

Logan, C.A. 1987. Fluctuations in fall and winter territory size in the Northern Mockingbird (*Mimus polyglottos*). *J. Field Ornithol.* 58(3):297–305.

Lombardo, M.P., L.C. Romagnano, P.C. Stouffer, A.S. Hoffenberg, and H.W. Power. 1989. The use of nest boxes as night roosts during the nonbreeding season by European Starlings in New Jersey. *Condor* 91:744–747.

Long, J.A., and P.C. Stouffer. 2003. Diet and preparation for spring migration in captive Hermit Thrushes (*Catharus guttatus*). *Auk* 120(2):323–330.

Long, J.L. 1981. *Introduced Birds of the World.* New York: Universe Books.

Longcore, T., C. Rich, P. Mineau, B. MacDonald, D.G. Bert, L.M. Sullivan, E. Mutrie, S.A. Gauthreaux Jr., M.L. Avery, R.L. Crawford, A.M. Manville II, E.R. Travis, and D. Drake. 2013. Avian mortality at communication towers in the United States and Canada: Which species, how many, and where? *Biological Conservation* 158:410–419.

Longcore, T., C. Rich, and L.M. Sullivan. 2009. Critical assessment of claims regarding management of feral cats by trap-neuter-return. *Conservation Biology* 23(4):887–894.

Lopez Ornat, A., and R. Greenberg. 1990. Sexual segregation by habitat in migratory warblers in Quintana Roo, Mexico. *Auk* 107:539–543.

Loss, S.R., T. Will, and P.P. Marra. 2013. The impact of free-ranging domestic cats on wildlife of the United States. *Nature Communications* 4, doi:10.1038/ncomms2380.

Lovette, I.J., and R.T. Holmes. 1995. Foraging behavior of American Redstarts in breeding and wintering habitats: Implications for relative food availability. *Condor* 97:782–791.

Lundberg, A. 1979. Residency, migration and a compromise: Adaptations to nest-site scarcity and food specialization in three Fennoscandian owl species. *Oecologia* 41:273–281.

Lynch, J. F. 1992. Distribution of overwintering Nearctic mirants in the Yucatan Peninsula, II: Use of native and human-modified vegetation. Pp. 178–196 in J.M. Hagan III and D.W. Johnston, eds., *Ecology and Conservation of Neotropical Migrant Landbirds*. Washington, DC: Smithsonian Institution Press.

Machin, P., J. Fernández-Elipe, M. Flores, J.W. Fox, J.I. Aiguirre, and R.H.G. Klaasen. 2015. Individual migration patterns of Eurasian Golden Plovers *Pluvialis apricaria* breeding in Swedish Lapland: Examples of cold spell-induced winter movements. *J. Avian Biol.* 46(6):634–642.

Macleod, R., A.G. Gosler, and W. Cresswell. 2005. Diurnal mass gain strategies and perceived predation risk in the Great Tit *Parus major*. *J. Anim. Ecol.* 74(5):956–964.

Mahecha, L., N. Villabona, L. Sierra, D. Ocampo, and O. Laverde-R. 2018. The Andean Cock-of-the-rock (*Rupicola peruvianus*) is a frugivorous bird predator. *Wilson J. Ornithol.* 130(2):558–560.

Mammen, D.L., and S. Nowicki. 1981. Individual differences and within-flock convergence in chickadee calls. *Behavioral Ecology and Sociobiology* 9:179–186.

Marantz, C.A., and J.V. Remsen Jr. 1991. Seasonal distribution of the Slaty Elaenia, a little-known austral migrant of South America. *J. Field Ornithol.* 62(2):162–172.

Marchant, S., and P.J. Higgins, eds. 1993. *Handbook of Australian, New Zealand and Antarctic Birds*. Vol. 2, *Raptors to Lapwings*. Melbourne: Oxford University Press.

Marks, J.S. 1993. Molt of Bristle-thighed Curlews in the northwestern Hawaiian Islands. *Auk* 110(3):573–587.

Marks, J.S., and R.L. Redmond. 1994. Migration of Bristle-thighed Curlews on Laysan Island: Timing, behavior and estimated flight range. *Condor* 96:316–330.

———. 1996. Demography of Bristle-thighed Curlews *Numenius tahitiensis* wintering on Laysan Island. *Ibis* 138:438–447.

Maron, J.L., and J.P. Myers. 1985. Seasonal changes in feeding success, activity patterns, and weights of nonbreeding Sanderlings (*Calidris alba*). *Auk* 102:580–586.

Marra, P.P. 2000. The role of behavioral dominance in structuring patterns of habitat occupancy in a migrant bird during the nonbreeding season. *Behavioral Ecology* 11(3):299–308.

Marra, P.P., and R.T. Holmes. 2001. Consequences of dominance-mediated habitat segregation in American Redstarts during the nonbreeding season. *Auk* 118(1):92–104.

Martin, S.G., and T.A. Gavin. 1995. Bobolink (*Dolichonyx oryzivorus*). In *The Birds of North America*, no. 176 (A. Poole and F. Gill, eds.). Academy of Natural Sciences, Philadelphia, and American Ornithologists' Union, Washington, DC.

Martin, T.E. 1985. Selection of second-growth woodlands by frugivorous migrating birds in Panama: An effect of fruit size and plant diversity. *J. Trop. Ecol.* 1:157–170.

Martinez-Curci, N.S., A.B. Aspiroz, J.P. Isacch, and R. Elias. 2015. Dietary relationships among Nearctic and Neotropical shorebirds in a key coastal wetland of South America. *Emu* 115:326–334.

Marzluff, J.M., B. Heinrich, and C.S. Marzluff. 1996. Raven roosts are mobile information centres. *Animal Behaviour* 51:89–103.

Masero, J.A. 2003. Assessing alternative anthropogenic habitats for conserving waterbirds: Salinas as buffer areas against the impact of natural habitat loss for shorebirds. *Biodiversity and Conservation* 12(6):1157–1173.

Mathot, K.J., B.D. Smith, and R.W. Elner. 2007. Latitudinal clines in food distribution correlate with differential migration in Western Sandpiper. *Ecology* 88(3):781–791.

Matthysen, E. 1990. Nonbreeding social organization in *Parus*. Pp. 209–249 in D.M. Power, ed., *Current Ornithology*, vol. 7. New York: Plenum Press.

Mayhew, P., and D. Houston. 1999. Effects of winter and early spring grazing by Wigeon *Anas penelope* on their food supply. *Ibis* 141:80–84.

Mazerolle, D.F., K.W. Dufour, K.A. Hobson, and H.E. den Haan. 2005. Effects of large-scale climatic fluctuations on survival and production of young in a Neotropical migrant songbird, the Yellow Warbler *Dendroica petechia*. *J. Avian Biol.* 36(2):155–163.

McCaw, J.H., III, P.J. Zwank, and R.J. Steiner. 1996. Abundance, distribution, and behavior of Common Mergansers wintering on a reservoir in southern New Mexico. *J. Field. Ornithol.* 67(4):669–679.

McDermott, M.E., and A.D. Rodewald. 2014. Conservation value of silvopastures to Neotropical migrants in Andean forest flocks. *Biological Conservation* 175:140–147.

McGrath, S. 2015. How young Chileans are saving the Hudsonian Godwit. *Audubon*, November–December.

McIntyre, J.W. 1978. Wintering behavior of Common Loons. *Auk* 95:396–403.

McKinney, R.A., S.R. McWilliams, and M.A. Charpentier. 2006. Waterfowl-habitat assocations during winter in an urban North Atlantic estuary. *Biological Conservation* 132(2):239–249.

McLandress, M.R., and D.G. Raveling. 1981a. Changes in diet and body composition of Canada Geese before spring migration. *Auk* 98:65–79.

———. 1981b. Hyperphagia and social behavior of Canada Geese prior to spring migration. *Wilson Bull.* 93(3):310–324.

McNamara, J.M., A.I. Houston, and S.L. Lima. 1994. Foraging routines of small birds in winter: A theoretical investigation. *J. Avian Biol.* 25:287–302.

McNeil, R. 1982. Winter resident repeats and returns of austral and boreal migrant birds banded in Venezuela. *J. Field Ornithol.* 53(2):125–132.

McNeil, R., M.T. Diaz, and A. Villeneuve. 1993. The mystery of shorebird over-summering: A new hypothesis. *Ardea* 82:143–152.

McNeil, R., and M. Robert. 1988. Nocturnal feeding strategies of some shorebird species in a tropical environment. Pp. 2328–2336 in H. Ouellet, ed., *Acta XIX Congressus Inernationalis Ornithologici*, vol. 2. Ottawa: University of Ottawa Press.

McNeil, R., and G. Rompré. 1995. Day and night feeding territoriality in Willets *Catoptrophorus semipalmatus* and Whimbrel *Numenius phaeopus* during the non-breeding season in the tropics. *Ibis* 137:169–176.

Meanley, B. 1965. The roosting behavior of the Red-winged Blackbird in the southern United States. *Wilson Bull.* 77(3):217–228.

Meijer, T., and R. Drent. 1999. Re-examination of the capital and income dichotomy in breeding birds. *Ibis* 141:399–414.

Middleton, A.L.A. 1993. American Goldfinch (*Carduelis tristis*). In *The Birds of North America*, no. 80 (A. Poole and F. Gill, eds.). Academy of Natural Sciences, Philadelphia, and American Ornithologists' Union, Washington, DC.

Midtgard, U. 1989. Circulatory adaptations to cold in birds. Pp. 211–222 in C. Bech and R.E. Reinertsen, eds., *Physiology of Cold Adaptation in Birds*. New York: Plenum Press.

Miller, J.B. 1996. Red-breasted Mergansers in an urban winter habitat. *J. Field Ornithol.* 67(3):477–483.

Miller, M.R., J. Beam, and D.P. Connelly. 1988. Dabbling duck harvest dynamics in the Central Valley of California—Implications for recruitment. Pp. 553–569 in M.W. Weller, ed., *Waterfowl in Winter*. Minneapolis: University of Minnesota Press.

Møller, A.P., E. Flensted-Jensen, K. Klarborg, W. Mardal, and J.T. Nielsen. 2010. Climate change affects the duration of the reproductive season in birds. *J. Anim. Ecol.* 79(4):777–784.

Møller, A.P., D. Rubolini, and E. Lehikoinen. 2008. Populations of migratory bird species that did not show a phenological response to climate change are declining. *Proc. Nat. Acad. Sci.* 105(42):16195–16200.

Møller, A.P., and T. Szép. 2005. Rapid evolutionary change in a secondary sexual character linked to climate change. *J. Evolutionary Biology* 18(2):481–495.

Moore, F.R. 1977. Flocking behavior and territorial competitors. *Animal Behaviour* 25:1063–1066.

———. 1990. Prothonotary Warblers cross the Gulf of Mexico together. *J. Field Ornithol.* 61(3):285–287.

Moreau, R.E. 1966. *The Bird Faunas of Africa and Its Islands.* New York: Academic Press.

———. 1972. *The Palaearctic-African Bird Migration Systems.* London: Academic Press.

Moreau, R.E., and W.M. Moreau. 1928. Some notes on the habits of Palaearctic migrants while in Egypt. *Ibis* series 12, 4(2):233–252.

Moreno, J., A. Lundberg, and A. Carlson. 1981. Hoarding of individual nuthatches (*Sitta europaea*) and Marsh Tits (*Parus palustris*). *Holarctic Ecol.* 4:263–269.

Morley, A. 1943. Sexual behaviour in British birds from October to January. *Ibis* 85:132–158.

Morrison, R.I.G., and K.A. Hobson. 2004. Use of body stores in shorebirds after arrival on high-Arctic breeding grounds. *Auk* 121(2):333–344.

Morton, E.S. 1971. Food and migration habits of the Eastern Kingbird in Panama. *Auk* 88:925–926.

———. 1976. The adaptive significance of dull coloration in Yellow Warblers. *Condor* 78:423.

———. 1977. Intratropical migration in the Yellow-green Vireo and Piratic Flycatcher. *Auk* 94:97–106.

———. 1980. Adaptations to seasonal changes by migrant land birds in the Panama Canal zone. Pp. 437–453 in A. Keast and E.S. Morton, eds., *Migrant Birds in the Neotropics: Ecology, Behavior, Distribution, and Conservation.* Washington, DC: Smithsonian Institution Press.

———. 1990. Habitat segregation by sex in the Hooded Warbler: Experiments on proximate causation and discussion of its evolution. *Am. Naturalist* 135(3):319–333.

Moskovits, D. 1978. Winter territorial and foraging behavior of Red-headed Woodpeckers in Florida. *Wilson Bull.* 90(4):521–535.

Moulton, D.W., C.D. Frentress, D.C. Stutzenbaker, D.S. Lobpries, and W.C. Brownlee. 1988. Ingestion of shotshell pellets by waterfowl wintering in Texas. Pp. 597–607 in M.W. Weller, ed., *Waterfowl in Winter.* Minneapolis: University of Minnesota Press.

Mouritsen, K.N. 1994. Day and night feeding in Dunlins *Calidris alpina*: Choice of habitat, foraging technique and prey. *J. Avian Biol.* 25:55–62.

Mowbray, T.B. 1999. Scarlet Tanager (*Piranga olivacea*). In *The Birds of North America*, no. 479 (A. Poole and F. Gill, eds.). The Birds of North America, Inc., Philadelphia.

Mowbray, T.B., F. Cooke, and B. Gantner. 2000. Snow Goose (*Chen caerulescens*). In *The Birds of North America*, no. 514 (A. Poole and F. Gill, eds.). The Birds of North America, Inc., Philadelphia.

Mueller, H.C., N.S. Mueller, D.D. Berger, G. Allez, W. Robichaud, and J.L. Kaspar. 2000. Age and sex differences in the timing of fall migration of hawks and falcons. *Wilson Bull.* 112(2):214–224.

Mugica, L., M. Acosta, D. Denis, A. Jiménez, A. Rodriguez, and X. Ruiz. 2007. Rice culture in Cuba as an important wintering site for migrant waterbirds from North America. Pp. 172–176 in G.C. Boere, C.A. Galbraith, and D.A. Stroud, eds., *Waterbirds around the World.* Edinburgh: Stationary Office.

Munn, C.A. 1985. Permanent canopy and understory flocks in Amazonia: Species composition and population density. Pp. 683–712 in P.A. Buckley, M.S. Foster, E.S. Morton, R.S. Ridgely, and F.G. Buckley, eds., *Neotropical Ornithology.* Washington, DC: American Ornithologists' Union.

Myers, J.P. 1980a. The pampas shorebird community: Interactions between breeding and nonbreeding members. Pp. 37–50 in S. Keast and E.S. Morton, eds., *Migrant Birds in the Neotropics: Ecology, Behavior, Distribution, and Conservation.* Washington, DC: Smithsonian Institution Press.

———. 1980b. Territoriality and flocking by Buff-breasted Sandpipers: Variations in non-breeding dispersion. *Condor* 82:241–240.

———. 1981. A test of three hypotheses for latitudinal segregation of the sexes in wintering birds. *Canadian J. Zool.* 59:1527–1534.

———. 1984. Spacing behavior of nonbreeding shorebirds. Pp. 271–321 in J. Burger and B.L. Olla, eds., *Behavior of Marine Animals: Current Perspectives in Research.* Vol. 6., *Shorebirds: Migration and Foraging Behavior.* New York: Plenum Press.

Myers, J.P., P.G. Connors, and F.A. Pitelka. 1979a. Territoriality in non-breeding shorebirds. Pp. 231–246 in F.A. Pitelka, *Studies in Avian Biology No. 2.* Los Angeles: Cooper Ornithological Society.

———. 1979b. Territory size in wintering Sanderlings: The effects of prey abundance and intruder density. *Auk* 96:551–561.

Myers, J.P., and L.P. Myers. 1979. Shorebirds of coastal Buenos Aires Province, Argentina. *Ibis* 121:186–200.

Myers, J.P., C.T. Schick, and G. Castro. 1988. Structure in Sanderling (*Calidris alba*) populations: The magnitude of intra- and interyear dispersal during the nonbreeding season. Pp. 604–615 in H. Ouellet, ed., *Acta XIX Congressus Internationalis Ornithologici*, vol. 1. Ottawa: University of Ottawa Press.

Nebel, S. 2006. Latitudinal clines in sex ratio, bill, and wing length in Least Sandpipers. *J. Field. Ornithol.* 77(1):39–45.

Nebel, S., D.B. Lank, P.D. O'Hara, G. Fernandez, B. Haase, F. Delgado, F.A. Estela, L.J.E. Ogden, B. Harrington, B.E. Kus, J.E. Lyons, F. Mercier, B. Ortego, J.Y. Takekawa, N. Warnock, and S.E. Warnock. 2002. Western Sandpipers (*Calidris mauri*) during the nonbreeding season: Spatial segregation on a hemispheric scale. *Auk* 119(4):922–928.

Nebel, S., J.L. Porter, and R.T. Kingsford. 2008. Long-term trends of shorebird populations in eastern Australia and impacts of freshwater extraction. *Biological Conservation* 141(4):971–980.

Nebel, S., K.G. Rogers, C.D.T. Minton, and D.I. Rogers. 2013. Is geographical variation in the size of Australian shorebirds consistent with hypotheses on differential migration? *Emu* 113:99–111.

Newton, I. 1995. Relationship between breeding and wintering ranges in Palaearctic-African migrants. *Ibis* 137:241–249.

———. 2004. Population limitation in migrants. *Ibis* 146:197–226.

———. 2008. *The Migration Ecology of Birds.* Amsterdam: Elsevier.

Newton, I., I. Wyllie, and P. Rothery. 1993. Annual survival of Sparrowhawks *Accipiter nisus* breeding in three areas of Britain. *Ibis* 135:49–60.

Nichols, J.D., and G.M. Haramis. 1980. Sex-specific differences in winter distribution patterns of Canvasbacks. *Condor* 82:406–416.

Nilsson, L. 1974. The behaviour of wintering Smew in southern Sweden. *Wildfowl* 25:84–88.

———. 1979. Variation in the production of young swans wintering in Sweden. *Wildfowl* 30:129–134.

———. 2014. Long-term trends in the number of Whooper Swans (*Cygnus cygnus*) breeding and wintering in Sweden. *Wildfowl* 64:197–206.

Nisbet, I.C.T., and Lord Medway. 1972. Dispersion, population ecology and migration of Eastern Great Reed Warblers *Acrocephalus orientalis* wintering in Malaysia. *Ibis* 114(4):451–494.

Nolan, V., Jr. and E.D. Ketterson. 1983. An analysis of body mass, wing length, and visible fat deposits of Dark-eyed Juncos wintering at different latitudes. *Wilson Bull.* 95(4):603–620.

Nolan, V., Jr., E.D. Ketterson, D.A. Cristol, C.M. Rogers, E.D. Clotfelter, R.C. Titus, S.J. Schoech, and E. Snajdr. 2002. Dark-eyed Junco (*Junco hyemalis*). In *The Birds of North America*, no. 716 (A. Poole and F. Gill, eds.). Academy of Natural Sciences, Philadelphia, and American Ornithologists' Union, Washington, DC.

Norris, D.R., P.P. Marra, G.J. Bowen, L.M. Ratcliffe, J.A. Royle, and T.K. Kyser. 2006a. Migratory connectivity of a widely distributed songbird, the American Redstart (*Setophaga ruticilla*). *Ornithological Monographs*, no. 61:14–28.

Norris, D.R., P.P. Marra, T.K. Kyser, T.W. Sherry, and L.M. Ratcliffe. 2003. Tropical winter habitat limits reproductive success on the temperate breeding grounds of a migratory bird. *Proc. Roy. Soc. Lond. B* 271:59–64.

Norris, D.R., M.B. Wunder, and M. Boulet. 2006b. Perspecitives on migratory connectivity. *Ornithological Monographs*, no. 61:79–88.

O'Connell, M. 2000. Threats to waterbirds and wetlands: Implications for conservation, inventory and research. *Wildfowl* 51:1–15.

O'Donnell, S., A. Kumar, and C. Logan. 2010. Army ant raid attendance and bivouac-checking behavior by Neotropical montane forest birds. *Wilson J. Ornithol.* 122(3):503–512.

———. 2014. Do Nearctic migrant birds compete with residents at army ant raids? A geographic and seasonal analysis. *Wilson J. Ornithol.* 126(3):474–487.

Odum, E.P., and F.A. Pitelka. 1939. Storm mortality in a winter starling roost. *Auk* 56:451–455.

Oglesby, R.T., and C.R. Smith. 1995. Climate change in the Northeast. Pp. 390–391 in E.T. LaRoe, ed., *Our Living Resources*. Washington, DC: US Department of the Interior, National Biological Service.

Oring, L.W., E.M. Gray, and J.M. Reed. 1997. Spotted Sandpiper (*Actitis macularia*). In *The Birds of North America*, no. 289 (A. Poole and F. Gill, eds.). Academy of Natural Sciences, Philadelphia, and American Ornithologists' Union, Washington, DC.

Oring, L.W., and D.B. Lank. 1982. Sexual selection, arrival times, philopatry and site fidelity in the polyandrous Spotted Sandpiper. *Behavioral Ecology and Sociobiology* 10:185–191.

Owen, M. 1991. Nocturnal feeding in waterfowl. Pp. 1105–1112 in B.D. Bell et al., eds., *Acta XX Congressus Internationalis Ornithologici*, vol. 2. Wellington: New Zealand Ornithological Trust Board.

Page, G., and D.F. Whitacre. 1975. Raptor predation on wintering shorebirds. *Condor* 77:73–83.

Page, G.W., N. Warnock, T.L. Tibbitts, D. Jorgensen, C.A. Hartman, and L.E. Stenzel. 2014. Annual migration patterns of Long-billed Curlew in the American West. *Condor* 116(1):50–61.

Parmelee, D.F. 1992 Snowy Owl (*Bubo scandiacus*). In *The Birds of North America*, no. 10 (A. Poole, P. Stettenheim, and F. Gill, eds.). Academy of Natural Sciences, Philadelphia, and American Ornithologists' Union, Washington, DC.

Parmesan, C., and G. Yohe. 2003. A globally coherent fingerprint of climate change across natural ecosystems. *Nature* 421:37–42.

Paruk, J.D., M.D. Chickering, D. Long IV, H. Uher-Koch, A. East, D. Poleschook, V. Gumm, W. Hanson, E.M. Adams, K.A. Kovach, and D.C. Evers. 2015. Winter site fidelity and winter movements in Common Loons (*Gavia immer*) across North America. *Condor* 117(4):485–517.

Patthey, P., S. Wirthner, N. Signorell, and R. Arlettaz. 2008. Impact of outdoor winter sports on the abundance of a key indicator species of alpine ecosystems. *J. Applied Ecology* 45(6):1704–1711.

Paulus, S.L. 1983. Dominance relations, resource use, and pairing chronology of Gadwalls in winter. *Auk* 100:947–952.

Paxton, E.H., S.L. Durst, M.K. Sogge, T.J. Koronkiewicz, and K.L. Paxton. 2017. Survivorship across the annual cycle of a migratory passerine, the Willow Flycatcher. *J. Avian Biol.* 48(8):1126–1131.

Peach, W., C. du Feu, and J. McMeeking. 1995. Site tenacity and survival rates of Wrens *Troglodytes troglodytes* and Treecreepers *Certhia familiaris* in a Nottinghamshire wood. *Ibis* 137:497–507.

Pearce-Higgins, J.W., and R.E. Green. 2014. *Birds and Climate Change*. Cambridge: Cambridge University Press.

Pearson, D.L. 1980. Bird migration in Amazonian Ecuador, Peru, and Bolivia. Pp. 273–283 in A. Keast and E.S. Morton, eds., *Migrant Birds in the Neotropics: Ecology, Behavior, Distribution, and Conservation.* Washington, DC: Smithsonian Institution Press.

Perfecto, I., R.A. Rice, R. Greenberg, and M.E. van der Voort. 1996. Shade coffee: A disappearing refuge for biodiversity. *BioScience* 46(8):598–608.

Perrins, C.M., ed. 1994. *Handbook of the Birds of Europe, the Middle East, and North Africa,* vol. 8. Oxford: Oxford University Press.

Peters, K.A., and D.L. Otis. 2007. Shorebird roost-site selection at two temporal scales: Is human disturbance a factor? *J. Applied Ecology* 44(1):196–209.

Petersen, M.R., and D.C. Douglas. 2004. Winter ecology of Spectacled Eiders: Environmental characteristics and population change. *Condor* 106:79–94.

Peterson, C.H., S.D. Rice, J.W. Short, D. Esler, J.L. Bodkin, B.E. Ballachey, and D.B. Irons. 2003. Long-term ecosystem response to the Exxon Valdez oil spill. *Science* 302(5653):2082–2086.

Petit, L.J. 1999. Prothonotary Warbler (*Protonotaria citrea*). In *The Birds of North America,* no. 408 (A. Poole and F. Gill, eds.). The Birds of North America, Inc., Philadelphia.

Petit, L.J., and D.R. Petit. 2003. Evaluating the importance of human-modified lands for Neotropical bird conservation. *Conservation Biology* 17(3):687–694.

Petracci, P.F., and K. Delhey. 2004. Nesting attempts of the Cliff Swallow *Petrochelidon pyrrhonota* in Buenos Aires Province, Argentina. *Ibis* 146:522–525.

Pierotti, R.J., and T.P. Good. 1994. Herring Gull (*Larus argentatus*). In *The Birds of North America,* no. 124 (A. Poole and F. Gill, eds.) Academy of Natural Sciences, Philadelphia, and American Ornithologists' Union, Washington, DC.

Piersma, T., R. Hoekstra, A. Dekinga, A. Koolhaas, P. Wolf, P. Battley, and P. Wiersma. 1993. Scale and intensity of intertidal habitat use by knots *Calidris cannutus* in the western Wadden Sea in relation to food, friends and foes. *Netherlands J. of Sea Research* 31:331–357.

Piersma, T., M. Klaassen, J.H. Bruggemann, A-M. Blomert, A. Gueye, Y. Ntiamoa-Baidu, and N.E. van Brederode. 1990a. Seasonal timing of the spring departure of waders from the Banc d'Arguin, Mauritania. *Ardea* 78:123–134.

Piersma, T., L. Zwarts, and J.H. Bruggemann. 1990b. Behavioral aspects of the departure of waders before long-distance flights: Flocking, vocalizations, flight paths and diurnal timing. *Ardea* 78:157–184.

Piper, W.H., and R. H. Wiley. 1989. Correlates of dominance in wintering White-throated Sparrows: Age, sex and location. *Animal Behaviour* 37:298–310.

Pitelka, F.A., J.P. Myers, and P.G. Connors. 1980. Spatial and resource-use patterns in wintering shorebirds: The Sanderling in central coastal California. Pp. 1041–1044 in *Acta XVII Congressus Internationalis Ornithologici,* vol. 2. Berlin: Deutschen Ornithologen-Gesellschaft.

Place, A.R., and E.W. Stiles. 1992. Living off the wax of the land: Bayberries and Yellow-rumped Warblers. *Auk* 109(2):334–345.

Plumpton, D.L., and D.E. Andersen. 1997. Habitat use and time budgeting by wintering Ferruginous Hawks. *Condor* 99:888–893.

Pool, D.B., A.O. Panjabi, A. Macias-Duarte, and D.M. Solhjem. 2014. Rapid expansion of croplands in Chihuahua, Mexico threatens declining North American grassland bird species. *Biological Conservation* 170:274–281.

Poole, A.F. 1989. *Ospreys: A Natural and Unnatural History.* Cambridge: Cambridge University Press.

Poole, A.F., R.O. Bierregaard, and M.S. Martell. 2002. Osprey (*Pandion haliaetus*). In *The Birds of North America,* no. 683 (A. Poole and F. Gill, eds.). The Birds of North America, Inc., Philadelphia.

Pott, C., and D.A. Wiedenfeld. 2017. Information gaps limit our understanding of seabird bycatch in global fisheries. *Biological Conservation* 210 (Part A):192–204.

Potvin, D.A., K. Välimäki, and A. Lehikoinen. 2016. Differences in shifts of wintering and breeding ranges lead to changing migration distances in European birds. *J. Avian Biol.* 47(5):619–628.

Poulin, B., and G. Lefebvre. 1996. Dietary relationships of migrant and resident birds from a humid forest in central Panama. *Auk* 113(2):277–287.

Powell, G.V.N. 1980. Migrant participation in Neotropical mixed species flocks. Pp. 477–483 in A. Keast and E.S. Morton, eds., *Migrant Birds in the Neotropics: Ecology, Behavior, Distribution, and Conservation.* Washington, DC: Smithsonian Institution Press.

Powell, G.V.N., J.H. Rappole, and S.A. Sader. 1992. Neotropical landbird use of lowland Atlantic habitats in Costa Rica: A test of remote sensing for identification of habitat. Pp. 287–298 in J.M. Hagan III and D.W. Johnston, eds., *Ecology and Conservation of Neotropical Migrant Landbirds.* Washington, DC: Smithsonian Institution Press.

Powell, L.L., R.C. Dobbs, and P.P. Marra. 2015. Habitat and body condition influence American Redstart foraging behavior during the non-breeding season. *J. Field Ornithol.* 86(3):229–237.

Pravosudov, V.V., and T.C. Grubb Jr. 1997. Energy management in passerine birds during the nonbreeding season: A review. Pp. 189–234 in V. Nolan Jr. et al., eds., *Current Ornithology*, vol. 14. New York: Plenum Press.

———. 1999. Effects of dominance on vigilance in avian social groups. *Auk* 116(1):241–246.

Pravosudov, V.V., T.C. Grubb Jr., P.F. Doherty Jr., C.L. Bronson, E.V. Pravosudova, and A.S. Dolby. 1999. Social dominance and energy reserves in wintering woodland birds. *Condor* 101:880–884.

Pravosudov, V.V., and J.R. Lucas. 2000. The costs of being cool: A dynamic model of nocturnal hypothermia by small food-caching birds in winter. *J. Avian Biol.* 31(4):463–472.

Pravosudova, E.V., T.C. Grubb Jr., and P.G. Parker. 2001. The influence of kinship on nutritional condition and aggression levels in winter social groups of Tufted Titmice. *Condor* 103:821–828.

Preston, C.R. 1981. Environmental influence on soaring in wintering Red-tailed Hawks. *Wilson Bull.* 93(3):350–356.

Pulliam, H.R., and G.C. Millikan. 1982. Social organization in the nonreproductive season. Pp. 169–197 in D.S. Farner, J.R. King, and K.C. Parkes, eds., *Avian Biology*, vol. 6. New York: Academic Press.

Rabenold, K.N. 1980. The Black-throated Green Warbler in Panama: Geographic and seasonal comparison of foraging. Pp. 297–307 in A. Keast and E.S. Morton, eds., *Migrant Birds in the Neotropics: Ecology, Behavior, Distribution, and Conservation.* Washington, DC: Smithsonian Institution Press.

Rabenold, K.N., and P.P. Rabenold. 1985. Variation in altitudinal migration, winter segregation, and site tenacity in two subspecies of Dark-eyed Juncos in the southern Appalachians. *Auk* 102(4):805–819.

Rabenold, P.P. 1986. Family associations in communally roosting Black Vultures. *Auk* 103:32–41.

Ralph, C.J., and L.R. Mewaldt. 1975. Timing of site fixation upon the wintering grounds in sparrows. *Auk* 92:698–705.

Rappole, J.H. 1995. *The Ecology of Migrant Birds: A Neotropical Perspective.* Washington, DC: Smithsonian Institution Press.

———. 2013. *The Avian Migrant: The Biology of Bird Migration.* New York: Columbia University Press.

Rappole, J.H., M.A. Ramos, and K. Winker. 1989. Wintering Wood Thrush movements and mortality in southern Veracruz. *Auk* 106:402–410.

Rappole, J.H. and A.R. Tipton. 1992. The evolution of avian migration in the Neotropics. *Ornitologia Neotropical* 3:45–55.

Rappole, J.H., and D.W. Warner. 1980. Ecological aspects of migrant bird behavior in Veracruz, Mexico. Pp. 353–393 in A. Keast and E.S. Morton, eds., *Migrant Birds in the Neotropics: Ecology, Behavior, Distribution, and Conservation.* Washington, DC: Smithsonian Institution Press.

Rautenberg, W. 1980. Temperature regulation in cold environment. Pp. 321–325 in R. Nohring, ed., *Acta XVII Congressus Internationalis Ornithologici*, vol. 1. Berlin: Verlag der Deutschen Ornithologen-Gesellschaft.

Raveling, D.G., W.E. Crews, and W.D. Klimstra. 1972. Activity patterns of Canada Geese during winter. *Wilson Bull.* 84(3):278–295.

Regehr, H.M., C.M. Smith, B. Arquilla, and F. Cooke. 2001. Post-fledging broods of migratory Harlequin Ducks accompany females to wintering areas. *Condor* 103:408–412.

Rehfisch, M.M., G.E. Austin, S.N. Freeman, M.J.S. Armitage, and N.H.K. Burton. 2004. The possible impact of climate change on the future distributions and numbers of waders on Britain's non-estuarine coast. *Ibis* 146 suppl. 1:70–81.

Reinertsen, R.E. 1983. Nocturnal hypothermia and its energetic significance for small birds living in the Arctic and subarctic regions: A review. *Polar Research* 1(3):269–284.

———. 1988. Behavioral thermoregulation in the cold: The energetic significance of microclimate selection. Pp. 2681–2689 in H. Ouellet, ed., *Acta XIX Congressus Internationalis Ornithologici*, vol. 2. Ottawa: University of Ottawa Press.

Reinertsen, R.E., and S. Haftorn. 1986. Different metabolic strategies of northern birds for nocturnal survival. *J. Comp. Physiology B* 156:655–663.

Renfrew, R.B., S.J.K. Frey, and J. Klavins. 2011. Phenology and sequence of the complete prealternate molt of Bobolinks in South America. *J. Field Ornithol.* 82(1):101–113.

Renfrew, R.B., D. Kim, N. Perlut, J. Smith, J. Fox, and P.P. Marra. 2013. Phenological matching across hemispheres in a long-distance migratory bird. *Diversity and Distributions* 19(8):1008–1019.

Reudink, M.W., P.P. Marra, T.K. Kyser, P.T. Boag, K.M. Langin, and L.M. Ratcliffe. 2009. Non-breeding season events influence sexual selection in a long-distance migratory bird. *Proc. Roy. Soc. Lond. B* 276:1619–1626.

Ridgely, R.S., and G. Tudor. 1994. *The Birds of South America*, vol. 2. Austin: University of Texas Press.

Rimmer, C.C., and C.H. Darmstadt. 1996. Non-breeding site fidelity in Northern Shrikes. *J. Field Ornithol.* 67(3):360–366.

Rizzolo, D.J., E. Esler, D.D. Roby, and R.L. Jarvis. 2005. Do wintering Harlequin Ducks forage nocturnally at high latitudes? *Condor* 107:173–177.

Robb, G.N., R.A. McDonald, D.E. Chamberlain, S.J. Reynolds, T.J.E. Harrison, and S. Bearhop. 2008. Winter feeding of birds increases productivity in the subsequent breeding season. *Biology Letters* 4(2):220–223.

Robbins, C.S., B.A. Dowell, D.K. Dawson, J.A. Colon, R. Estrada, A. Sutton, R. Sutton, and D. Weyer. 1992. Comparison of Neotropical and migrant landbird populations wintering in tropical forest, isolated forest fragments, and agricultural habitats. Pp. 207–220 in J.M. Hagan III and D.W. Johnston, eds., *Ecology and Conservation of Neotropical Migrant Landbirds*. Washington, DC: Smithsonian Institution Press.

Robert, M., R. McNeil, and A. Leduc. 1989. Conditions and significance of night feeding in shorebirds and other water birds in a tropical lagoon. *Auk* 106:94–101.

Robertson, G., M. McNeill, N. Smith, B. Wienecke, S. Candy, and F. Olivier. 2006. Fast sinking (integrated weight) longlines reduce mortality of White-chinned Petrels (*Procellaria aequinoctialis*) and Sooty Shearwaters (*Puffinus griseus*) in demersal longline fisheries. *Biological Conservation* 132(4):458–471.

Robillard, A., J.F. Therrien, G. Gauthier, K.M. Clark, and J. Bêty. 2016. Pulsed resources at tundra breeding sites affect winter irruptions at temperate latitudes of a top predator, the Snowy Owl. *Oecologia* 181:423–433.

Robin, F., T. Piersma, F. Meunier, and P. Bocher. 2013. Expansion into an herbivorous niche by a customary carnivore: Black-tailed Godwits feeding on rhizomes of *Zostera* at a newly established wintering site. *Condor* 115(2):340–347.

Robinson, S.K., J. Terborgh, and J.W. Fitzpatrick. 1988. Habitat selection and relative abundance of migrants in southeastern Peru. Pp. 2298–2307 in H. Ouellet, ed., *Acta XIX Congressus Internationalis Ornithologici*, vol 2. Ottawa: University of Ottawa Press.

Robinson, W.D. 1996. Summer Tanager (*Piranga rubra*). In *The Birds of North America*, no. 248 (A. Poole and F. Gill, eds.). Academy of Natural Sciences, Philadelphia, and American Ornithologists' Union, Washington, DC.

Rockwell, S.M., C.I. Bocetti, and P.P. Marra. 2012. Carry-over effects of winter climate on spring arrival date and reproductive success in an endangered migratory bird, Kirtland's Warbler (*Setophaga kirtlandii*). *Auk* 129(4):744–752.

Rodrigues, A.A.F., A.T.L. Lopes, E.C. Goncalves, A. Silva, and M.P.C. Schneider. 2007. Philopatry of the Semipalmated Sandpiper (*Calidris pusilla*) on the Brazilian Amazonian coast. *Ornitologia Neotropical* 18:285–291.

Rogers, C.M., and R. Heath-Coss. 2003. Effects of experimentally altered food abundance on fat reserves in wintering birds. *J. Anim. Ecol.* 72(5):822–830.

Rogers, D.I., T. Piersma, and C.J. Hassell. 2006. Roost availability may constrain shorebird distribution: Exploring the energetic costs of roosting and disturbance around a tropical bay. *Biological Conservation* 133(2):225–235.

Root, T. 1988. Energy constraints on avian distributions and abundance. *Ecology* 69:330–339.

Rohwer, F.C., and M.G. Anderson. 1988. Female-biased philopatry, monogamy, and the timing of pair formation in migratory waterfowl. Pp. 187–221 in R.F. Johnston, ed., *Current Ornithology*, vol. 5. New York: Plenum Press.

Rohwer, S. 1975. The social significance of avian winter plumage variability. *Evolution* 29(4):593–610.

Rohwer, S., L.K. Butler, and D.R. Froehlich. 2005. Ecology and demography of east-west differences in molt scheduling of Neotropical migrant passerines. Pp. 87–105 in R. Greenberg and P.P. Marra, eds., *Birds of Two Worlds: The Ecology and Evolution of Migration*. Baltimore: Johns Hopkins University Press.

Rojas de Azuaje, L.M., S. Tai, and R. McNeil. 1993. Comparisons of rod/cone ratio in three species of shorebirds having different nocturnal foraging strategies. *Auk* 110(1):141–145.

Root, T. 1988. Environmental factors associated with avian distributional boundaries. *J. Biogeography* 15(3):489–505.

Root, T.L., and J.D. Weckstein. 1995. Changes in winter ranges of selected birds, 1901–1989. Pp. 386–389 in E.T. LaRoe, ed., *Our Living Resources*. Washington, DC: US Department of the Interior, National Biological Service.

Roth, T.C., II, S.L. Lima, and W.E. Vetter. 2005. Survival and causes of mortality in wintering Sharp-shinned Hawks and Cooper's Hawks. *Wilson Bull.* 117(3):237–244.

Ruiz, G.M., P.G. Connors, S.E. Griffin, and F.A. Pitelka. 1989. Structure of a winter Dunlin population. *Condor* 91:562–570.

Rushing, C.S., P.P. Marra, and M.R. Dudash. 2016. Winter habitat quality but not long-distance dispersal influences apparent reproductive success in a migratory bird. *Ecology* 97(5):1218–1227.

Saalfeld, D.T., S.T. Saalfeld, W.C. Conway, and K.M. Hartke. 2016. Wintering grassland bird responses to vegetation structure, exotic invasive plant composition, and disturbance regime in coastal prairies of Texas. *Wilson J. Ornithol.* 128(2):290–305.

Saarela, S., B. Klapper, and G. Heldmaier. 1989. Thermogenic capacity of Greenfinches and Siskins in winter and summer. Pp. 115–122 in C. Bech and R.E. Reinertsen, eds., *Physiology of Cold Adaptation in Birds*. New York: Plenum Press.

Sabine, W.S. 1959. The winter society of the Oregon Junco: Intolerance, dominance, and the pecking order. *Condor* 61:110–135.

Safina, C., and J.M. Utter. 1989. Food and winter territories of Northern Mockingbirds. *Wilson Bull.* 101:97–101.

Saino, N., R. Ambrosini, M. Carioli, A. Romano, M. Romano, D. Rubolini, C. Scandolara, and F. Liechti. 2017. Sex-dependent carry-over effects on timing of reproduction and fecundity of a migratory bird. *J. Anim. Ecol.* 86(2):239–249.

Saino, N., T. Szép, R. Ambrosini, M. Romano, and A.P. Møller. 2004. Ecological conditions during winter affect sexual selection and breeding in a migratory bird. *Proc. Roy. Soc. Lond. B* 271:681–686.

Salewski, V., F. Bairlein, and B. Leisler. 2003. Niche partitioning of two Palearctic passerine migrants with Afrotropical residents in their West African winter quarters. *Behavioral Ecology* 14(4):493–502.

Salewski, V., and P. Jones. 2006. Palearctic passerines in Afrotropical environments: A review. *J. Ornithol.* 147:192–201.

Salomonsen, F. 1955. The evolutionary significance of bird-migration. *Dan. Biol. Medd.* 22(6):1–62.

Salomonson, M.G., and R.P. Balda. 1977. Winter territoriality of Townsend's Solitaires (*Myadestes townsendi*) in a pinon-juniper-ponderosa pine ecotone. *Condor* 79:148–161.

Salvador, S.A., L.A. Salvador, F.A. Gandoy, and J.I. Areta. 2016. La golondrina rabadilla canela (*Petrochelidon pyrrhonota*) cria in Sudamérica. *Ornitologia Neotropical* 27:163–168.

Sanchez, L.E., D.M. Buehler, and A.I. Castillo. 2006. Shorebird monitoring in the Upper Bay of Panama. Pp. 166–171 in G.C. Boere, C.A. Galbraith, and D.A. Stroud, eds., *Waterbirds around the World*. Edinburgh: Stationary Office.

Sander, L. 2015. Grassland Revival. *Bird Conservation*, Fall 2015, 24–29.

Sanderson, F.J., P.F. Donald, D.J. Pain, I.J. Burfield, and F.P.J. van Bommel. 2006. Long-term population declines in Afro-Palearctic migrant birds. *Biological Conservation* 131(1):93–105.

Sandvik, H., K.E. Erikstad, R.T. Barrett, and N.G. Yoccoz. 2005. The effects of climate on adult survival of North Atlantic seabirds. *J. Anim. Ecol.* 74(5):817–831.

Sauer, E.G.F. 1963. Migration habits of Golden Plovers. Pp. 454–467 in C.G. Sibley, ed., *Proceedings XIII International Ornithological Congress*. Baton Rouge, LA: American Ornithologists' Union.

Savard, J.-P.L. 1985. Evidence of long-term pair bonds in Barrow's Goldeneye (*Bucephala islandica*). *Auk* 102:389–391.

———. 1988. Winter, spring and summer territoriality in Barrow's Goldeneye: Characteristics and benefits. *Ornis Scandinav.* 19:119–128.

Schaefer, R.R., D.C. Rudolph, and J.F. Fagan. 2007. Winter prey caching by Northern Hawk Owls in Minnesota. *Wilson J. Ornithol.* 119(4):755–758.

Schamber, J.L., J.S. Sedinger, and D.H. Ward. 2012. Carry-over effects of winter location contribute to variation in timing of nest initiation and clutch size in Black Brant (*Branta bernicla nigricans*). *Auk* 129(2):205–210.

Schlaich, A.E., R.H.G. Klaassen, W. Bouten, V. Bretagnolle, B.J. Koks, A. Villers, and C. Both. 2016. How individual Montagu's Harriers cope with Moreau's Paradox during the Sahelian winter. *J. Anim. Ecol.* 85(6):1491–1501.

Schmidt, N.M., J.B. Mosbacher, E.J. Veterinen, T. Roslin, and A. Michelsen. 2018. Limited dietary overlap amongst resident Arctic herbivores in winter: Complementary insights from complementary methods. *Oecologia* 187:689–699.

Schuetz, J.G., G.M. Langham, C.U. Soykan, C.B. Wilsey, T. Auer, and C.C. Sanchez. 2015. Making spatial prioritizations robust to climate change uncertainties: A case study with North American birds. *Ecological Applications* 25(7):1819–1831.

Schwabl, H. 1992. Winter and breeding territorial behaviour and levels of reproductive hormones of migratory European Robins. *Ornis Scandinav.* 23:271–276.

Schwartz, P. 1963. Orientation experiments with Northern Waterthrushes wintering in Venezuela. Pp. 481–484 in C.G. Sibley, ed., *Proceedings XIII International Ornithological Congress*. Baton Rouge, LA: American Ornithologists' Union.

Şekercioğlu, Ç.H., R.B. Primack, and J. Wormworth. 2012. The effects of climate change on tropical birds. *Biological Conservation* 148(1):1–18.

Şekercioğlu, Ç.H., S.H. Schneider, J.P. Fay, and S.R. Loarie. 2008. Climate change, elevational range shifts, and bird extinctions. *Conservation Biology* 22(1):140–150.

Senner, N.R. 2012. One species but two patterns: Populations of the Hudsonian Godwit (*Limosa haemastica*) differ in spring migration timing. *Auk* 129(4):670–682.

Shaffer, S.A., Y. Tremblay, H. Weimerskirsch, D. Scott, D.R. Thompson, P.M. Sagar, H. Moller, G.A. Taylor, D.G. Foley, B.A. Block, and D.P. Costa. 2006. Migratory shearwaters integrate oceanic resources across the Pacific Ocean in an endless summer. *Proc. Nat. Acad. Sci.* 103(34):12799–12802.

Sherry, D.F. 1984. Food storage by Black-capped Chickadees: Memory for the location and contents of caches. *Animal Behaviour* 29:1260–1266.

———. 1989. Food storing in the Paridae. *Wilson Bull.* 101(2):289–304.

Sherry, T.W., and R.T. Holmes. 1995. Summer versus winter limitation of populations: What are the issues and what is the evidence? Pp. 85–120 in T.E. Martin and D. M. Finch, eds., *Ecology and Migration of Neotropical Migratory Birds: A Synthesis and Review of Critical Issues*. New York: Oxford University Press.

Sillett, T.S., and R.T. Holmes. 2002. Variation in survivorship of a migratory songbird throughout its annual cycle. *J. Anim. Ecol.* 71(2):296–308.

Simmons, K.E.L. 1951. Interspecific territorialism. *Ibis* 93:407–413.

Simmons, R.E., H. Kolberg, R. Braby, and B. Erni. 2015. Declines in migrant shorebird populations from a winter-quarter perspective. *Conservation Biology* 29(3):877–887.

Siriwardena, G.M., D.K. Stevens, G.Q.A. Anderson, J.A. Vickery, N.A. Calbrade, and S. Dodd. 2007. The effect of supplementary winter seed food on breeding populations of farmland birds: Evidence from two large-scale experiments. *J. Applied Ecology* 44(5):920–932.

Sitters, H.P., P.M. Gonzalez, T. Piersma, A.J. Baker, and D.J. Price. 2001. Day and night feeding habitat of Red Knots in Patagonia: Profitability versus safety? *J. Field Ornithol.* 72(1):86–95.

Skokkan, K.-A. 1992. Energetics and adaptations to cold in ptarmigan in winter. *Ornis Scandinav.* 23:366–370.

Smallwood, J.A. 1988. A mechanism of sexual segregation by habitat in American Kestrels (*Falco sparverius*) wintering in south-central Florida. *Auk* 105:36–46.

Smallwood, J.A., and D.M. Bird. 2002. American Kestrel (*Falco sparverius*). In *The Birds of North America*, no. 602 (A. Poole and F. Gill, eds.). The Birds of North America, Inc., Philadelphia.

Smith, C.J., M.D. Johnson, B.R. Campos, and C.M. Bishop. 2012. Variation in aggression of Black-throated Blue Warblers wintering in Jamaica. *Condor* 114(4):831–839.

Smith, J.A.M., L.R. Reitsma, L.L. Rockwood, and P.P. Marra. 2008. Roosting behavior of a Neotropical migrant songbird, the Northern Waterthrush *Seiurus noveboracensis*, during the non-breeding season. *J. Avian Biol.* 39:460–465.

Smith, J.N.M., R.D. Montgomerie, M.J. Tait, and Y. Yom-Tov. 1980. A winter feeding experiment on an island Song Sparrow population. *Oecologia* 47:164–170.

Smith, K.G, W.M. Davis, T.E. Kienzle, W. Post, and R.W. Chinn. 1999. Additional records of fall and winter nesting by Killdeer in southern United States. *Wilson Bull.* 111(3):424–426.

Snow, D.W. 1955. The abnormal breeding of birds in the winter 1953/54. *Brit. Birds* 48:120–126.

Solheim, R. 1984. Caching behavior, prey choice and surplus killing by Pygmy Owls *Glaucidium passerinum* during winter, a functional response of a generalist predator. *Ann. Zool. Fenn.* 21:301–308.

Somershoe, S.G., C.R.D. Brown, and R.T. Poole. 2009. Winter site fidelity and over-winter site persistence of passerines in Florida. *Auk* 121(1):119–125.

Spear, L.B., and D.G. Ainley. 1999. Migration routes of Sooty Shearwaters in the Pacific Ocean. *Condor* 101:205–218.

Spottiswoode, C.N., A.P. Tottrup, and T. Coppack. 2006. Sexual selection predicts advancement of avian spring migration in response to climate change. *Proc. Roy. Soc. Lond. B* 273:3023–3029.

Spurr, E., and H. Milne. 1976. Adaptive significance of autumn pair formation in the Common Eider *Somateria mollissima* (L.). *Ornis Scandinav.* 7:85–89.

Staicer, C.A. 1992. Social behavior of the Northern Parula, Cape May Warbler, and Prairie Warbler wintering in second-growth forest in southwestern Puerto Rico. Pp. 308–320 in J.M. Hagan III and D.W. Johnston, eds., *Ecology and Conservation of Neotropical Migrant Landbirds*. Washington, DC: Smithsonian Institution Press.

Stanley, C.Q., E.A. McKinnon, K.C. Fraser, M.P. Macpherson, G. Casbourn, L. Friesen, P.P. Marra, C. Studds, T.B. Ryder, N.E. Diggs, and B.J.M. Stutchbury. 2015. Connectivity of Wood Thrush breeding, wintering, and migration sites based on range-wide tracking. *Conservation Biology* 29(1):164–174.

Steadman, D.W. 2005. The paleoecology and fossil history of migratory landbirds. Pp. 5–17 in R. Greenberg and P.P. Marra, eds., *Birds of Two Worlds: The Ecology and Evolution of Migration*. Baltimore: Johns Hopkins University Press.

Stenhouse, I.J., C. Egevang, and R.A. Phillips. 2012. Trans-equatorial migration, staging sites and wintering area of Sabine's Gull *Larus sabini* in the Atlantic Ocean. *Ibis* 154:42–51.

Steube, M.M., and E.D. Ketterson. 1982. A study of fasting in Tree Sparrows (*Spizella arborea*) and Dark-eyed Juncos (*Junco hyemalis*): Ecological implications. *Auk* 99:299–308.

Stevens, T.K., A.M. Hale, K.B. Karsten, and V.J. Bennett. 2013. An analysis of displacement from wind turbines in a wintering grassland bird community. *Biodiversity and Conservation* 22(8):1755–1767.

Stewart, R.E., Jr., G.L. Krapu, B. Conant, H.F. Percival, and D.L. Hall. 1988. Pp. 613–617 in M.W. Weller, ed., *Waterfowl in Winter*. Minneapolis: University of Minnesota Press.

Stirnemann, R.L., J. O'Halloran, M. Ridgway, and A. Donnelly. 2012. Temperature-related increases in grass growth and greater competition for food drive earlier migrational departure of wintering Whooper Swans. *Ibis* 154:542–553.

Stokes, D.L., P.D. Boersma, J. Lopez de Casenave, and P. Garcia-Borboroglu. 2014. Conservation of migratory Magellanic Penguins requires marine zoning. *Biological Conservation* 170(1):151–161.

Stotz, D.F., J.W. Fitzpatrick, T.A. Parker III, and D.K. Moskovits. 1996. *Neotropical Birds: Ecology and Conservation*. Chicago: University of Chicago Press.

Stouffer, P.C., and G.M. Dwyer. 2003. Sex-biased winter distribution and timing of migration of Hermit Thrushes (*Catharus guttatus*) in eastern North America. *Auk* 120(3):836–847.

Studds, C.E., T.K. Kyser, and P.P. Marra. 2008. Natal dispersal driven by environmental conditions interacting across the annual cycle of a migratory songbird. *Proc. Nat. Acad. Sci.* 105:2929–2933.

Studds, C.E., and P.P. Marra. 2005. Nonbreeding habitat occupancy and population processes: An upgrade experiment with a migratory bird. *Ecology* 86(9):2380–2385.

———. 2007. Linking fluctuations in rainfall to nonbreeding season performance in a long-distance migratory bird, *Setophaga ruticilla*. *Climate Research* 35(1–2):115–122.

Stutchbury, B.J. 1994. Competition for winter territories in a Neotropical migrant: The role of age, sex, and color. *Auk* 111(1):63–69.

Stutchbury, B.J.M., R. Siddiqui, K. Applegate, G.T. Hvenegaard, P. Mammenga, N. Mickle, M. Pearman, J.D. Ray, A. Savage, T. Shaheen, and K.C. Fraser. 2016. Ecological causes and consequences of intratropical migration in temperate-breeding migratory birds. *Am. Naturalist* 188(1):S28–S40.

Suhonen, J., R.V. Alatalo, A. Carlson, and J. Hogland. 1992. Food resource distribution and the organization of the *Parus* guild in a spruce forest. *Ornis Scandinav.* 23:467–474.

Suhonen, J., M. Halonen, T. Mappes, and E. Korpimaki. 2007. Interspecific competition limits larders of Pygmy Owls *Glaucidium passerinum*. *J. Avian Biol.* 36:630–634.

Summers, R.W., L.G. Underhill, C.F. Clinning, and M. Nicoll. 1989. Populations, migrations, biometrics and moult of the Turnstone *Arenaria i. interpres* on the east Atlantic coastline, with special reference to the Siberian population. *Ardea* 77:145–168.

Sutherland, W.J., J.A. Alves, T. Amano, C.H. Chang, N.C. Davidson, G.M. Finlayson, J.A. Gill, R.E. Gill Jr., P.M. Gonzalez, T.G. Gunnarsson, D. Kleijn, C.J. Spray, T. Székely, and D.B.A. Thompson.

2012. A horizon scanning assessment of current and potential future threats to migratory shorebirds. *Ibis* 154:663–679.

Swennen, C. 1984. Differences in quality of roosting flocks of Oystercatchers. Pp. 177–189 in P.R. Evans, J.D. Goss-Custard, and W.G. Hale, eds., *Coastal Waders and Wildfowl in Winter*. Cambridge: Cambridge University Press.

Swenson, J.E., A.V. Andreev, and S.V. Drovetskii. 1995. Factors shaping winter social organization in Hazel Grouse *Bonasa bonasia*: A comparative study in the eastern and western Palearctic. *J. Avian Biol.* 26(1):4–12.

Systad, G.H., J.O. Bustnes, and K.E. Erikstad. 2000. Behavioral responses to decreasing day length in wintering sea ducks. *Auk* 117(1):33–40.

Szép, T. 1995. Relationship between west African rainfall and the survival of central European Sand Martins *Riparia riparia*. *Ibis* 137:162–168.

Tarboton, W.R. 1981. Cooperative breeding and group territoriality in the Black Tit. *Ostrich* 52(1):216–225.

Tejeda-Cruz, C., and W.J. Sutherland. 2004. Bird responses to shade coffee production. *Animal Conservation* 7(2):169–179.

Telleria, J.L., and J. Perez-Tris. 2004. Consequences of the settlement of migrant European Robins *Erithacus rubecula* in wintering habitats occupied by conspecific residents. *Ibis* 146:258–268.

———. 2007. Habitat effects on resource tracking ability: Do wintering Blackcaps *Sylvia atricapilla* track fruit availability? *Ibis* 149:18–25.

Terborgh, J. 1989. *Where Have All the Birds Gone?* Princeton, NJ: Princeton University Press.

Terborgh, J.W., and J.R. Faaborg. 1980. Factors affecting the distribution and abundance of North American migrants in the eastern Caribbean region. Pp. 145–155 in A. Keast and E.S. Morton, eds., *Migrant Birds in the Neotropics: Ecology, Behavior, Distribution, and Conservation*. Washington, DC: Smithsonian Institution Press.

Terrill, S.B. 1990. Ecophysiological aspects of movements by migrants in the winter quarters. Pp. 130–143 in E. Gwinner, ed., *Bird Migration: Physiology and Ecophysiology*. Berlin: Springer-Verlag.

Terrill, S.B., and R.L. Crawford. 1988. Additional evidence of nocturnal migration by Yellow-rumped Warblers in winter. *Condor* 90:261–263.

Terrill, S.B., and R.D. Ohmart. 1984. Facultative extension of fall migration by Yellow-rumped Warblers (*Dendroica coronata*). *Auk* 101(3):427–438.

Thaler, E. 1991. Survival strategies in Goldcrest and Firecrest (*Regulus regulus*, *R. ignicapillus*) during winter. Pp. 1791–1798 in B.D. Bell et al., eds., *Acta XX Congressus Internationalis Ornithologici*, vol. 3. Wellington: New Zealand Ornithological Trust Board.

Therrien, J.-F., N. Lecomte, T. Zgirski, M. Jaffré, A. Beardsell, L.J. Goodrich, J. Bêty, A. Franke, E. Zlonis, and K.L. Bildstein. 2017. Long-term phenological shifts in migration and breeding-area residency in eastern North American raptors. *Auk* 134(4):871–881.

Thiel, D., S. Jenni-Eiermann, V. Braunsich, R. Palme, and L. Jenni. 2008. Ski tourism affects habitat use and evokes a physiological stress response in Capercaillie *Tetrao urogallus*: A new methodological approach. *J. Applied Ecology* 45(3):845–853.

Thomas, D.C., and J.J. Lennon. 1999. Birds extend their ranges northward. *Nature* 399:213.

Thomas, K., R.G. Kvitek, and C. Bretz. 2003. Effects of human activity on the foraging behavior of sanderlings *Calidris alba*. *Biological Conservation* 109(1):67–71.

Thuermer, A.M., Jr. 2015. Planned gas field risks 2000 Wyoming sage-grouse. *WyoFile*, March 10.

Thurber, W.A. 1980. Hurricane Fifi and the 1974 autumn migration in El Salvador. *Condor* 82:212–218.

Tomback, D.F. 1977. Foraging strategies of Clark's Nutcracker. *Living Bird* 16:123–161.

Tordoff, H.B., and W.R. Dawson. 1965. The influence of daylength on reproductive timing in the Red Crossbill. *Condor* 67:416–422.

Tøttrup, A.P., R.H.G. Klaassen, R. Strandberg, K. Thorup, M.W. Kristensen, P.S. Jørgensen, J. Fox, V. Afanasyev, C. Rahbek, and T. Alerstam. 2012. The annual cycle of a trans-equatorial Eurasian-African passerine migrant: Different spatio-temporal strategies for autumn and spring migration. *Proc. Roy. Soc. Lond. B* 279(1730):1008–1016.

Townsend, J.M., C.C. Rimmer, K.P. McFarland, and J.E. Goetz. 2012. Site-specific variation in food resouces, sex ratios, and body condition of an overwintering migrant songbird. *Auk* 129(4):683–690.

Townshend, D.J., P.J. Dugan, and M.W. Pienkowski. 1984. The unsociable plover—uses of intertidal areas by Grey Plovers. Pp. 140–159 in P.R. Evans, J.D. Goss-Custard, and W.G. Hale, eds., *Coastal Waders and Wildlife in Winter*. Cambridge: Cambridge University Press.

Tramer, E.J., and T.R. Kemp. 1980. Foraging ecology of migrant and resident warblers and vireos in the highlands of Costa Rica. Pp. 285–307 in A. Keast and E.S. Morton, eds., *Migrant Birds in the Neotropics: Ecology, Behavior, Distribution, and Conservation*. Washington, DC: Smithsonian Institution Press.

Travers, S.E., B. Marquardt, N.J. Zerr, J.B. Finch, M.J. Boche, R. Wilk, and S.C. Burdick. 2015. Climate change and shifting arrival date of migratory birds over a century in the northern Great Plains. *Wilson J. Ornithol.* 127(1):43–51.

Trierweiler, C., W.C. Mullié, R.H. Drent, K.-M. Exo, J. Komdeur, F. Bairlein, A. Harouna, M. de Bakker, and B.J. Koks. 2013. A Palaearctic migratory raptor species tracks shifting prey availability within its wintering range in the Sahel. *J. Anim. Ecol.* 82(1):107–120.

Troy, D.M. 1983. Recaptures of redpolls: Movements of an irruptive species. *J. Field Ornithol.* 54(2):146–151.

Turnbull, R.E., and G.A. Baldassarre. 1987. Activity budgets of Mallards and American Wigeon wintering in east-central Alabama. *Wilson Bull.* 99(3):457–463.

Uhlmann, S., D. Fletcher, and H. Moller. 2005. Estimating incidental takes of shearwaters in driftnet fisheries: Lessons for the conservation of seabirds. *Biological Conservation* 123(2):151–163.

Underhill, L.G. 1987. Waders (Charadrii) and other waterbirds at Langebaan Lagoon, South Africa 1975–1986. *Ostrich* 58(4):145–155.

Valéry, L., and V. Schricke. 2013. Trends in abundance and wintering phenology of the Dark-bellied Brent Goose *Branta b. bernicla* in France between 1982 and 2012. *Wildfowl* Special Issue 3:57–73.

Van Buskirk, J. 2012. Changes in the annual cycle of North American raptors associated with recent shifts in migration timing. *Auk* 129(4):691–698.

Van Buskirk, J., R.S. Mulvihill, and R.C. Leberman. 2010. Declining body sizes in North American birds associated with climate change. *Oikos* 119(6):1047–1055.

van de Pol, M., Y. Vindenes, B.-E. Saether, S. Engen, B.J. Ens, K. Oosterbeek, and J.M. Tinbergen. 2010. Effects of climate change and variability on population dynamics in a long-lived shorebird. *Ecology* 91(4):1192–1204.

Vander Wall, S.B. 1990. *Food Hoarding in Animals*. Chicago: University of Chicago Press.

van Dijk, A.J., F.E. de Roder, E.C.L. Marteijn, and H. Spiekman. 1990. Summering waders on the Banc d'Arguin, Mauritania: A census in June 1988. *Ardea* 78:145–156.

van Eerden, M.R. 1984. Waterfowl movements in relation to food stocks. Pp. 84–100 in P.R. Evans, J.D. Goss-Custard, and W.G. Hale, eds., *Coastal Waders and Wildfowl in Winter*. Cambridge: Cambridge University Press.

Vézina, F., A. Dekinga, and T. Piersma. 2010. Phenotypic compromise in the face of conflicting ecological demands: An example in red knots *Calidris canutus*. *J. Avian Biol.* 41(1):88–93.

von Haartman, L. 1968. The evolution of resident versus migrant habits in birds: Some considerations. *Ornis Fennica* 45(1):1–7.

Voous, K.H. 1960. *Atlas of European Birds*. London: Nelson.

Votier, S.C., T.R. Birkhead, D. Oro, M. Trinder, M.J. Grantham, J.A. Clark, R.H. McCleery, and B.J. Hatchwell. 2008. Recruitment and survival of immature seabirds in relation to oil spills and climate variability. *J. Anim. Ecol.* 77(5):974–983.

Votier, S.C., B.J. Hatchwell, A. Beckerman, R.H. McCleery, F.M. Hunter, J. Pellatt, M. Trinder, and T.R. Birkhead. 2005. Oil pollution and climate have wide-scale impacts on seabird demographics. *Ecology Letters* 8:1157–1164.

Waide, R.B. 1981. Interactions between resident and migrant birds in Campeche, Mexico. *J. Trop. Ecol.* 22(1):134–154.

Waite, T.A. 1991. Nocturnal hypothermia in Gray Jays *Perisoreus canadensis* wintering in interior Alaska. *Ornis Scandinav.* 22:107–110.

———. 1992. Winter fattening in Gray Jays: Seasonal, diurnal and climatic correlates. *Ornis Scandinav.* 23(4):499–503.

Waite, T.A., and T.C. Grubb Jr. 1987. Dominance, foraging and predation risk in the Tufted Titmouse. *Condor* 89:936–940.

Waite, T.A., and D. Strickland. 2006. Climate change and the demographic demise of a hoarding bird living on the edge. *Proc. Roy. Soc. Lond. B* 273:2809–2813.

Walk, J.W., T.L. Esker, and S.A. Simpson. 1999. Continuous nesting of Barn Owls in Illinois. *Wilson Bull.* 111(4):572–573.

Walsberg, G.E. 1990. Communal roosting in a very small bird: Consequences for the thermal and respiratory gas environments. *Condor* 92:795–798.

Walter, H. 1979. *Eleonora's Falcon: Adaptations to Prey and Habitat in a Social Raptor*. Chicago: University of Chicago Press.

Wang, L.C.H. 1988. Comparative aspects of hypothermia and torpor in birds and mammals. Pp. 2702–2707 in H. Ouellet, ed., *Acta XIX Congressus Internationalis Ornithologici*, vol. 2. Ottawa: University of Ottawa Press.

Ward, D.H., J. Helmericks, J.W. Hupp, L. McManus, M. Budde, D.C. Douglas, and K.D. Tape. 2016. Multi-decadal trends in spring arrival of avian migrants to the central Arctic coast of Alaska: Effects of environmental and ecological factors. *J. Avian Biol.* 47(2):197–207.

Ward, P. 1963. Lipid levels in birds preparing to cross the Sahara. *Ibis* 105:109–111.

Warkentin, I.G., and E.S. Morton. 2000. Flocking and foraging behavior of wintering Prothonotary Warblers. *Wilson Bull.* 112(1):88–98.

Warnock, N., G.W. Page, and L.E. Stenzel. 1995. Non-migratory movements of Dunlins on their California wintering grounds. *Wilson Bull.* 107(1):131–139.

Webb, D.R., and C.M. Rogers. 1988. Nocturnal energy expenditure of Dark-eyed Juncos roosting in Indiana during winter. *Condor* 90:107–112.

Webster, M.D. 1999. Verdin (*Auriparus flaviceps*). In *The Birds of North America*, no. 470 (A. Poole and F. Gill, eds.). The Birds of North America, Inc., Philadelphia.

Webster, M.S., and P.P. Marra. 2005. The importance of understanding migratory connectivity and season interactions. Pp. 199–209 in R. Greenberg and P.P. Marra, eds., *Birds of Two Worlds: The Ecology and Evolution of Migration*. Baltimore: Johns Hopkins University Press.

Wellbrock, A.H.J., C. Bauch, J. Rozman, and K. Witte. 2017. "Same procedure as last year?" Repeatedly tracked swifts show individual consistency in migration pattern in successive years. *J. Avian Biol.* 48(6):897–903.

Wetmore, A., R.F. Pasquier, and S.L. Olson. 1984. *The Birds of the Republic of Panama*, part 4. Washington, DC: Smithsonian Institution Press.

Whittington, P.A., B.D. Dyer, R.J.M. Crawford, and A.J. Williams. 1999. First recorded breeding of Leach's Storm Petrel *Oceanodroma leucorhoa* in the Southern Hemisphere, at Dyer Island, South Africa. *Ibis* 141:327–330.

Wiedenfeld, D.A. 1992. Foraging in temperate- and tropical-breeding and wintering male Yellow Warblers. Pp. 321–328 in J.M. Hagan III and D.W. Johnston, eds., *Ecology and Conservation of Neotropical Migrant Landbirds*. Washington, DC: Smithsonian Institution Press.

Willis, E.O. 1966. The role of migrant birds at swarms of army ants. *Living Bird* 5:187–231.

Wilson, E.O. 2016. *Half-Earth: Our Planet's Fight for Life*. New York: Liveright.

Wilson, H.B., B.E. Kendall, R.A. Fuller, D.A. Milton, and H.P. Possingham. 2011. Analyzing variability and the rate of decline of migratory shorebirds at Moreton Bay, Australia. *Conservation Biology* 25(4):758–766.

Wilson, J., and W. Peach. 2006. Impact of an exceptional winter flood on the population dynamics of bearded tits (*Panurus biarmicus*). *Animal Conservation* 9(4):463–473.

Wilson, S., S.L. LaDeau, A.P. Tottrup, and P.P. Marra. 2011. Range-wide effects of breeding- and nonbreeding-season climate on the abundance of a Neotropical migrant songbird. *Ecology* 92(9):1789–1798.

Winger, B.M., F.K. Barker, and R.H. Ree. 2014. Temperate origins of long-distance seasonal migration in New World songbirds. *Proc. Nat. Acad. Sci.* 111(33):12115–12120.

Winiarski, K.J., D.L. Miller, P.W.C. Paton, and S.R. McWilliams. 2014. A spatial conservation prioritization approach for protecting marine birds given proposed offshore wind energy development. *Biological Conservation* 169:79–88.

Winker, K. 1998. The concept of floater. *Ornitologia Neotropical* 9(2):111–119.

Winker, K., J.H. Rapple, and M.A. Ramos. 1990. Population dynamics of the Wood Thrush in southern Veracruz, Mexico. *Condor* 92:444–460.

Winstanley, D., R. Spencer, and K. Williamson. 1974. Where have all the whitethroats gone? *Bird Study* 21(1):1–14.

With, K.A. 1994. McCown's Longspur (*Calcarius mccownii*). In *The Birds of North America*, no. 96 (A. Poole and F. Gill, eds.). Academy of Natural Sciences, Philadelphia, and American Ornithologists' Union, Washington, DC.

Withers, P.C. 1977. Respiration, metabolism, and heat exchange of euthermic and torpid poorwills and hummingbirds. *Physiological Zoology* 50:43–52.

Wojtowicz, S. 2016. The search is on for Piping Plovers. US Fish and Wildlife Service. March 9. https://usfwsnortheast.wordpress.com/2016/03/09/the-search-is-on-for-piping-plovers/.

Wolfson, A. 1945. The role of the pituitary, fat deposition, and body weight in bird migration. *Condor* 46:95–127.

Wood, B. 1979. Changes in numbers of over-wintering Yellow Wagtails *Motacilla flava* and their food supplies in a West African savanna. *Ibis* 121:228–231.

———. 1992. Yellow Wagtail *Motacilla flava* migration from West Africa to Europe: Pointers towards a conservation strategy for migrants on passage. *Ibis* 134 suppl. 1:66–76.

Woodin, M.C., and T.C. Michot. 2002. Redhead (*Aythya americana*). In *The Birds of North America*, no. 695 (A. Poole and F. Gill, eds.). The Birds of North America, Inc. Philadelphia.

Woodrey, M.S., and C.R. Chandler. 1997. Age-related timing of migration: Geographic and interspecific patterns. *Auk* 109(1):52–67.

Wormworth, J., and Ç. Şekercioğlu. 2011. *Winged Sentinels: Birds and Climate Change*. Cambridge: Cambridge University Press.

Wunderle, J.M., Jr. 1995. Population characteristics of Black-throated Blue Warblers wintering in three sites on Puerto Rico. *Auk* 112(4):931–946.

Wunderle, J.M., Jr., and S.C. Latta. 2000. Winter site fidelity of Nearctic migrants in shade coffee plantations of different sizes in the Dominican Republic. *Auk* 117(3):596–614.

Ydenberg, R.C., J. Barrett, D.B. Lank, C. Xu, and M. Faber. 2017. The redistribution of non-breeding Dunlins in response to the post-DDT recovery of falcons. *Oecologia* 183(4):1101–1110.

Ydenberg, R.C., and H.H.T. Prins. 1984. Why do birds roost communally? Pp. 123–139 in P.R. Evans, J.D. Goss-Custard, and W.G. Hale, eds., *Coastal Waders and Wildfowl in Winter*. Cambridge: Cambridge University Press.

Ydenberg, R.C., H.H.T. Prins, and J. van Dijk. 1983. The post-roost gatherings of wintering Barnacle Geese: Information centres? *Ardea* 71:125–131.

Yom-Tov, Y. 2001. Global warming and body mass decline in Israeli passerine birds. *Proc. Roy. Soc. Lond. B* 268:947–952.

Yom-Tov, Y., S. Yom-Tov, J. Wright, C.J.R. Thorne, and R. Du Feu. 2006. Recent changes in body weight and wing length among some British passerine birds. *Oikos* 112(1):91–101.

Zahavi, A. 1971. The social behaviour of the White Wagtail *Motacilla alba alba* wintering in Israel. *Ibis* 113:203–211.

Zimmerman, J.L. 1966. Effects of extended tropical photoperiod and temperature on the Dickcissel. *Condor* 68:377–387.

Zwarts, L., J. van der Kamp, E. Klop, M. Sikkema, and E. Wymenga. 2014. West African mangroves harbor millions of wintering European warblers. *Ardea* 102(2):121–130.

Zydelis, R., C. Small, and G. French. 2013. The incidental catch of seabirds in gillnet fisheries: A global review. *Biological Conservation* 162:76–88.

INDEX